T0392734

Machine Learning in Quantum Sciences

Artificial intelligence is reshaping the world, including scientific research. It plays an essential role in scientific discovery by enhancing and accelerating research across multiple fields. This book dives into the interplay between artificial intelligence and the quantum sciences, the outcome of a collaborative effort from the world's leading experts. After presenting the key concepts and foundations of machine learning, a subfield of artificial intelligence, its applications in quantum chemistry and physics are presented in an accessible way, enabling readers to engage with emerging literature on machine learning in science. By examining its state-of-the-art applications, readers will discover how machine learning is currently being applied in their own field and appreciate its broader impact on science and technology.

This book is accessible to undergraduates and more advanced readers from physics, chemistry, engineering, and computer science. Online resources include coding exercises as Jupyter notebooks for self-study of some key topics introduced in the book.

Machine Learning in Quantum Sciences

Anna Dawid
University of Warsaw
ICFO – The Institute of Photonic Sciences
Flatiron Institute

Julian Arnold
University of Basel

Borja Requena
ICFO – The Institute of Photonic Sciences

Alexander Gresch
Heinrich Heine University Düsseldorf
Hamburg University of Technology

Marcin Płodzień
ICFO – The Institute of Photonic Sciences

Kaelan Donatella
University of Paris

Kim A. Nicoli
Technical University of Berlin
Berlin Institute for the Foundations of Learning and Data

Paolo Stornati
ICFO – The Institute of Photonic Sciences

Rouven Koch
Aalto University

Miriam Büttner
Albert-Ludwig University of Freiburg

Robert Okuła
University of Gdańsk
Gdańsk University of Technology

Gorka Muñoz-Gil
University of Innsbruck

Rodrigo A. Vargas-Hernández
University of Toronto
Vector Institute for Artificial Intelligence
McMaster University

Alba Cervera-Lierta
Barcelona Supercomputing Center

Juan Carrasquilla
Vector Institute for Artificial Intelligence

Vedran Dunjko
Leiden University

Marylou Gabrié
Polytechnic Institute of Paris

Patrick Huembeli
Swiss Federal Institute of Technology in Lausanne

Evert van Nieuwenburg
Leiden University
Niels Bohr Institute

Filippo Vicentini
Swiss Federal Institute of Technology in Lausanne
Polytechnic Institute of Paris

Lei Wang
Chinese Academy of Sciences
Songshan Lake Materials Laboratory

Sebastian J. Wetzel
Perimeter Institute for Theoretical Physics

Giuseppe Carleo
Swiss Federal Institute of Technology in Lausanne

Eliška Greplová
Delft University of Technology

Roman Krems
University of British Columbia

Florian Marquardt
Max Planck Institute for the Science of Light
Friedrich-Alexander University of Erlangen-Nuremberg

Michał Tomza
University of Warsaw

Maciej Lewenstein
ICFO – The Institute of Photonic Sciences
ICREA

Alexandre Dauphin
ICFO – The Institute of Photonic Sciences
PASQAL SAS

Shaftesbury Road, Cambridge CB2 8EA, United Kingdom

One Liberty Plaza, 20th Floor, New York, NY 10006, USA

477 Williamstown Road, Port Melbourne, VIC 3207, Australia

314–321, 3rd Floor, Plot 3, Splendor Forum, Jasola District Centre, New Delhi – 110025, India

103 Penang Road, #05–06/07, Visioncrest Commercial, Singapore 238467

Cambridge University Press is part of Cambridge University Press & Assessment, a department of the University of Cambridge.

We share the University's mission to contribute to society through the pursuit of education, learning and research at the highest international levels of excellence.

www.cambridge.org
Information on this title: www.cambridge.org/9781009504935

DOI: 10.1017/9781009504942

When citing this work, please include a reference to the DOI 10.1017/9781009504942

First published 2025

A catalogue record for this publication is available from the British Library

A Cataloging-in-Publication data record for this book is available from the Library of Congress

ISBN 978-1-009-50493-5 Hardback

In memory of Peter Wittek

Contents

Contributors

Julian Arnold is a theoretical physicist working at the interface between the quantum sciences, information theory, and machine learning (ML). His research includes the design of methods for the automated detection of phase transitions and the application of differentiable programming to solve inverse design problems in quantum many-body physics.

Miriam Büttner earned her MSc in molecular science at the Friedrich-Alexander University of Erlangen-Nuremberg. In 2017, she went to Shenzhen, China, for an elective Master's project on ML in quantum chemistry and has since then been growing her ML knowledge. She is currently doing her PhD in many-body physics.

Giuseppe Carleo is a computational physicist, best known for pioneering ML tools for quantum systems. He holds a PhD from the International School for Advanced Studies in Italy (SISSA, 2011) and worked in National Centre for Scientific Research (France), ETH Zürich (Switzerland), and Flatiron Institute (USA). Since 2018, he is a professor at Swiss Federal Institute of Technology in Lausanne (EPFL), Switzerland, leading the Computational Quantum Science Lab.

Juan Carrasquilla's research interests are at the intersection of condensed matter physics, quantum computing, and ML. He completed his PhD in Physics at SISSA. Juan has been recently appointed an associate professor of computational physics at ETH Zürich.

Alba Cervera-Lierta is Senior Researcher at the Barcelona Supercomputing Center. She earned her PhD in Physics at the University of Barcelona. She works on near-term quantum algorithms and their applications. Since October 2021, she is the coordinator of the Quantum Spain project, an initiative to boost the Spanish quantum computing ecosystem.

Alexandre Dauphin is VP quantum simulation at PASQAL, a neutral-atom quantum computing company. During his career, he has worked on a broad range of topics going from quantum simulation of many-body phases of matter to ML applied to physics and quantum machine learning. He is the recipient of the *New Journal of Physics* (NJP) Early Career Award 2019.He is a member of the editorial board of NJP since 2020 and a member of European Laboratory for Learning and Intelligent Systems since 2021.

Anna Dawid just transitioned from a research fellow at the Flatiron Institute, New York, to an assistant professor at Leiden University. She holds a PhD in quantum physics awarded by the University of Warsaw and the Institute of Photonic Sciences (ICFO). Her research spans interpretable ML for scientific discovery, quantum simulations, and foundations of deep learning.

Kaelan Donatella is a Franco-Irish physicist trained at Ecole Normale Supérieure and the University of Paris. His interests range from quantum computing to the history and philosophy of science, with recent work being focused on analog computing for artificial intelligence (AI).

Vedran Dunjko's research interest lies in the intersection of computer science and quantum physics, including quantum computing and quantum cryptography. For over 10 years, he has been focusing on the interplay between quantum computing, ML, and AI, publishing over 50 research papers in this area.

Marylou Gabrié is Assistant Professor at Ecole Polytechnique since 2022. She was awarded a fellowship from L'Oréal-UNESCO for Women in Science program during her PhD in École Normale Supérieure in 2019. Her research lies at the boundary of ML and statistical physics.

Eliška Greplová is Associate Professor at the Kavli Institute of Nanoscience at Delft University of Technology in the Netherlands. Eliška leads the QMAI group working at the intersection of quantum technologies, AI, and condensed matter physics.

Alexander Gresch (PhD student at the universities of Düsseldorf and Hamburg) is a theoretical physicist specializing in mathematical and ML methods in the context of quantum technologies. This includes, in particular, the efficient and accurate read-out of hybrid quantum algorithms and the role of quantum data for ML.

Patrick Huembeli earned his PhD at ICFO, bridging ML with quantum information. He was a postdoctoral researcher at EPFL, combining classical ML and quantum computing. Currently, he is a staff scientist at a stealth startup focusing on probabilistic ML software and hardware.

Rouven Koch is a doctoral researcher at Aalto University, working in the intersection of condensed matter theory and ML. His research focuses on the combination of theory and experiments with the help of AI. Personally, he is interested in daily-life applications of AI.

Roman Krems is a professor of chemistry and Distinguished University Scholar at the University of British Columbia in Vancouver, Canada. He is also a member of the computer science department and a principal investigator at the Stewart Blusson Quantum Matter Institute.

Maciej Lewenstein studied in Warsaw, was on the faculty of Center for Theoretical Physics, Polish Academy of Sciences, French Alternative Energies and Atomic Energy Commission Paris-Saclay center, Universität Hannover, and is presently at ICFO. He is a Fellow of American Physical Society and Optica, has an H-index of 107 (WoS) and 124 (Google).

Florian Marquardt is a theoretical physicist working at the intersection between ML and physics, as applied to nanophysics and quantum optics. Since 2016,he is a scientific director in the Max Planck Society, leading the theory division at the Max Planck Institute for the Science of Light. His long-term vision is true artificial scientific discovery.

Gorka Muñoz-Gil is a Marie Skłodowska-Curie fellow in Innsbruck University. Before, he received his PhD at ICFO (Spain) for the study of stochastic processes and their connection to ML. His research focuses on the application of ML techniques to a variety of topics, with a special interest in interpretable solutions.

Kim A. Nicoli is a postdoctoral researcher at the Helmholtz Institute for Radiation and Nuclear Physics and the University of Bonn. He got his PhD in ML from Technical University of Berlin in 2023. His research interests extend across probabilistic modeling, quantum computing, generative models, lattice quantum field theory, and neuromorphic computing.

Evert van Nieuwenburg might be known from the "confusion scheme" or perhaps from TiqTaqToe. He researches ML for quantum systems and often tries to gamify quantum challenges so that AI can learn to play and solve them.

Robert Okuła is a PhD student interested in all things quantum, especially quantum cryptography and quantum Darwinism. He considers ML to be a useful tool in that regard.

Marcin Płodzień (Ph.D., 2014, Jagiellonian University) is Research Fellow at ICFO, specializing in theoretical physics with a focus on many-body quantum systems and quantum information theory. His research interests include quantum simulators and quantum metrology, entanglement and Bell correlations, quantum reservoir computing, and applications of ML to quantum mechanics.

Borja Requena develops ML algorithms for scientific applications. His contributions span multiple fields, from quantum to statistical and biophysics. Additionally, Borja has worked in high-tech companies such as Xanadu Quantum Technologies and Telefonica R&D, and he has been high ranked in ML and quantum computing competitions.

Paolo Stornati is a postdoctoral researcher in quantum simulation and quantum many-body theory. I have deep interest in the development of novel numerical tools to study exotic phases of matter and lattice Gauge theories.

Michał Tomza is a professor of theoretical physics at the University of Warsaw. He specializes in physics of ultracold quantum matter, including interactions and collisions of ultracold atoms, ions, and molecules. He won the ERC Starting Grant and National Science Center Award in Physical Sciences and Engineering and is a member of the Polish Young Academy.

Rodrigo A. Vargas-Hernández's main research interest is the development of numerical tools that could help us crack the exponential wall to simulate quantum systems.

Filippo Vicentini is a computational quantum physicist who develops innovative ML-inspired algorithms. He has been awarded the Atos – Joseph Fourier Prize for AI in sciences in 2019 and leads the NetKet open-software collaboration since.

Lei Wang is a computational quantum physicist. His Erdős number is 2.

Sebastian J. Wetzel's research interest is AI in the physical sciences where his most important contributions are related to using ML to calculate phase diagrams and artificial scientific discovery through the interpretation of neural networks.

Preface

We live in fascinating times where scientists are starting to incorporate artificial intelligence (AI) algorithms for knowledge discovery. Advances in this booming field have led to a rapid increase in the interest and confidence of the scientific community in these methods. This trend can be observed by tracking the percentage of machine learning (ML)-based publications in physics, chemistry, and material science, shown in Fig. 0.1. As the number of ML applications grows, keeping track of all advances becomes challenging. Moreover, it is difficult to find reliable intermediate-level learning material that allows one to efficiently bridge the gap between the rapidly developing field of ML and scientists interested in incorporating ML tools into their own research.

The idea of creating this book was born out of *Summer School: Machine Learning in Quantum Physics and Chemistry* that took place between August 23 and September 03, 2021, in Warsaw, Poland. As such, its aim is to give an educational and self-contained overview of modern applications of ML in quantum sciences. The scientific content of this work is inspired by the topics covered by the lecturers and invited speakers of the school. We invite the reader to take a look at the school tutorials in [2] and to reuse the figures prepared for this book, which are available in [3].

The target audience of this book is quantum scientists who want to familiarize themselves with ML methods. Therefore, we assume a basic knowledge of linear algebra, probability theory, and quantum information theory. We also expect familiarity with concepts such as Lagrange multipliers, Hilbert space, and Monte Carlo methods. We also assume that the reader is familiar with quantum mechanics and has a basic grasp of the current challenges in quantum sciences.

Our book is roughly divided into three parts. The first part is devoted to establishing a solid foundation of basic ML concepts needed for understanding its applications in natural sciences. In the second part, we dive into four core application areas of ML in quantum sciences. This covers the use of deep learning and kernel methods in supervised, unsupervised, and reinforcement learning algorithms for phase

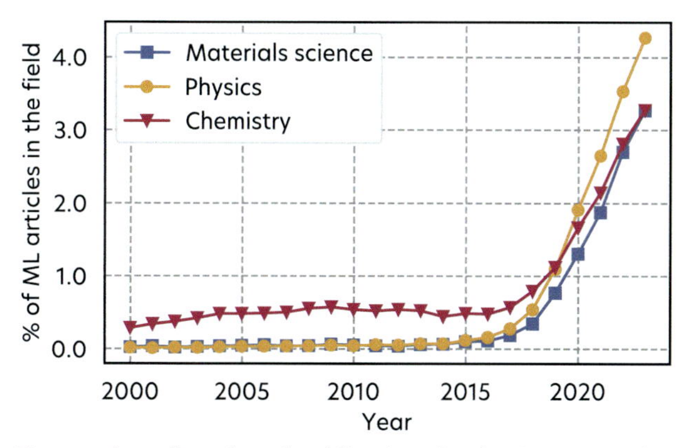

Figure 0.1 The number of ML-based publications in physics, materials science, and chemistry is growing exponentially. Adapted from [1] under the MIT License.

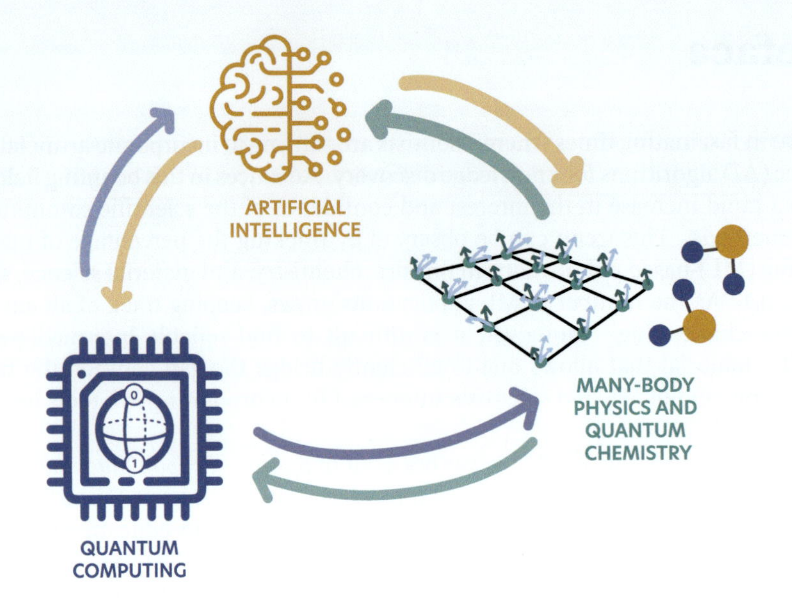

Figure 0.2 Interplay of AI and quantum sciences, in particular quantum computing, many-body physics, and quantum chemistry. Within this book, we focus not only on the influence of AI on quantum sciences but also cover the reverse impact of statistical physics and quantum computing on ML.

classification, representation of many-body quantum states, quantum feedback control, and quantum circuit optimization among other applications. In the third part, we introduce and discuss more specialized topics such as differentiable programming, generative models, statistical physics approaches to ML, and quantum machine learning. All in all, this book discusses the fruitful interplay of AI and quantum sciences, presented schematically in Fig. 0.2.

We do not aim to provide an exhaustive list of ML applications in quantum sciences. Such reviews already exist and nicely summarize the latest achievements [4–6]. Instead, our objective is to provide the reader with enough knowledge, intuition, and tricks of the trade to start implementing ML methods of choice in their own research. As such, we selected the ML applications presented in this work that, we believe, are pedagogically appealing while keeping a broad overview of the field. To this end, we focus on what a reader could do and not only on what has been done. To fulfill this ambition, we conclude each chapter with an outlook and open problems that we recognize as important and promising.

Online resources accompanying this book include coding exercises as Jupyter notebooks for self-study, and focus on key topics introduced in the book. These are available for download at www.cambridge.org/dawidQML.

Acknowledgments

We thank Hans J. Briegel, Lorenzo Cardarelli, Kacper Cybiński, and Mario Krenn for useful discussions and Fesido Studio Graficzne for the graphical design of the book.

Author contributions This manuscript is a result of a unique collaboration between the participants and lecturers of *Summer School: Machine Learning in Quantum Physics and Chemistry* which took place in Warsaw, Poland, during August–September 2021 and which was organized by M. Tomza, A. Dauphin, A. Dawid, and M. Lewenstein. All authors of this manuscript participated in the reading and improvement of its content. In particular:

- *"Introduction"* was written by A. Dawid with the help of M. Płodzień, M. Lewenstein, A. Gresch, R. Koch, B. Requena, and G. Muñoz-Gil.
- *"Basics of machine learning"* was written by A. Dawid, A. Gresch, and J. Arnold with the help of K. Nicoli, K. Donatella, and M. Płodzień.
- *"Phase classification"* was written by J. Arnold, A. Dawid, A. Gresch, R. Koch, M. Płodzień, and S. Wetzel based on the scientific content provided by E. Greplová, P. Huembeli, and S. Wetzel.
- *"Gaussian processes and other kernel methods"* was written by A. Gresch, A. Dawid, K. Nicoli, J. Arnold, R. Krems, and R. A. Vargas-Hernández based on the scientific content provided by R. Krems.
- *"Neural-network quantum states"* was written by K. Donatella, B. Requena, and P. Stornati with the help of R. Okuła and M. Płodzień based on the scientific content provided by G. Carleo, J. Carrasquilla, and F. Vicentini.
- *"Reinforcement learning"* was written by B. Requena, M. Płodzień, and A. Gresch with the help of R. Okuła and G. Muñoz-Gil based on the scientific content provided by V. Dunjko, F. Marquardt, and E. van Nieuwenburg.
- *"Differentiable programming"* was written by J. Arnold, *"Generative models in many-body physics"* – by K. A. Nicoli and M. Gabrié with the help of M. Płodzień, K. Donatella, and A. Dawid, *"Machine learning for experiments"* – by M. Büttner, R. Koch, and A. Dawid, based on the scientific content provided by J. Carrasquilla, E. Greplová, and L. Wang.
- *"Statistical physics for machine learning"* was written by A. Dawid with the help of M. Płodzień based on the scientific content of M. Gabrié, and *"Quantum machine learning"* was written by P. Stornati, A. Dauphin, M. Płodzień, and R. Koch with the help of R. Okuła, based on the lectures of A. Cervera-Lierta and V. Dunjko.
- Finally, *"Conclusions"* was written by A. Dauphin, B. Requena, M. Płodzień, and A. Dawid with the help of all the coauthors.

The project was led by A. Dawid and supervised by A. Dauphin with the help of M. Lewenstein and M. Tomza.

Funding information An. D. acknowledges the financial support from the National Science Centre, Poland, within the Preludium Grant No. 2019/33/N/ST2/03123 and the Etiuda Grant No. 2020/36/T/ST2/00588 as well as from the Foundation for Polish Science. The Flatiron Institute is a division of the Simons Foundation. J. A. acknowledges financial support from the Swiss National Science Foundation individual grant (Grant No. 200020 200481). A. G. acknowledges financial support from the Deutsche Forschungsgemeinschaft (DFG, German Research Foundation) – Project No. 441423094. M. P. acknowledges the support of the Polish National Agency for Academic Exchange, the Bekker Programme No. PPN/BEK/2020/1/00317. K. A. N. acknowledges support by the Federal Ministry of Education and Research (BMBF) for the Berlin Institute for the Foundations of Learning and Data (BIFOLD) (01IS180-37A). R. K. acknowledges financial support from the Academy of Finland Projects No. 331342 and No. 336243. G. M-G. acknowledges support from the Austrian Science Fund (FWF) through SFB BeyondC F7102. A. C-L. acknowledges the support by the Ministry of Economic Affairs and Digital Transformation of the Spanish Government through the QUANTUM ENIA project call – QUANTUM SPAIN project, and by the European Union through the Recovery, Transformation and Resilience Plan – NextGenerationEU within the framework of the Digital Spain 2025 Agenda. M. G. acknowledges funding as an Hi!Paris Chair Holder. L. W. is supported by the Strategic Priority Research Program of the Chinese Academy of Sciences under Grant No. XDB30000000 and National Natural Science Foundation of China under Grant No. T2121001. M. T. acknowledges the financial support from the Foundation for Polish Science within the First Team programme cofinanced by the EU Regional Development Fund. Al. D. acknowledges the financial support from a fellowship granted by la Caixa Foundation (ID 100010434, fellowship code LCF/BQ/PR20/11770012). This project has received funding from the European Union's Horizon 2020 research and innovation program under the Marie Skłodowksa-Curie Grant Agreement No. 895439 "ConQuER."

ICFO group acknowledges support from: ERC AdG NOQIA; MICIN/AEI (PGC2018-0910.13039/501100011033, CEX2019-000910-S/10.13039/501100011033, Plan National FIDEUA PID2019-106901GB-I00, FPI; MICIIN with funding from European Union NextGenerationEU (PRTR-C17.I1): QUANTERA MAQS PCI2019-111828-2); MCIN/AEI/10.13039/501100011033 and by the "European Union NextGeneration EU/PRTR" QUANTERA DYNAMITE PCI2022-132919 (QuantERA II Programme co-funded by European Union's Horizon 2020 programme under Grant Agreement No. 101017733), Ministry of Economic Affairs and Digital Transformation of the Spanish Government through the QUANTUM ENIA project call – Quantum Spain project, and by the European Union through the Recovery, Transformation and Resilience Plan – NextGenerationEU within the framework of the Digital Spain 2026 Agenda. Fundació Cellex; Fundació Mir-Puig; Generalitat de Catalunya (European Social Fund FEDER and CERCA program, AGAUR Grant No. 2021 SGR 01452, QuantumCAT U16-011424, co-funded by ERDF Operational Program of Catalonia 2014–2020); Barcelona Supercomputing Center MareNostrum (FI-2023-1-0013); EU Quantum Flagship (PASQuanS2.1, 101113690); EU Horizon 2020 FET-OPEN OPTOlogic (Grant No. 899794); EU Horizon Europe Program (Grant Agreement No. 101080086 — NeQST), National Science Centre, Poland (Symfonia Grant No. 2016/20/W/ST4/00314); ICFO Internal "QuantumGaudi" project;

"La Caixa" Junior Leaders fellowships ID100010434: LCF/BQ/PI19/11690013, LCF/BQ/PI20/11760031, LCF/BQ/PR20/11770012, and LCF/BQ/PR21/11840013. Views and opinions expressed are, however, those of the author(s) only and do not necessarily reflect those of the European Union, European Commission, European Climate, Infrastructure and Environment Executive Agency (CINEA), nor any other granting authority. Neither the European Union nor any granting authority can be held responsible for them.

Research at Perimeter Institute is supported in part by the Government of Canada through the Department of Innovation, Science and Economic Development Canada and by the Province of Ontario through the Ministry of Economic Development, Job Creation and Trade. We thank the National Research Council of Canada for its partnership with Perimeter on the PIQuIL.

Note on the text

Numbers and arrays

$\boldsymbol{A}$	matrix
A	tensor
$\boldsymbol{a}$	vector
A	random variable
a	scalar

Physical constants and quantities

β	$1/k_\mathrm{B}T$
δ	Kronecker delta
$\langle x\rangle_p$ or $\mathbb{E}[x \mid p]$	estimator of quantity x with respect to distribution p
$\hat{\sigma}$	Pauli matrix
$\hat{H}$	quantum Hamiltonian
$\mathcal{H}$	Hilbert space
σ	spin variable
H	classical Hamiltonian
k_B	Boltzmann constant
m	magnetization
Z	partition function

Machine learning quantities

$\boldsymbol{b}$	model biases
K_i or K	ith class or number of classes in a classification problem
$\mathcal{D}$	dataset
n	size of $\mathcal{D}$, that is, the number of training examples
η	learning rate
ϕ	feature map
m	dimensionality of data point $\boldsymbol{x}$, that is, the number of data features
$\hat{f}$	model with converged $\boldsymbol{\theta}$
$\mathcal{L}$	loss (or cost/error) function
$\boldsymbol{H}$	Hessian matrix
d	size of $\boldsymbol{\theta}$, that is, number of model parameters
$\boldsymbol{\theta}$	model parameters
$\boldsymbol{\theta}^*$	converged $\boldsymbol{\theta}$
π	policy
π^*	optimal policy
ℓ_n	L(n) regularization
ς	activation function
$\boldsymbol{w}$ or $\boldsymbol{W}$	vector or matrix of model weights
a	action
D_KL	Kullback–Leibler divergence
G	return
r	reward
s	state

Abbreviations

AD automatic differentiation

AE autoencoder

AI artificial intelligence

ANN artificial neural network

AR autoregressive

ARNN autoregressive neural network

BIC Bayesian information criterion

BO Bayesian optimization

CPU central processing unit

CE cross-entropy

CNN convolutional neural network

∂P differentiable programming

DL deep learning

DNN deep neural network

DQN deep Q-network

ECM episodic and compositional memory

GAMP generalized approximate message passing

GAN generative adversarial network

GNS generative neural sampler

GP Gaussian process

GPR Gaussian process regression

GPU graphics processing unit

IGT Ising gauge theory

KRR kernel ridge regression

KL Kullback–Leibler

L-BFGS limited-memory Broyden–Fletcher–Goldfarb–Shanno algorithm

LASSO least absolute shrinkage and selection operator

LE local ensemble

MAE mean absolute error

MAP maximum a posteriori

MCMC Markov chain Monte Carlo

MDP Markov decision process

ML machine learning

MLE maximum likelihood estimation

MPS matrix product state

MSE mean-squared error

NF normalizing flow

NIS neural importance sampling

NISQ noisy intermediate-scale quantum

NMCMC neural Markov chain Monte Carlo

NN neural network

NQS neural quantum state

ODE ordinary differential equation

PC principal component

PCA principal component analysis

PES potential energy surface

POVM positive operator-valued measure

PPT positive under partial transposition

PQC parametrized quantum circuit

PS projective simulation

QAOA quantum approximate optimization algorithm

QD quantum dot

QML quantum machine learning

RBM restricted Boltzmann machine

RKHS reproducing kernel Hilbert space

RL reinforcement learning

RNN recurrent neural network

RUE resampling uncertainty estimation

SGD stochastic gradient descent

SE state evolution

SVM support vector machine

t-SNE t-distributed stochastic neighbor embedding

t-VMC time-dependent variational Monte Carlo

TD temporal difference

TN tensor network

TNS tensor network state

VAE variational autoencoder

VQE variational quantum eigensolver

1 Introduction

Making *intelligent* machines, that is, machines capable of learning and utilizing the gathered knowledge in thinking and reasoning, is a long-lived dream of human civilization. The more we know about the human brain, intelligence, and psychology, the more challenging it seems. However, despite the many obstacles and challenges in creating artificial intelligence (AI), the joint effort of researchers working in the natural, cognitive, mathematical, and computer sciences has produced impressive machinery that is already revolutionizing our daily life, industry, and science.

1.1 How do computers learn?

The ultimate goal of AI is to endow machines with the ability to conceptualize and create abstractions. Both of these features are mechanisms that underlie learning representations of knowledge and reasoning based on experience in humans. We have multiple ways of representing ideas. For example, we can encode a piece of music in a digital format on a computer, in an analog format on a vinyl disk, or we can write it down in a music score. Although the representations are entirely different, the piece of music is the same. Therefore, the properties of abstract ideas do not depend on the data source.

Furthermore, conceptualization and abstraction bring the possibility of considering various levels of details within a particular representation or the ability to switch from one level to another while preserving the relevant information [7–11]. Our brain excels at extracting abstract ideas from different representations of knowledge. In our daily lives, we constantly process information from multiple sources that represent the same concept in completely different ways. For example, we can identify the concept of a dog by seeing one, hearing or smelling it, reading the word "dog," painting a snout on someone's face, or even casting shadows with our hands that resemble the shade of a dog. This level of abstraction and conceptualization enables us to reason, connecting high-level ideas. All the properties of our brain mentioned above form what we call intelligence. Conferring these properties to a computer would result in a general problem-solving machine.

Today, we are at a point in our technological advances at which the human brain and computers have a disjoint set of tasks in which they naturally excel.[1] Some tasks are easy for computers but difficult for humans. These are problems that can be described by a list of formal, mathematical rules. Therefore, computers excel at solving logic, algebra, geometry, and optimization problems, which we can tackle with hard-coded solutions or knowledge-based AI. However, we would like to tackle problems that are not easy to present in a formal mathematical way, such as face

[1]This observation was first made in the 1980s, and it is called Moravec's paradox. As Moravec wrote in 1988 [12], "It is comparatively easy to make computers exhibit adult level performance on intelligence tests or playing checkers, and difficult or impossible to give them the skills of a one-year-old when it comes to perception and mobility" (p. 15).

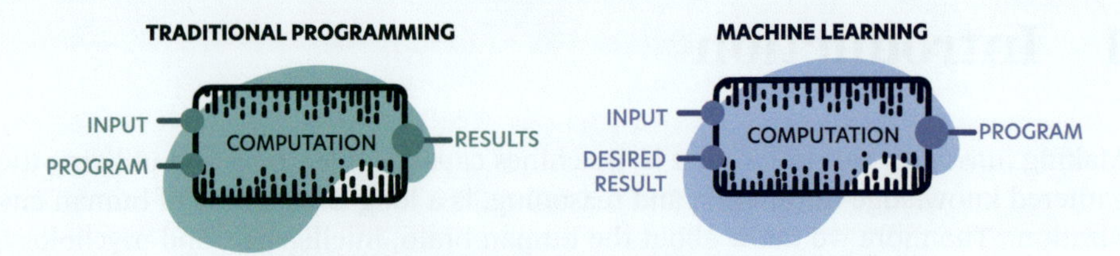

Figure 1.1 Schematic representation of the difference between traditional programming, based on the algorithmic approach, and the experience-based/data-driven approach, which is the backbone of the ML paradigm. The ML paradigm is the first step toward learning abstractions by computers through the extraction of common features from data.

recognition, or whose exact mathematical formulation is not yet known, such as detecting new quantum phases.

A particularly exciting direction is the development of algorithms that are not explicitly programmed. The main principle is to enable computers to learn from experience (or data). The shift toward this data-driven paradigm led to the birth of machine learning (ML), schematically depicted in Fig. 1.1. This field leverages fundamental concepts of applied statistics, emphasizing the use of computers to estimate complicated functions and with a decreased emphasis on proving confidence intervals around them [13]. This trend has accelerated with the rise of deep learning (DL), where enormous and heavily parametrized hierarchical models are used to deal with complex patterns from real-world data and do this with unprecedented accuracy. Interestingly, many DL architectures are designed to mimic some of the properties of the human thinking process, such as understanding correlations in visual patterns or recurrence in sound signals. We present a schematic representation of the relationship between these three fields (AI, ML, and DL) in Fig. 1.2.

To make a computer learn, we need three main ingredients:

1. A *task* to solve (Section 1.4).
2. *Data* that can be considered as an equivalent of *experience*. The latter can be provided in the form of, for example, an interacting environment, and allows for solving the task (Section 1.5).
3. A *model* that learns how to solve the task (Section 2.4).

To check whether a computer successfully learns how to solve a task, we need to define a performance measure that can be as simple as the comparison between the prediction of the model and the expected answer. In these terms, the learning process can be described as the iterative minimization of the model error or maximization of the model performance on the given task and data.

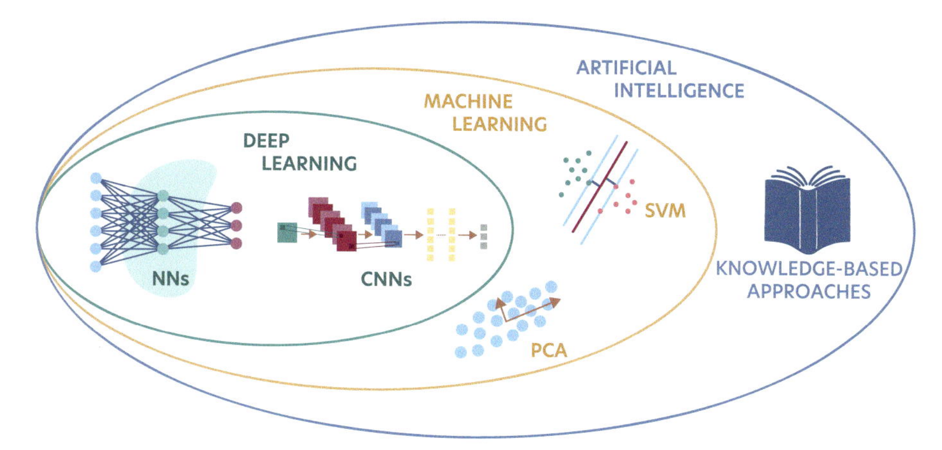

Figure 1.2 Sketch of the relation between AI, ML, and DL with examples from each field, including support vector machines (SVMs), principal component analysis (PCA), neural networks (NNs), and convolutional neural networks (CNNs).

1.2 Historical view on learning machines

The foundations of the theory of learning were established already in the 1940s. Its development has followed two parallel paths: a knowledge-based approach, which dominated the AI research field for decades, and a data-based one, which is currently on the rise. Throughout the years, ML has gone under various names (such as cybernetics or connectionism) and experienced a few cycles of intense popularity,[2] followed by criticism and disappointment, followed by funding cuts, followed by renewed interest years or decades later [13]. To give the reader some insights into the giants on whose shoulders we stand, we briefly present milestones in the development of ML following [13, 15–17]:

- 1943 – Walter Pitts and Warren McCulloch create a computer model inspired by the neural networks (NNs) of the human brain called the *threshold logic*. Their field of expertise is called *cybernetics*.
- 1949 – Donald Hebb hypothesizes how learning in biological systems works and formulates *Hebbian learning*. For example, if certain neurons "fire together, they wire together."
- 1957 – Frank Rosenblatt introduces a *Rosenblatt perceptron* modeling a single neuron. A perceptron is also called "an artificial neuron" and, after modifications in 1969 by Marvin Minsky and Seymour Papert, to this day, remains widely used as a building block of artificial neural networks (ANNs).
- 1962 – David Hubel and Torsten Wiesel present, for the first time, the response properties of single biological neurons recorded with a microelectrode.

[2]Some argue that the "AI winter" is upon us unless we rethink AI or combine it with knowledge-based approaches [14]. It is also important to remember that such hype cycles are frequent with emerging new technologies.

- 1969 – Marvin Minsky and Seymour Papert point out the computational limitations and disadvantages of linear models, including a single artificial neuron, contributing to the first "AI winter."
- 1986 – David Rumelhart, Geoffrey Hinton, and Ronald Williams use *backpropagation* to train an NN with one or two hidden layers, which, next to the revival of Hebb's ideas, causes renewed interest in the field that at this time is called *connectionism*. In the same year, David Rummelhart *et al.* publish a widely discussed two-volume book *Parallel Distributed Processing*, which collected original contributions from the field, including backpropagation and Boltzmann machines.
- The mid-1990s – Second AI winter whose appearance is ascribed [13] to exceedingly ambitious claims of the community, which led to the disappointment of investors, and the simultaneous progress of kernel methods, which require less computational resources.

Interestingly, we can see how closely the development of AI was intertwined with neuroscience. This makes sense, as the human brain provides proof by example that intelligent behavior is possible. A natural approach to AI would be to try to reverse engineer the brain to reproduce its functionality. However, while the perceptron was inspired by biological neurons and some ML models are loosely inspired by neurological discoveries, there is nowadays a consensus that models should not be designed to be realistic simulators of biological functions [13].[3] Instead, scientists attempt to solve the mysteries of the human brain using ML.

Since 2006, DL has been thriving again thanks to a breakthrough in the efficient training of deep NNs [18] via backpropagation, followed by multiple analyses confirming the importance of its depth. At the same time, there has been a rapid improvement in computational power in recent decades, which has allowed the exploration of larger ML models. Here, the development of graphics processing units (GPUs) [19, 20] has played a particularly important role: highly parallelizable algorithms, such as NNs, which are based on matrix and vector operations, can profit immensely from the parallel architecture of GPUs, enabling them to process large amounts of data more efficiently than central processing units (CPUs). Furthermore, we have started to produce and store large amounts of easily accessible electronic data throughout the world [21–23], enabling data-driven programming approaches. Since then, progress in the field has enabled realizations of concepts known, so far, only in science-fiction literature, such as self-driving cars or robots mimicking human emotions on their artificial faces (even if we are still far from human-like intelligence [24]). DL has dominated the field of computer vision for years and has found great success in time series analysis, with applications such as stock market and weather forecasting [25]. Another fruitful direction is natural language processing, where sequence-to-sequence models have achieved great feats, even combining text with images [26, 27]. The DL-based algorithms obtained superhuman performance in video games [28, 29] and complex board games, such as Go [30].

[3] Interestingly, we know that actual biological neurons compute very different functions than the perceptrons constituting our modern NNs, but greater realism has not yet led to any improvement in model performance [13].

Overall, the continuous progress in the field of ML is supported by the steady increase in computational power and its easy applicability to real-world problems. The increasing amount of data produced by our society and the monetary benefit of its processing have made the largest technological companies focus enormous economic efforts on the development of ML models. It is, hence, not a coincidence that the most important research groups in the field are associated with such companies. Importantly, one should understand the extent to which the trends of the field are dictated by the thirst for scientific discovery or by the particular needs of one or another technological giant. In summary, ML has become a day-to-day tool, acting in the shades of multiple technological tools we use today [24], with the potential to solve some of the most important problems of the modern world and thus contribute to improving the quality of life of people around the world.

1.3 Learning machines viewed by a statistical physics

It is also worth noting that the above-sketched developments of AI, data science, cognitive science, and neuroscience, related to ML and NN, were also intertwined with the development of the statistical physics of spin glasses and NN. A wonderful retrospective of these developments can be found in the lecture of the late Naftali Tishby, "Statistical physics and ML: A 30-year perspective." Therefore, here we present a similar list of historical milestones as in Section 1.2 but focused on statistical physics achievements:

- 1975 – Philipp W. Anderson and Samuel F. Edwards formulate the Edwards–Anderson spin glass model with short-range random interactions between Ising spins.
- 1975 – A little later, David Sherrington and Scott Kirkpatrick formulate the Sherrington–Kirkpatrick spin-glass model with infinite-range interactions, for which the mean-field solution should be exact. They propose to solve it using the replica trick, but this approximate solution turns out to be clearly incorrect at low temperatures.
- 1979 – Giorgio Parisi proposes an ingenious replica symmetry-breaking solution of the Sherrington–Kirkpatrick model.
- 1982 – John J. Hopfield publishes his seminal paper on attractor NNs, where, by assuming the symmetry of interneuron coupling, he relates the model to a disordered Ising model of N spins, very much analogous to spin glasses. The maximal storage capacity is found to be $0.14N$.
- 1985 – Daniel Amit, Hannoch Gutfreund, and Haim Sompolinski formulate the statistical physics of the Hopfield model and relate limited storage capacity to the spin-glass transition.
- 1987 – Marc Mezard, Giorgio Parisi, and Miguel Angel Virasoro publish the book *Spin Glass Theory and Beyond: An Introduction to the Replica Method and Its Applications*. Interestingly, it is one of the first works bringing together statistical physics and NNs but also putting them in a more general context of complex systems like optimization and protein folding.

- 1988 – Elisabeth Gardner formulates the so-called Gardner's program to ML, where learning abilities are related to the relative volume in the space of those NNs that realize learning tasks and teacher–student scenarios (see Section 8.1.1).
- 1989 – Daniel Amit publishes the book *Modeling Brain Function: The World of Attractor NNs* where he brings closer neurophysiology and artificial NNs by introducing dynamical patterns whose temporal sequence encodes the information.
- 1990 – Géza Györgyi shows that sharp phase transitions from bad to good generalization can occur in learning using Gardner's program on the perceptron.
- 1995 – David Saad, Sara Solla, Michael Biehl, and Holm Schwarze adapt Gardner's idea to study the dynamics of gradient descent in perceptrons and simple two-layer NNs called committee machines.
- Late 2010s – The statistical mechanics predictions for the perceptron and the committee machine start being made mathematically rigorous by Nicolas Macris, Jean Barbier, Lenka Zdeborová, and Florent Krzakala.
- 2010s–today – With the explosion of DLs, interest in the statistical mechanics approach to learning is rekindled. Analyses are developed for increasingly complex models beginning to bridge the gap from perceptrons to deep NNs.
- 2021 – Giorgio Parisi receives the Nobel Prize in Physics "for the discovery of the interplay of disorder and fluctuations in physical systems from atomic to planetary scales."
- 2024 – John J. Hopfield and Geoffrey Hinton receive the Nobel Prize in Physics "for foundational discoveries and inventions that enable machine learning with artificial neural networks."

We discuss the intersection of statistical physics and ML in more detail in Section 8.1.

1.4 Examples of tasks

As stated in Section 1.1, the first ingredient needed for a computer to learn is the notion of a *learning task*. The archetypical ML task is the study of a response variable, $y(x)$, influenced by an explanatory variable x. In principle, there is no restriction on whether y or x or both are continuous, discrete, or even categorical.[4] Throughout the book, we restrict both variables, possibly encoded accordingly, to be of quantitative nature. That is, we can treat variables straightforwardly from a numerical perspective and easily adjust them to fit our needs.

[4]When the inputs are, for example, words in a sentence as they are in the field of natural language processing, we can still process them by representing words by a suitable encoding, which can be either continuous or discrete.

Regression. We start by considering *regression* tasks. In this setting, we typically assume an immediate relationship between the two variables x and y, which is often deterministic. More precisely, we seek to express the variable $\boldsymbol{y}$, also known as (a.k.a.) the output or target, in terms of the variable $\boldsymbol{x}$, a.k.a. the input. In general, both variables can be multidimensional, as indicated by our notation. The objective of regression is to find the function f that yields the mapping $\boldsymbol{y} = f(\boldsymbol{x})$ for all possible tuples of $(\boldsymbol{x}, \boldsymbol{y})$. Of course, from a practical point of view, we can neither optimize over the set of all possible functions nor over the entire domain of $\boldsymbol{x}$. Instead, we resort to a finite dataset for which we opt to find a model that maps every input $\boldsymbol{x}$ to its corresponding target $\boldsymbol{y}$. Usually, the model is predefined up to some parameters,[5] which are tuned to fit the dataset. The most simple model assumes a linear relationship between the input and the output. We give more details of this model archetype in Section 2.4.1. From here, there is a multitude of ways to extend the model by incorporating nonlinear dependencies on both the model parameters and the input $\boldsymbol{x}$. We find interesting regression problems in a large range of study fields, such as sociology (e.g., annual salary as a function of years of work experience), psychology (e.g., perceived happiness relative to wealth), finance (e.g., housing market prices depending on socioeconomic factors), and, of course (quantum) physics and chemistry. We cover some examples in this book, for instance, the prediction of potential energy surfaces (PESs) in quantum chemistry in Section 4.5, or the estimation of Hamiltonian's parameters given the measurement data in Section 7.3.

Classification. Another large class of tasks is *classification*. In this case, our goal is to use an algorithm to assign *discrete* class labels to examples. In contrast to regression, we are optimizing a model to find a mapping from an input vector $\boldsymbol{x}$ to a target $\boldsymbol{y}$, which encodes a representation of the different possible classes. The simplest example of this kind of task is binary classification, in which an algorithm has to distinguish between two classes, for example, true or false. When the task involves more than two classes, we speak of multi-class classification. A canonical example for such a task is the classification of the images of handwritten digits contained in the famous MNIST [31] dataset (named after the Modified National Institute of Standards and Technology) over ten classes, one for each number from zero to nine. Other famous ML classification datasets are Iris [32], CIFAR-10 and 100 [33], and ImageNet [34].[6] A popular example from physics is the classification of different classical and quantum phases of matter, described in Chapter 3. Another set of examples is provided by the classification subroutines in the automation of (quantum) experiments highlighted in Section 7.3.

Both regression and classification tasks require a training dataset consisting of examples of inputs $\boldsymbol{x}$ together with their corresponding labels $\boldsymbol{y}$. Nonetheless, there are also tasks that do not require explicit labels. An example of such is *density estimation*, where the aim is to infer the probability density function of the dataset. This is

[5]There are also nonparametric approaches, for example, see Sections 4.4.2 and 7.2.2.

[6]The Iris database contains 150 data points with four features of three species of iris. The CIFAR-10 dataset consists of 60 000 32×32 color images in 10 classes and was named after the Canadian Institute for Advanced Research. Finally, the ImageNet is a gigantic project with over 10 million labeled images whose most popular subset spans 1 000 object classes.

directly related to the field of *generative problems*, where the goal is to *generate* new data instances that resemble some given input data. The distinction between the two fields is that the latter does not require explicit knowledge or reconstruction of the underlying data distribution to sample new instances. We present more details on density estimation in Section 7.2.

In all the previous cases, we try to infer properties of a given predefined dataset. However, there are other tasks that involve starting from scratch and building a dataset on the fly, from which we can then learn. A paradigmatic example of such a task is learning how to play a game. In this case, we start tabula rasa and progressively build a dataset with the experience gathered as we play the game. From these data (or during their retrieval), our goal is to learn a function that chooses the best possible action or move according to the current state of the game. In this example, we can periodically alternate between collecting experience and learning, or we can do both at the same time.

This list of tasks is, of course, not exhaustive. Other examples that do not directly fall into the previous categories include text translation, imputation of missing values, anomaly detection, and data denoising, to name a few.

1.5 Types of learning

The second learning ingredient is *data*, whose accessibility also often determines the type of learning we have to consider. It is clear, of course, that the notions of task, as presented in Section 1.4, and data are intertwined: Certain tasks can only be solved if sufficient data are available and, in turn, a richer dataset allows us to transfer from one task to another with seemingly low effort. Although the term data is often used for a variety of concepts across many fields, there is a precise definition of it in the ML community. We usually refer to *data* in terms of a dataset $\mathcal{D}$, containing a finite amount of data instances often called *data points* $\boldsymbol{x}_i$, which may be presented as is, that is, $\mathcal{D} = \{\boldsymbol{x}_i\}$ or may be accompanied by predefined labels or targets $\boldsymbol{y}_i$, that is, $\mathcal{D} = \{(\boldsymbol{x}_i, \boldsymbol{y}_i)\}$. To shorten the notation, we also represent the input data points $\{\boldsymbol{x}_i\}$ by a matrix $\boldsymbol{X}$, which can either be stacked row- or column-wise.

Although the notation is clear, there is much less convention and an even lesser understanding of how the data should be *represented*. This is because, on the one hand, the data can be arbitrarily preprocessed (e.g., the data mean is often subtracted prior to any further analysis), which already provides some degree of freedom. On the other hand, even choosing the right descriptors to characterize our object of interest is challenging: Too few might not capture all relevant aspects of the object, whereas too many can lead to spurious correlations that can interfere with the conclusions that we want to draw from data. We refer to each element at each data point $\boldsymbol{x}_i$ as a *feature*. As stated in Section 1.1, a central problem in ML relates to the correct representation of the data and its features. This is the core of the field of representation learning, which we only touch upon, for example, by means of principal component analysis (PCA) and autoencoders (AEs) in Chapter 3 and Section 7.2, respectively.

Finally, we emphasize that data can, loosely speaking, be identified with experience: Data can be produced as a result of a repeated interaction with an entity (such as an experiment or a simulation) that then leaves us with a certain amount of experience about its underlying mechanism. In some cases, this experience may be used to further interact with such an entity and learn from it. To this end, we set up a model. In summary, the type of data to which we have access effectively defines the types of learning with which our model can be faced. These are usually divided into three: supervised, unsupervised, and reinforcement learning (RL).

Supervised learning. *Supervised learning* can be seen as a generalized notion of regression and classification, introduced in Section 1.4, and describes ML algorithms that learn from *labeled* data, that is, $\mathcal{D} = \{(\boldsymbol{x}_i, \boldsymbol{y}_i)\}$. There exist various approaches to supervised learning, ranging from statistical methods to classical ML and DL, both introduced in Section 2.4. The concept of supervised learning appears repeatedly in this book and forms the basis of many chapters, including phase classification (Chapter 3), Gaussian processes (Chapter 4), as well as the selected topics of DL for quantum sciences (Chapter 8). Importantly, some of the latter are specially suited to deal with experimental data, as, for instance, in the efficient readout of quantum dots or the identification of Hamiltonian parameters describing quantum experimental setups. In most of these examples (but there are notable exceptions), large amounts of data are required for the training process. On top of the data, as stated above, supervised learning requires correctly labeled data. This is usually considered one of its most prominent downsides, as perfectly matching labels are not always accessible or have to be added manually by humans.

Unsupervised learning. Supervised learning is not always the best option: the scarcity of labeled data is an example in which a classical input–output design might fail. Instead, we often have access to data where no prior information, for example, in terms of labels, is given (i.e., $\mathcal{D} = \{\boldsymbol{x}_i\}$). In this case, we can employ *unsupervised learning*. Unsupervised learning can either be used for preliminary preprocessing steps, such as dimensionality reduction, or for representation learning, such as in clustering. In contrast, dimensionality can also be increased by adding features via generative models. In this book, we discuss the application of unsupervised learning for phase classification in Chapter 3 and density estimation in Section 7.2. This example is particularly interesting because it demonstrates how the choice of unsupervised learning over supervised learning can aid in the automated discovery of new physics when the interpretation of a process, for example, the nature of two different phases in a transition is unknown.

Reinforcement learning. In contrast to the two previous types of learning, in RL, we usually do not have a dataset available *at all*. Instead, we have an *environment* with which we have to interact to achieve a certain task. This interaction is augmented with feedback, that is, some extra information on whether the action has been beneficial or harmful in achieving the task at hand. The collection of visited environment states, actions taken, and rewards or penalizations received take the role of a dataset. Feedback is very important in RL because we do not have a clear-cut route in achieving our task. In fact, initially, we typically do not even know the necessary ingredients for achieving the task. Often, we only know that we *achieved* a specific goal but not *why* we did it. The field of RL is precisely concerned with tackling the issue of how. To introduce it properly, we devote Chapter 6 to it.

Other types of learning. While supervised, unsupervised, and RL are the most common learning schemes in the ML applications in quantum sciences, there are ML approaches that go beyond this classification. An interesting example is active learning. This field includes selection strategies that allow for an iterative construction of a model's training set in interaction with a human expert or environment. The aim of active learning is to select the most informative examples and minimize the cost of labeling [35, 36]. We only touch on this topic by means of BO[7] in Section 4.3 and local ensembles (LEs) in Section 3.5.3. Another example of learning is semi-supervised learning in which a large amount of unlabeled data is explored to get better feature representations and improve the models trained on the small number of labeled data [37].

1.6 How to read this book

This book aims at providing an educational and self-contained overview of modern applications of ML in quantum sciences. As such, Chapter 2 is devoted to the ML prerequisites that are necessary to fully enjoy all the more advanced and further contents of this book. We discuss in detail four main ML paradigms that have been successfully explored in quantum physics and chemistry: In Chapter 3, we describe how supervised and unsupervised learning can be utilized to classify phases of matter. In Chapter 4, we introduce kernel methods with a special focus on Gaussian processes (GPs) and BO. Chapter 5 presents an overview of various representations of quantum states based on NNs. Finally, in Chapter 6, we dive into the foundations of RL and how it can be applied to quantum experiments.

In addition to these four pillars of ML in quantum sciences, there exists an exciting two-way interplay between the natural sciences and AI. Chapter 7 focuses on more specialized examples of how ML-related methods revolutionize quantum science. In particular, we introduce the paradigm of differentiable programming (∂P) and describe how it is becoming an important numerical research tool. Moreover, we discuss how ML methods assist researchers in tasks related to density estimation, as well as optimizations and speedup of scientific experiments. There exists a vibrant reverse influence on ML coming from statistical physics (which we discuss in Section 8.1) and finally quantum computing. We describe the promises of quantum machine learning (QML) in Section 8.2. All in all, this book discusses the fruitful interplay of AI and physical sciences. Its content with references to relevant sections is illustrated in Fig. 1.3.

We encourage the reader to start with Chapters 1 and 2, that is, "Introduction" and "Basics of Machine Learning." Then, the reader is free to wander into any of the independent Chapters 3–6 covering the four main paradigms, Section 7.1 on

[7] Bayesian optimization (BO) and active learning, while similar, are not the same. Active learning aims to determine optimal sampling, while BO aims to find an extremum of a black-box function with as few function evaluations as possible.

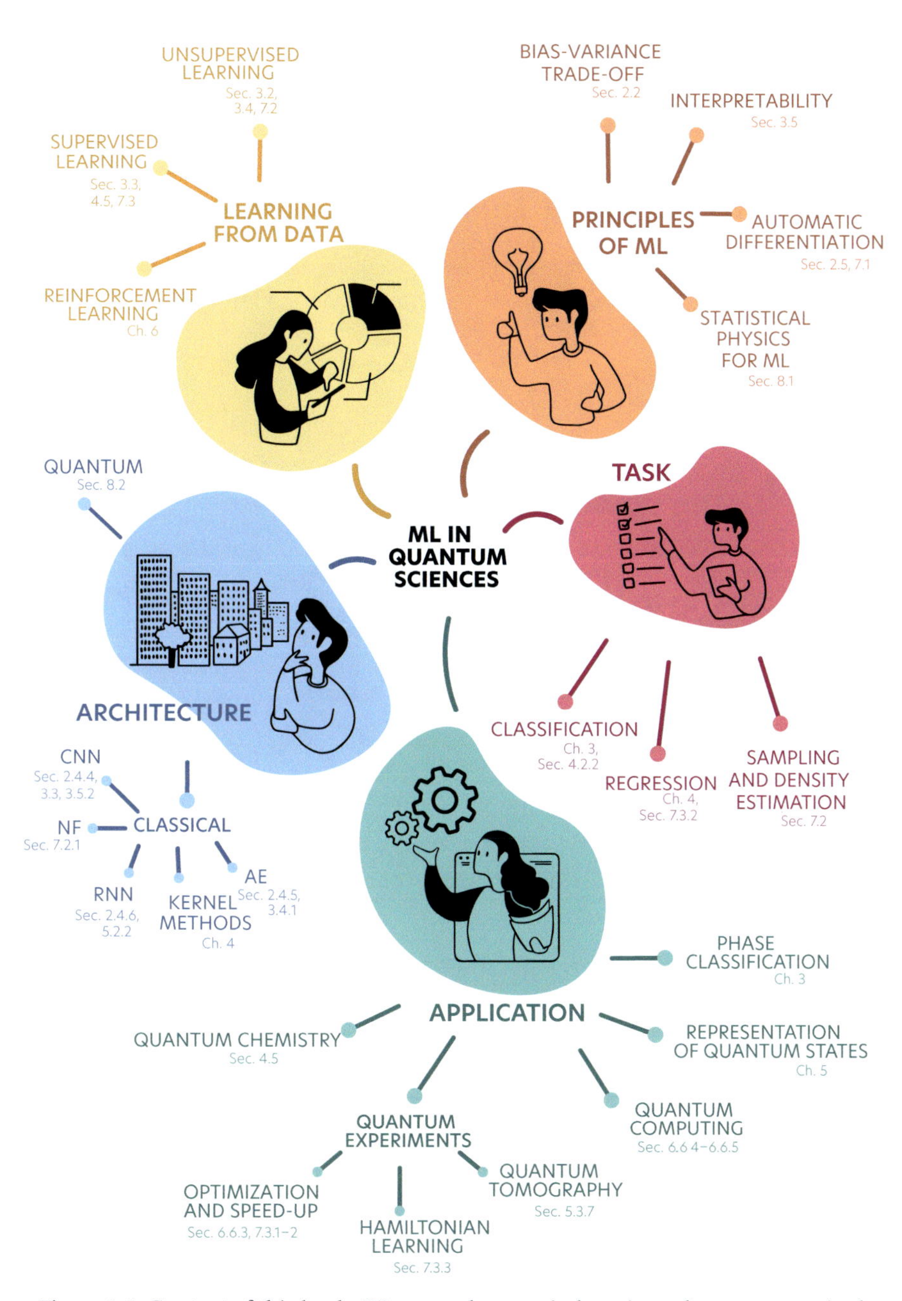

Figure 1.3 Content of this book. We cover three main learning schemes: supervised, unsupervised, and reinforcement learning (RL), examples of ML tasks like classification, regression, and density estimation, various applications in quantum sciences, quantum and classical ML architectures. We also dive into the principles of ML.

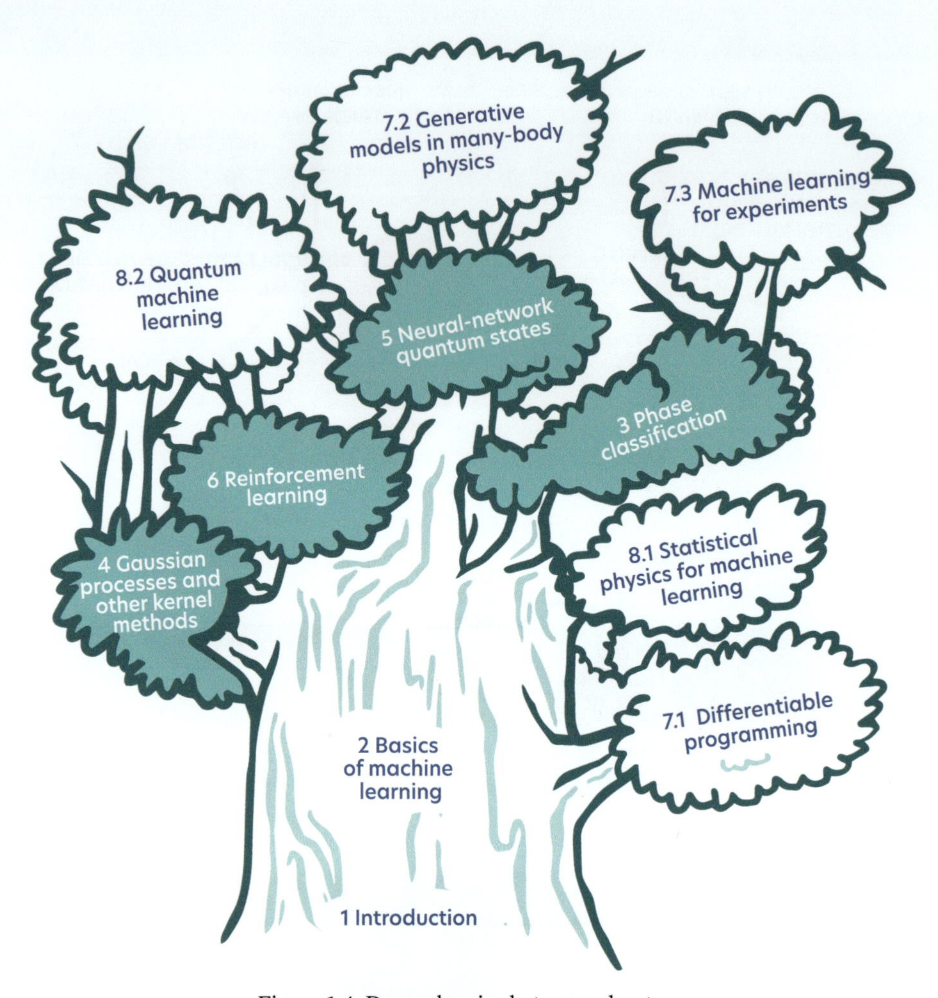

Figure 1.4 Dependencies between chapters.

differentiable programming or Section 8.1 on how statistical physics tackles the puzzles of ML. Section 7.2 on generative models builds on Chapter 5 on neural quantum states (NQSs), while Section 7.3 on ML for experiments requires knowledge of the methods used for phase classification presented in Chapter 3. Finally, Section 8.2 on QML utilizes concepts introduced in Chapters 4–6. The dependencies between chapters are visualized as a tree in Fig. 1.4.

Further reading

- Carleo, G. *et al.* (2019). *Machine learning and the physical sciences.* Rev. Mod. Phys. 91, 045002. This detailed review summarizes the development of ML in physics and its achievements till 2019 [5].

- Carrasquilla, J. (2020). *Machine learning for quantum matter.* Adv. Phys. X 5, 1. A concise review focusing on phase classification and quantum state representation [6].
- Krenn, M. *et al.* (2023). *Artificial intelligence and machine learning for quantum technologies.* Phys. Rev. A 107(1), 010101. A perspective focusing on how quantum computing, quantum communication, and quantum simulation benefit from the ML revolution [38].
- Chollet, F. (2019). *On the measure of intelligence.* The review of different measures used to quantify *intelligence*, which provides a perspective on AI development [39].
- Krenn, M. *et al.* (2022) *On scientific understanding with artificial intelligence.* Nat. Rev. Phys. 4, 761–769 [40]. A beautiful paper discussing ways in which AI could contribute to scientific discovery. It touches upon the philosophy of understanding and draws conclusions from dozens of anecdotes from scientists on their computer-guided discoveries.
- Recordings of lectures of the Summer School: Machine Learning in Quantum Physics and Chemistry which took place between August 23 and September 03, 2021, in Warsaw, Poland.
- Jupyter notebooks prepared as tutorials for the Summer School: Machine Learning in Quantum Physics and Chemistry [2].

2 Basics of machine learning

In this section, we describe basic machine learning (ML) concepts connected to optimization and generalization. Moreover, we present a probabilistic view on ML that enables us to deal with uncertainty in the predictions we make. Finally, we discuss various ML models. Together, these topics form the ML preliminaries needed for understanding the contents of Chapters 3–9.

2.1 Learning as an optimization problem

We have already discussed that ML can solve various tasks (e.g., classification or regression) and that there are different ways for the machine to access the data. The final ingredient is a model that learns how to solve the given task with the data at hand. In general, it is a function of the input data, $f(\boldsymbol{x})$, whose output is interpreted as a prediction made for the input data. The form of the output depends on the task. It can be, for example, a class from a discrete set of possible classes in the classification task or a tensor from a continuous target distribution in the regression task. Finding the function that provides the best mapping between the data and the desired outcome for a specific task is at the heart of ML. We start with declaring a certain parametrization of a model (function), for example, $f(\boldsymbol{x}) = \boldsymbol{w}^{\intercal}\boldsymbol{x} + \boldsymbol{b}$ with $\boldsymbol{\theta} \supset \{\boldsymbol{w}, \boldsymbol{b}\}$. Then, all possible parametrizations of this function form the set of functions, that is, the hypothesis class. Section 2.4 presents specific examples of the hypothesis classes (or spaces), but for now, we focus on the learning process itself.

The mentioned learning schemes, that is, supervised, unsupervised, or reinforcement learning, have the same underlying process of learning: finding an optimal model $\hat{f} \equiv f_{\boldsymbol{\theta}^*}$ with optimal parameters $\boldsymbol{\theta}^*$ in the hypothesis space, which minimizes the target loss function or maximizes a model performance. For the remainder of this section, for clarity, we focus on minimizing the loss function, $\mathcal{L}$, which intuitively plays a role of a penalty for errors of a model.

> Machines "learn" by minimizing the loss function of the training data, that is, all the data accessible to the ML model during the learning process. The minimization is done by tuning the parameters of the model. The loss function formula varies between tasks, and there is a certain freedom in how it can be chosen. In general, the loss function compares model predictions or a developed solution against the reality or expectations. Therefore, learning becomes an optimization problem.

In this book, we use the terms of loss, error, and cost functions[1] interchangeably following [13]. Popular examples of loss functions include the mean-squared error

[1]The literature also uses the terms of criterion or cost, error, or objective functions. Their definitions are not very strict. Following [13]: "The function we want to minimize or maximize is called the objective function, or criterion. When we are minimizing it, we may also call it the cost function, loss function, or

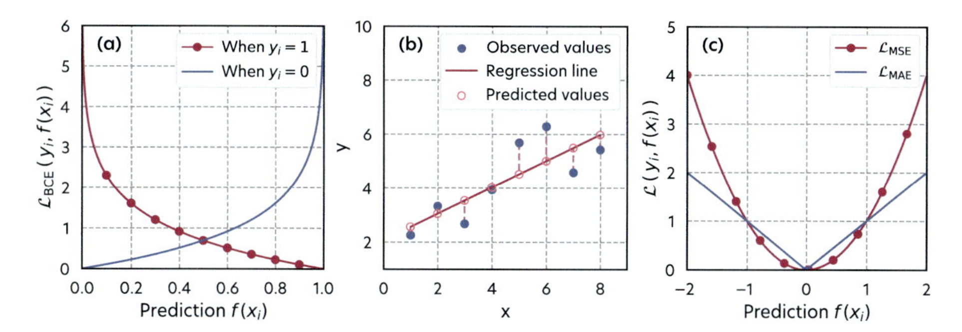

Figure 2.1 Examples of loss functions. (a) Plot of the binary cross-entropy (BCE) for a single data point, x_i, when the ground-truth label is $y_i = 0$ (blue) or 1 (purple). (b) The intuition behind loss functions used in regression problems. Dashed lines are differences between the labels, y_i, and values predicted by a model, $f(x_i)$. (c) Plots of the MSE (purple) and MAE (blue) for a single data point, x_i, when the ground-truth label $y_i = 0$.

(MSE) and the cross-entropy (CE), used for supervised regression and classification[2] problems. The output of the loss function depends on the model (which enters into formulas via predictions) and the dataset. They are also normalized by the number of data points n to compare their values between problems with different dataset sizes. The MSE is a popular loss function inherited from linear regression problems and is defined as

$$\mathcal{L}_{\text{MSE}} = \frac{1}{n}\sum_{i=1}^{n}(y_i - f(\boldsymbol{x}_i))^2 . \tag{2.1}$$

It has an information-theoretic justification discussed in more detail in Sections 2.2 and 2.4.1. In the former, we also introduce the mean absolute error (MAE) as another viable loss function, which is more sensitive to small errors than the MSE as shown in Fig. 2.1(b) and (c). CE is also a concept drawn from information theory and has connections to probability theory (see Section 2.3). In the binary classification task, we can use the binary CE (BCE), also known as the log loss (Eq. (2.2) and panel (a) in Fig. 2.1), while for the multiclass classification, we use the categorical CE (CCE). They are defined as

$$\mathcal{L}_{\text{BCE}} = -\frac{1}{n}\sum_{i=1}^{n} y_i \cdot \log(f(\boldsymbol{x}_i)) + (1 - y_i) \cdot \log(1 - f(\boldsymbol{x}_i)) , \tag{2.2}$$

$$\mathcal{L}_{\text{CCE}} = -\frac{1}{n}\sum_{i=1}^{n}\sum_{c=1}^{K} y_{i,c} \cdot \log(f(\boldsymbol{x}_i)) , \tag{2.3}$$

error function. In this book, we use these terms interchangeably, though some ML publications assign special meaning to some of these terms." For example, the loss function may be defined for a single data point, the cost or error function may be a sum of loss functions, so check the definitions used in each paper.

[2]For classification, a more intuitive measure of the performance could be, for example, accuracy, which is the ratio between the number of correctly classified examples and the dataset size. Note, however, that gradient-based optimization requires smooth and differentiable performance measures. These conditions distinguish loss functions from evaluation metrics such as accuracy, recall, and precision.

where K is the number of classes. This formula requires representing labels in a way called *one-hot encoding*. For example, in a K-class problem, instead of having a label with K possible values such as $y_i = 1, 2, \ldots, K$, each label is encoded as a K-element vector with all-zero elements except for one at the index corresponding to the class. For example, $\boldsymbol{y}_i = [0, 0, 1, \ldots, 0]$, means a sample i belongs to the third class, as only $y_{i,3}$ is nonzero.

Once we choose a loss function, we can minimize it by varying the parameters of the ML model, using any optimization method of our choice. In general, we can find the minimum of the loss function either via analytical construction or optimization methods that can be either gradient-based or gradient-free. A popular example of a gradient-based method is *gradient descent*. Optimization usually starts in a random place within the loss landscape (meaning with a model with randomly initialized parameters, $\boldsymbol{\theta} = \boldsymbol{\theta}_0$).[3] Using the model with $\boldsymbol{\theta}_0$, one makes predictions over the training data and from them computes the loss function. The next step consists of computing the gradients of the loss function with respect to each model parameter, θ_j. The final step is to update the parameters by subtracting the respective gradients multiplied by a learning rate, η, that is,

$$\theta_j := \theta_j - \eta \frac{\partial \mathcal{L}}{\partial \theta_j}. \tag{2.4}$$

These steps need to be repeated until the minimum is reached, and each repetition is called an *epoch*. The intuition is that gradient descent updates model parameters by taking steps toward the minimum of the function (so in the opposite direction than the gradient, which indicates where the function value grows). The learning rate controls the size of these steps. Figure 2.2 presents in a simplified way the importance of the η choice. Both too large and too small η make optimization more challenging, and only an optimal η promises efficient convergence to a minimum. There is rarely an obvious way of choosing η that, therefore, has to be found, for example, by trial and error. As such, the learning rate is one of the so-called *hyperparameters* of the learning process. Hyperparameters are parameters whose values control the learning process (especially the speed of convergence and the quality of the minimum) and are chosen by a user (in contrast to model parameters, which are derived through training). The total number of epochs or the choice of the loss function are hyperparameters, too. We encounter more examples of hyperparameters in this introductory chapter.

To find optimal hyperparameters, we should form (in addition to the training dataset) a separate *validation dataset*. These data are only used to *validate* the model and not for training. Then, we can set various hyperparameters and choose them in such a way that the error on the validation set is minimized.[4] Dividing the dataset into smaller subsets can be problematic in the case of a limited number of data. Alternative approaches for model validation exist, like k-fold *cross-validation* [13], which consists of splitting the dataset into k nonoverlapping subsets. The validation error

[3] In practice, parameters are usually initialized randomly, but with the constraint to have a mean at zero and constant variance across layers, otherwise, we may encounter problems with vanishing or exploding gradients [41].

[4] One can even use optimization methods to find optimal hyperparameters which minimize the validation error (a popular library is Optuna [42]), but a choice of hyperparameters guided by intuition may prove to be a faster and cheaper approach.

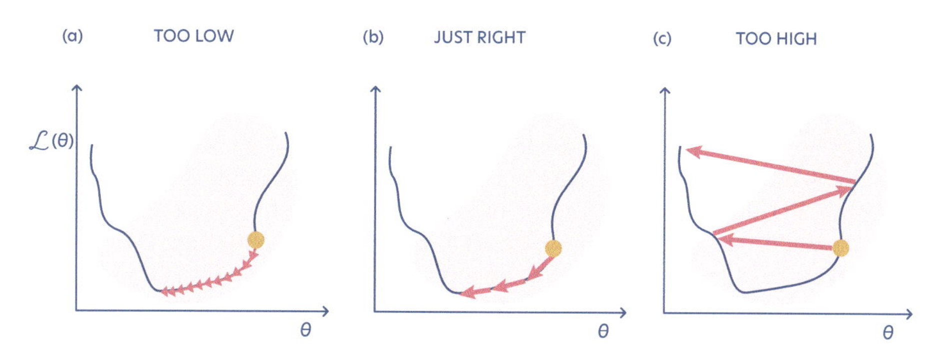

Figure 2.2 Choosing a learning rate has an impact on convergence to the minimum. (a) If η is too small, the training needs many epochs. (b) The right η allows for a fast convergence to a minimum and needs to be found. (c) If η is too large, optimization can take you away from the minimum (you *overshoot*). This figure suggests that the loss function is convex which is rarely true.

can then be estimated by taking the average error over k trials where the ith trial uses ith subset as a validation set and the rest as training data. Note that cross-validation comes at the price of increased computational cost.

Returning to the gradient descent, note that to perform it, we must first compute the gradient of the loss function with respect to the parameters to be tuned, $\nabla_\theta \mathcal{L}$, before each step, see Eq. (2.4). A priori, there exist several different approaches to compute these derivatives. For example, one could work out the analytical derivatives by hand or approximate them numerically based on finite differences. When we are concerned with the accurate numerical evaluation of derivatives and not their symbolic form, automatic differentiation (AD) is a good choice. AD makes use of the fact that computer programs that compute the corresponding loss function can be decomposed into a sequence of a handful of elementary arithmetic operations (e.g., additions or multiplications) and functions (e.g., exp or sin). Therefore, the numerical value of the derivative of the program, that is, the loss function, can be computed in an automated fashion by repeated applications of basic predefined differentiation rules, such as the chain rule,

$$\frac{df(g(x))}{dx} = f'(g(x))g'(x). \tag{2.5}$$

For more details on how to compute derivatives of computer programs, in particular AD, see Section 7.1.[5]

The optimization procedure that we have described in the previous paragraphs and in Fig. 2.2 is very efficient when the *loss landscape*, that is, the representation of the loss values around the parameter space of the model, is convex. However, especially for DL, loss landscapes are highly nonconvex and usually exhibit multiple local minima [43, 44]. Two immediate questions arise from this nonconvexity: First, how can one avoid getting stuck in local minima corresponding to large loss

[5]The special case where AD is applied in *reverse-mode* to NNs is known as backpropagation and constitutes the workhorse that enables efficient NN training.

function values or in saddle points of such landscapes? Second, are some minima better than others? Currently, such questions concerning learning dynamics are still being explored in various ongoing research directions, but some intuitions are already provided by statistical physics (see Section 8.1). A popular approach to deal with the aforementioned problems considers a slight modification of the gradient descent algorithm, so-called *stochastic gradient descent (SGD)* [45]. This optimization method (whose pseudocode is provided in Algorithm 1) consists of computing the loss function at each epoch on randomly selected mini-batches (subsets) of the training data. This means that during each epoch, the gradients may point in various directions. Effectively, the resulting stochasticity has been shown to help escape saddle points and narrow local minima [46].[6] Furthermore, computing the loss function and gradients only for a mini-batch of data instead of the whole dataset provides a nice computational speedup for large datasets.

Let us examine the minimum reached during the optimization of DL models in more detail. To do that and to describe the curvature around such a minimum, we use the Hessian of the training loss function, $\boldsymbol{H}_{\theta^*} = \frac{\partial^2}{\partial\theta_i\theta_j}\mathcal{L}_{\text{train}}|_{\theta=\theta^*}$, that is, the square matrix of second-order partial derivatives of $\mathcal{L}$ with respect to the model parameters, calculated at the minimum, $\theta = \theta^*$. The eigenvectors of $\boldsymbol{H}_{\theta^*}$ corresponding to the largest positive eigenvalues indicate the directions with the steepest ascent around the minimum. A high curvature implies that the training data strongly determine the model parameters along that direction. What may be surprising is that the training of an ML model leads to a local minimum or a saddle point[7] [49–51]: The vast

Algorithm 1 Minibatch stochastic gradient descent (SGD)

Require: Learning rate η
Initialize θ to random values
for epoch = 1 to no_epochs **do**
Shuffle $\mathcal{D}_{\text{train}}$
for i = 1 to m (where m is a minibatch size) **do**
$\boldsymbol{x}_i, y_i \sim \mathcal{D}_{\text{train}}$ ▷ Draw random data point from dataset without replacement
$\mathcal{L} \leftarrow \frac{1}{m}\sum_{i=1}^{m}\mathcal{L}(y_i, f(\boldsymbol{x}_i))$ ▷ Compute loss function on the minibatch
$(\nabla\mathcal{L})_j \leftarrow \frac{\partial\mathcal{L}}{\partial\theta_j}$ ▷ Compute gradients
$\theta_j \leftarrow \theta_j - \eta\frac{\partial}{\partial\theta_j}\mathcal{L}$ ▷ Update parameters
end for
end for
return θ

[6] In practice, stochasticity is helpful in avoiding saddle points but theoretical works show it is not a necessary condition for a proper convergence [47].

[7] One can wonder why we should trust a model that does not land in the global minimum. A series of empirical results as well as applying spin-glass theory to deep learning [48] indicate, among others, that for large networks, most local minima are equivalent and yield similar performance on a test set. Also, the probability of finding a *bad* (high value) local minimum is nonzero for small networks and decreases

majority of the eigenvalues are close to zero, indicating various flat directions, and some small negative eigenvalues are also present, indicating directions with negative curvature. We present more examples of what information one can gain from $\boldsymbol{H}_{\theta^*}$ in Section 3.5.3.

Up to this point, the only gradient-based optimization method we have described is SGD. Popular alterations to this scheme consist of, for example, including a momentum term that takes previous update directions into account [52, 53] or adaptive learning rates between epochs [54] or both, culminating in the celebrated Adam optimizer [55, 56]. Another different idea is to incorporate the second derivative in the update rule, as is accomplished by the limited-memory Broyden–Fletcher–Goldfarb–Shanno algorithm (L-BFGS) algorithm [57]. There are also gradient-free optimization approaches that are used, especially when the gradients or loss function itself are expensive or impossible to compute, for example, when optimizing experiments. Examples include genetic algorithms, particle swarm optimization, random search, and simulated annealing [58]. Another example we discuss in more detail in Section 4.3 is Bayesian optimization.

2.2 Generalization and regularization

So far, ML may seem like a function fitting in disguise. This changes when we go beyond simply trying to maximize the performance of a model on the available data.

> The heart of ML lies in *generalization*, which is the ability to make accurate predictions on new data, never seen during the training.

The ability to generalize can be quantified with the generalization error. The generalization error is the expected error of a model on new data drawn from the distribution of input/output pairs we expect the model to encounter in practice [13]. However, such a distribution is generally inaccessible. Therefore, we approximate the generalization error of an ML model by measuring its performance on an additional held-out dataset, commonly referred to as the *test set*, composed of data points that are not used either to optimize the model parameters or to search for the best hyperparameters characterizing the learning process. The error made on the test set, called the test error, serves as a tractable measure of the generalization ability of the model and is only used to report the final performance of the model.[8] Therefore, the original dataset needs to be separated into a training, a validation, and a test set.[9] One needs to be particularly careful in the preparation of these datasets to prevent *information leakage*, that is, the use of information in the training process that is not expected to

quickly with network size. Finally, attempting to find the global minimum on the training set (as opposed to one of the many good local ones) is not useful in practice and may lead to overfitting, that is, much better performance on the training set than on the test set, which is equivalent to bad generalization.

[8] We need the test set because the performance as evaluated on the validation set may be overestimated because we use it to find the best hyperparameters of the learning process.

[9] The ratio between the sizes of these sets depends on how much data is available in total, but we suggest starting with, for example, 8:1:1.

be available at prediction time.[10] Also note that we always assume all data points to be drawn independently from the same distribution.[11] This ensures two things: first, the samples in our dataset are uncorrelated; and second, we can split the dataset into smaller subsets.

A common feature of the training of an ML model is a higher test error than the training error. Their difference is a proxy for the generalization gap, which is the difference between the training error and the generalization error. This lower model performance on the test set compared to the training set persists even when all data points are generated by an identical probability distribution, and it only disappears in the infinite data limit. The main reason is the large capacity of DL models.[12]

The *capacity* can be loosely understood as the measure of a model's ability to fit a variety of functions. When the capacity of the model is much higher than the required to solve the task, the model tends to *overfit*, that is, memorize all possible properties of the training set, which may not be true for the general distribution (and particularly, the test set).

In particular, the model can even fit the noise in the training data. As a result, overfitting increases the test error while keeping the training error low (or even decreasing it). An optimal capacity provides the lowest generalization error, minimizing the gap between the test and the training error. However, a capacity that is too low results in an overly constrained model that can *underfit*, that is, have a high training error. The intuition behind the under- and overfitting is schematically shown in Fig. 2.3. Therefore, we can improve the generalization of the model by controlling its capacity.

Every modification of the model aiming to improve its generalization, even at the cost of increasing the training error, is called a *regularization* technique.

One can think of regularization in terms of the Occam razor.[13] The additional motivation to use regularization is the *no free lunch theorem*, which states that, when averaged over all possible data generating distributions, every classification algorithm has the same error rate when classifying previously unobserved points [13, 61]. Therefore, no ML model is universally better than another; and no regularization technique

[10] A common mistake is to normalize the whole dataset first and then separate it into a training, a validation, and a test set. Normalization contains information about the most extreme data points, which may not even be part of the training set. This information can be exploited by the model to achieve better performance on the available data. Thus, the reported test error may not be a faithful indicator of the performance of the model on unseen data.

[11] In practice, a trained model may be confronted with a sample from a distribution different than the one that generated training samples. In such a scenario, models are known to be overconfident [59], which makes them unreliable. As such, the detection of out-of-distribution samples is an essential challenge in the deployment of ML in safety-critical applications.

[12] DL models are even able to fit large datasets with random labels [60]!

[13] This principle states that among competing hypotheses that explain known observations equally well, one should choose the simplest one. It is sometimes summarized as "entities should not be multiplied beyond necessity."

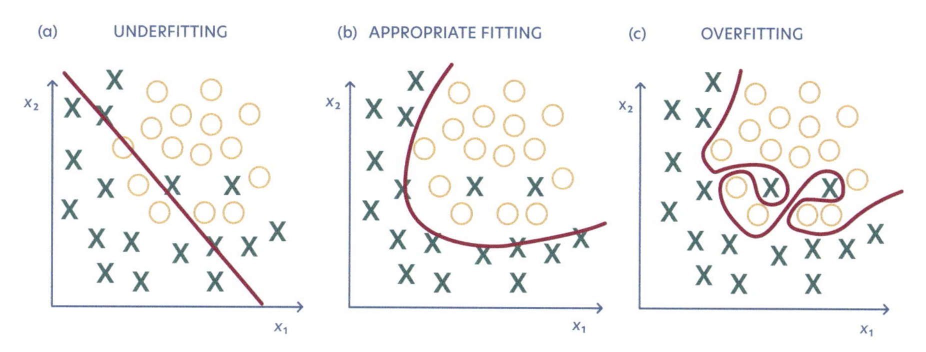

Figure 2.3 Scheme of under- and overfitting. (a) When the model capacity is too low, the model cannot fit the training data properly. (b) With the model capacity corresponding to the task complexity, the fitting is optimal. (c) When the model capacity exceeds the task complexity, the model tends to overfit, and the generalization error increases.

is universally better than another. This implies that we need to design our ML algorithms to perform well on a specific task, for example, by regularizing it in a way that is tailored to this task.

A straightforward way of restricting the model's capacity is to limit the magnitude of its trainable parameters, which effectively limits the hypothesis space of a parametrized model. This can be done by adding a penalizing term to the training loss function, which increases with the parameters' magnitude. Such an approach is used within the two popular regularization techniques, that is, ℓ_1 and ℓ_2 regularization. In particular, ℓ_2 regularization is described in more detail in Sections 2.4.1 and 4.2.1.

Until now, we have discussed the relationship between a model's complexity and its performance on the training and test set in intuitive terms. In the following, we formalize this intuition through the *bias-variance trade-off*. Consider the standard situation encountered in regression problems: We are given an ensemble of data points $\mathcal{D} = \{\boldsymbol{x}, f(\boldsymbol{x}) + \epsilon\}$ that derives from the function $f(\boldsymbol{x})$ and some noise ϵ inherent in the data. The function $f(\boldsymbol{x})$ is generally unknown, and our goal is to infer it. We do this by constructing a regression fit of the data $\hat{f}(\boldsymbol{x})$. What we are ultimately interested in is for the test error (or generalization error) to be as small as possible. The test error is given as an average of the loss function $\mathcal{L}$ evaluated over test points,

$$\mathrm{Err}_{\mathcal{T}} = \mathbb{E}\left[\mathcal{L}(\boldsymbol{y}, \hat{f}(\boldsymbol{x})) \mid \mathcal{T}\right], \tag{2.6}$$

where $\mathcal{T}$ is a fixed training set. This quantity is difficult to calculate, and we can instead resort to the expected prediction error obtained by averaging the generalization error over many training sets,

$$\mathrm{Err} = \mathbb{E}\left[\mathcal{L}(\boldsymbol{y}, \hat{f}(\boldsymbol{x}))\right] = \mathbb{E}\left[\mathrm{Err}_{\mathcal{T}}\right]. \tag{2.7}$$

Let us look at the expected prediction error at a given point $\boldsymbol{x}_0$

$$\mathrm{Err}(\boldsymbol{x}_0) = \mathbb{E}\left[\left(f(\boldsymbol{x}_0) + \epsilon - \hat{f}(\boldsymbol{x}_0)\right)^2\right], \tag{2.8}$$

where, for now, we consider an MSE as the loss function (Eq. (2.1)). Averaging in Eq. (2.8) is performed over all random variables inside the expression $\mathbb{E}[\cdot]$, namely, the noise ϵ as well as the model through the choice of different training sets. We can expand this expression as

$$\mathrm{Err}(\boldsymbol{x}_0) = \mathbb{E}\left[\left(f(\boldsymbol{x}_0) - \hat{f}(\boldsymbol{x}_0)\right)^2\right] + \mathbb{E}\left[2\epsilon(f(\boldsymbol{x}_0) - \hat{f}(\boldsymbol{x}_0))\right] + \mathbb{E}\left[\epsilon^2\right], \tag{2.9}$$

where $\mathbb{E}\left[\epsilon^2\right]$ is the (fixed) variance of the underlying noise in the data. Next, we use the property of independent random variables $\mathbb{E}[AB] = \mathbb{E}[A]\,\mathbb{E}[B]$ to obtain

$$\mathbb{E}\left[2\epsilon(f(\boldsymbol{x}_0) - \hat{f}(\boldsymbol{x}_0))\right] = 2\,\mathbb{E}[\epsilon]\,\mathbb{E}\left[f(\boldsymbol{x}_0) - \hat{f}(\boldsymbol{x}_0)\right] = 0, \tag{2.10}$$

where we assumed unbiased noise $\mathbb{E}[\epsilon] = 0$. Thus, we are left with

$$\mathbb{E}\left[\left(f(\boldsymbol{x}_0) - \hat{f}(\boldsymbol{x}_0)\right)^2\right] + \mathbb{E}\left[\epsilon^2\right] = \mathbb{E}\left[\left(\hat{f}(\boldsymbol{x}_0)\right)^2\right] - 2f(\boldsymbol{x}_0)\,\mathbb{E}\left[\hat{f}(\boldsymbol{x}_0)\right] + \left(f(\boldsymbol{x}_0)\right)^2 + \mathbb{E}\left[\epsilon^2\right]. \tag{2.11}$$

We modify Eq. (2.11) by adding and subtracting $\mathbb{E}\left[\hat{f}(\boldsymbol{x}_0)\right]\mathbb{E}\left[\hat{f}(\boldsymbol{x}_0)\right]$ to get

$$\mathrm{Err}(\boldsymbol{x}_0) = \left(\mathbb{E}\left[\hat{f}(\boldsymbol{x}_0)\right] - f(\boldsymbol{x}_0)\right)^2 + \left(\mathbb{E}\left[\left(\hat{f}(\boldsymbol{x}_0)\right)^2\right] - \mathbb{E}\left[\hat{f}(\boldsymbol{x}_0)\right]\mathbb{E}\left[\hat{f}(\boldsymbol{x}_0)\right]\right) + \mathbb{E}\left[\epsilon^2\right]. \tag{2.12}$$

We can identify the first term as the squared bias of our model

$$\mathrm{Bias}^2[\hat{f}(\boldsymbol{x}_0)] := \left(\mathbb{E}\left[\hat{f}(\boldsymbol{x}_0)\right] - f(\boldsymbol{x}_0)\right)^2, \tag{2.13}$$

and the second as its variance

$$\mathrm{Var}[\hat{f}(\boldsymbol{x}_0)] := \mathbb{E}\left[\left(\hat{f}(\boldsymbol{x}_0)\right)^2\right] - \mathbb{E}\left[\hat{f}(\boldsymbol{x}_0)\right]\mathbb{E}\left[\hat{f}(\boldsymbol{x}_0)\right]. \tag{2.14}$$

This results in

$$\mathrm{Err}(\boldsymbol{x}_0) = \mathrm{Bias}^2[\hat{f}(\boldsymbol{x}_0)] + \mathrm{Var}[\hat{f}(\boldsymbol{x}_0)] + \mathbb{E}\left[\epsilon^2\right]. \tag{2.15}$$

The average prediction error at a given unseen test point $\boldsymbol{x}_0$ can therefore be decomposed into the bias of our model, its variance as well as the variance of the noise underlying our data (which is irreducible from a model perspective).[14]

> The more complex a model $\hat{f}(\boldsymbol{x})$ is, the lower its bias is after training. However, the increased model complexity generally also results in larger fluctuations in capturing the data points, resulting in a larger variance – a situation we refer to as overfitting. This is the bias-variance trade-off.

Figure 2.4 shows an illustration of the bias-variance trade-off, which makes clear that the ideal model realizes an optimal trade-off between the training error and

[14]This does not only hold for an MSE loss as many variations of the bias-variance decomposition are known [62].

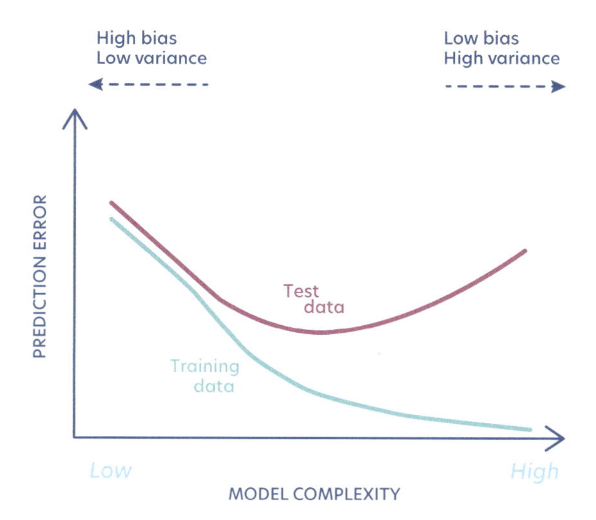

Figure 2.4 Illustration of the bias-variance tradeoff and its relation to the prediction error observed on training (green curve) and test sets (red curve). The ideal model, which results in the lowest test error, has both intermediate model complexity (e.g., capacity) and training error. Adapted from [63] with publisher permission.

the model complexity. Interestingly, empirical studies indicate that modern large DL models with enormous capacities can generalize very well [64]. How over-parametrized models can generalize so well remains a challenging puzzle of the field[15]. but some insight is provided with tools of statistical physics (see Section 8.1).

2.3 Probabilistic view on machine learning

The need for a probabilistic approach to ML becomes apparent when we consider that this field has to tackle three sources of uncertainty (following [13]). First, there may be an inherent stochasticity of the system that generates the data we have access to (especially when dealing with quantum data). Second, we need to account for a possible incomplete observability, that is, an unavoidable lack of information regarding all possible variables that influence the system.[16] In other words, we have only partial access (by means of the available data) to all relevant parts of the mechanism or distribution underlying the system. Finally, the models we use are rarely complete and need to discard some available information. An example of incomplete modeling may be a robot whose movement space we discretize. Such a discretization immediately makes the robot uncertain about *omitted* parts of the space. To mathematically

[15] Promising observations are provided by the lottery ticket hypothesis [65].

[16] This is, in fact, a feature and not a bug: For example, we easily understand the rotation of the earth around the sun due to its gravitational interaction. However, predicting the *exact* orbit of the earth would require us to take into account all other gravitational masses in the solar system as well. Unless we intend to send a satellite into space, we are very happy to neglect these other interactions in favor of only a small error in our predictions.

account for the uncertainty of a model, we can follow the so-called Bayesian approach to probability, which interprets the probability as an expectation or quantification of a *belief.*

In this section, we provide a concise reminder of basic concepts from probability theory which appear in the rest of this book:

- Discarding any mathematical rigor, random variables are variables taking on random values. If they are independent and identically distributed (i.e., drawn independently from the same probability distribution), they are called i.i.d. random variables.
- A *probability distribution* is a measure of how likely a random variable X is to take on each of its possible states x,[17] for example, $p(X = x)$. A probability distribution over discrete (continuous) variables is called a probability mass function (probability density function). A *joint* probability distribution is a probability distribution over many variables at the same time and is denoted, for example, as $p(X = x, Y = y)$. When the notation is clear, we typically also drop the random variable and just write $p(X = x) \equiv p(x)$ instead.
- Two random variables X and Y are *independent* if their joint probability distribution can be expressed as a product of two factors, one involving only X and one involving only Y:

$$\forall x, y, \quad p(X = x, Y = y) = p(X = x)p(Y = y). \tag{2.16}$$

  You can denote this independence by $X \perp Y$.
- A vector whose elements consists of random variables is called a *random vector*, and we denote it simply with $\boldsymbol{x}$.
- A *conditional* probability is a probability of one event given that some other event has happened. We denote the conditional probability with $p(Y = y \mid X = x)$, meaning the probability of $Y = y$ given the observation that $X = x$. It can be calculated as:

$$p(Y = y \mid X = x) = \frac{p(Y = y, X = x)}{p(X = x)}. \tag{2.17}$$

- Any joint probability distribution over many random variables may be decomposed into conditional distributions over only one variable each, which is called the *chain rule* or *product rule* of probability:

$$p\left(x^{(1)}, \ldots, x^{(n)}\right) = p\left(x^{(1)}\right) \prod_{i=2}^{n} p\left(x^{(i)} \mid x^{(1)}, \ldots, x^{(i-1)}\right). \tag{2.18}$$

- Finally, let us discuss a situation where we know the conditional probability $p(y \mid x)$ and need to know the opposite one, $p(x \mid y)$. Fortunately, if we also know $p(x)$, we can compute the desired quantity using *Bayes' rule*:

$$p(x \mid y) = \frac{p(y \mid x)p(x)}{p(y)}. \tag{2.19}$$

[17]This notation is easily generalized to vector-valued random variables.

Bayes' rule is a direct consequence of the definition of conditional probability in Eq. (2.17). If we do not know $p(y)$, we can compute it via $p(y) = \sum_x p(y \mid x)p(x)$, the *sum rule* of probabilities. Coming back to the notion of *belief* in Bayesian statistics, we can give the other terms of Eq. (2.19) a clear interpretation. In this theory, $p(x)$ encodes our prior knowledge about a proposition x, that is, modeled *without* any evidence y collected. Hence, $p(x)$ is called the *prior*. Consequently, Bayes' rule gives us the recipe for how to update our beliefs using the likelihood $p(y \mid x)$ to arrive at the *posterior* $p(x \mid y)$. The posterior now models our updated knowledge about the proposition x that takes the evidence y collected into account.

We are now armed with enough tools to look at ML models in a probabilistic way. In particular, we can reformulate the definition of supervised and unsupervised learning. Unsupervised learning consists of observing some outcomes of a random variable X, for example, $x_1, x_2, \ldots, x_n$, and then learning the probability distribution $p(X)$ or some of its properties.[18] Supervised learning is about observing instances of a random variable X and an associated variable Y, for example, $\{x_1, y_1\}, \{x_2, y_2\}, \ldots, \{x_n, y_n\}$, and learning to predict y from x, usually by estimating $p(Y = y \mid X = x)$ from data. The so-called *Bayes classifier* bases its predictions on the true conditional probability $p(Y = y \mid X = x)$, that is, predicts the label $y_{\text{Bayes}} = \arg\max_y p(y \mid X = x)$ given the sample x. It is *optimal* as there exists no other classifier that outperforms it in the classification task at hand (i.e., that achieves a lower misclassification probability) [66]. However, even a Bayes classifier can be wrong and may only achieve a nonzero misclassification probability. This irreducible error (which is achieved by a Bayes classifier) is called *Bayes error* and is inherent to the classification task under consideration, that is, is a fundamental limit independent of the choice of the predictive model. A classification problem has a nonzero Bayes error if there exist identical samples x that are given distinct labels y, resulting in the class-conditional probabilities $p(y \mid X = x)$ being different from 0 and 1.

In the following, we now seek to combine this probabilistic view with our notion of learning as an optimization task in Section 2.1. From our considerations above, we now understand that ML models are used to estimate probability distributions given data. Because ML models are typically parametrized, the concept of likelihood must enter the picture. The *likelihood function* is the joint probability of the observed data as a function of the parameters of the chosen model, $p(\mathcal{D} \mid \boldsymbol{\theta})$, estimating the data-generating probability distribution.[19] The likelihood provides us with the missing link between the given data for which we would like to infer, for example, $p(\mathcal{D})$, and the parameters of our model parameters that we want to learn. To this end, it is useful to consider how one can compare two probability distributions over the same random variable X, for example, $p(x)$ and $q(x)$ with each other. An example of a measure that one can use for such a comparison is a *relative entropy*, called the Kullback–Leibler (KL) divergence, $D_{\text{KL}}(p||q)$. To be precise, the KL divergence is a measure of how the probability distribution q differs from a reference probability distribution p. As

[18] Again, we can easily generalize this notion to random vectors.

[19] Do not confuse likelihood and probability! Intuitively, probability is a property of a sample coming from some distribution. Likelihood, on the other hand, is a property of a parametrized model. In particular, if you plot $p(\mathcal{D} \mid \boldsymbol{\theta})$ as a function of possible $\boldsymbol{\theta}$, it does not have to integrate to one.

we typically employ it in classification tasks where p and q are both distributions of a discrete variable, it is defined as:

$$D_{\mathrm{KL}}(p||q) = \left\langle \log \frac{p(x)}{q(x)} \right\rangle_p = \frac{1}{n}\sum_i^n p(x_i) \log \frac{p(x_i)}{q(x_i)}. \tag{2.20}$$

For continuous distributions, the sum has to be replaced by an integral. $D_{\mathrm{KL}}(p||q)$ has some properties of a distance, that is, is non-negative and zero if and only if p and q are equal.[20] But it is not a proper distance measure as it is not symmetric, $D_{\mathrm{KL}}(p||q) \neq D_{\mathrm{KL}}(q||p)$.[21] Using the properties of the logarithm, $D_{\mathrm{KL}}(p||q)$ can be expressed as

$$D_{\mathrm{KL}}(p||q) = \frac{1}{n}\sum_i^n p(x_i) \log p(x_i) - \frac{1}{n}\sum_i^n p(x_i) \log q(x_i) =: -\mathcal{S}(p) + \mathcal{L}_{\mathrm{CE}}(p,q), \tag{2.21}$$

where $\mathcal{S}(p)$ is the Shannon entropy of the reference probability distribution p, and as the second term we obtain the CE, which we have already introduced in Eqs. (2.2) and (2.3)! We rediscover it by noting that minimizing $D_{\mathrm{KL}}(p||q)$, that is, the difference of p with respect to q, is equivalent to minimizing the CE because q does not appear in $\mathcal{S}(p)$.

While the utility of comparing probability distributions is clear in the case of estimating an unknown probability distribution by a parametrized one, it may not be immediately obvious for arbitrary ML models. Let us discuss the case of supervised learning with a model f. Consider a labeled training dataset consisting of n tuples $\{x_i, y_i\}$, where x_i is a given sample with label y_i. Each label belongs to one out of K classes. Next, we can think of each one-hot-encoded label $y_i = k$ as a very specific probability distribution $y_i = q(x_i) = \delta_{k,j}$, where $k, j \in \{1, \ldots, K\}$ (one-hot encoding). Next, the training data x_i are fed to the model f, and as an output we obtain the probability distribution $p(x_i) = f(x_i)$, which gives us the probabilities of a given sample x_i belonging to each class. In the last step, we have to compare two probability distributions, p and q. Therefore, we rediscover the CCE from Eq. (2.3). Similarly, one can show that the MSE loss, Eq. (2.1), emerges naturally from a probabilistic viewpoint. We refer to the example of linear regression in Section 2.4.1 for this analysis. In conclusion, we have seen that there exists a deep connection between probability theory and our optimization perspective of ML.

2.4 Machine learning models

We have already described two out of three ingredients of ML: tasks (Section 1.4) and data (Section 1.5). The final element is a model that learns how to solve a task given some data. ML models can be broadly divided into two classes which are *standard* ML and DL. In Sections 2.4.1–2.4.3, we give an overview of the former, while the latter is explained in more depth in Sections 2.4.4–2.4.6. Let us start by stressing the following point:

[20] Equal in the case of discrete variables, and equal "almost everywhere," that is, throughout all of the relevant space except for on a set of measure zero, in the case of continuous variables.

[21] We recommend an illustrative discussion of the asymmetry of $D_{\mathrm{KL}}(p||q)$ in Fig. 3.6 of [13].

DL is a subfield of ML itself as depicted in Fig. 1.2. However, it is customary to distinguish between ML methods based on whether they use neural networks (NNs). Henceforth, in the remainder of the chapter, we refer to *standard* ML as any algorithm that does not make use of NNs.

The distinction here becomes more subtle: in a nutshell, what distinguishes traditional learning from DL is the level of abstraction and the flexibility the algorithm has in extracting the features. In other words, traditional ML requires very specific algorithms designed and tailored to the problem at hand. The choice of the model then often comes down to experience and further intuition of the task of interest. On the other hand, NNs are a very flexible yet general tool whose main objective is to reproduce a target function without any (or little) constraints on the functional class from which to search. As a down-side, they usually do not support an easy interpretation of their mapping (compared to traditional ML methods) and are often referred to as *black-box* functions. We explain to what extent this is actually the case in Section 3.5. The distinction we can infer is that DL does not require an explicit set of instructions on how to connect the input to the output. Traditional ML methods, on the other hand, are often constructed by geometric or information-theoretic arguments, which already provide intuition into the method by their very construction and, hence, provide them with immediate interpretability.

A natural question that arises now is: Which approach to choose, traditional ML or DL? As always: it depends. For instance, DL performs at the state-of-the-art level in the big-data regime where an NN has enough available information to infer and perform the feature extraction. The low-data limit is where traditional ML is still prevailing. Here, we need to incorporate as much information as possible into the algorithm of choice to make learning efficient. Thus, algorithms become much less general and more problem-dependent.

Let us now focus on the standard ML algorithms and leave the discussion about NNs for Section 2.4.4 and forward. Some prominent examples of traditional ML we encounter during the rest of this book are the following: Principal component analysis (PCA) is a very elegant approach for the task of dimensionality reduction, that is, for data compression. It takes multidimensional samples and compresses their feature space while maintaining only a few relevant features. The compressed data can undergo further ML routines (see Section 3.2.1). Gaussian processes (GPs) are another example of a traditional ML algorithm that deals well with learning tasks when only limited data are available. Together with BO, they represent one of the most powerful examples of variational inference (see Chapter 4). They are furthermore an instance of so-called *kernel methods* which are as powerful as widely used. The elegance comes from the efficient application of a feature transformation of the input data. In this way, data in a representation that is difficult to analyze, get mapped into a domain where they are easier to analyze.

The mentioned methods are discussed in more detail throughout the book. Sections 2.4.1–2.4.6 constitute a primer of the standard ML models, describing basic approaches such as linear and logistic regression, linear support vector machines (SVMs), and continue into the DL regime with description of NNs with focus on convolutional neural networks (CNNs) and autoregressive neural networks (ARNNs).

2.4.1 Linear (ridge) regression

Before diving into the details of the topic, let us restate the problem of regression sketched in Section 1.4. We encounter a labeled data set $\mathcal{D} = \{(\boldsymbol{x}_i, y_i)\}_{i=1}^n \equiv \{(\boldsymbol{X}, \boldsymbol{y})\}$ of observations that are derived from an underlying function f, possibly subject to some (stochastic) noise ϵ. The latter is often assumed to be sampled from an unknown noise distribution $\mathcal{E}$, that is, $\epsilon \sim \mathcal{E}$:

$$y_i = f(\boldsymbol{x}_i) + \epsilon_i \quad \forall\, (\boldsymbol{x}_i, y_i) \in \mathcal{D}. \tag{2.22}$$

The function f is generally unknown, and our goal is to infer it. To this end, we build a regression fit of the data $\hat{f}$ such that $\hat{f}(\boldsymbol{x}) \approx y$.[22]

Arguably, the simplest parametrized fitting method one can produce is a linear model, where we seek to find parameters $\boldsymbol{\theta} \in \mathbb{R}^d$ that linearly connect the input variable $\boldsymbol{x}$ with the prediction $\hat{y}$, that is,

$$\hat{y} = \sum_{i=1}^{d-1} \theta_i x_i + b \equiv \sum_{i=0}^{d-1} \theta_i x_i = \boldsymbol{x}^\intercal \boldsymbol{\theta}. \tag{2.23}$$

To shorten the notation, we have absorbed the constant $\theta_0 = b$, the so-called *bias*, in the definition of the input $\boldsymbol{x}$ via setting $x_0 = 1$. Up to now, the linear model aims to find a hyperplane[23] through the data points. We can extend the model by a nonlinear transformation ϕ of the input, that is, $\boldsymbol{x} \mapsto \phi(\boldsymbol{x})$. This is still linear regression as we maintain linearity in the parameters $\boldsymbol{\theta}$ that we seek to optimize. As an example of a nonlinear transformation, the map $\phi_p : x \mapsto (1, x, x^2, x^3, \dots, x^p)$ promotes our model to polynomial regression up to the pth degree. To simplify the notation in the rest of the section, we consider the case where no feature maps are applied. The inclusion of a feature map is a central element of Chapter 4 and is discussed there to a far greater extent.

Once a certain hyperplane is defined, by means of its parameters $\boldsymbol{\theta}$, we need to define a quality measure that compares our predictions to their corresponding ground-truth values. That is, we have to choose a suitable loss function $\mathcal{L}$. The most conventional choice for the loss is the MSE over the dataset $\mathcal{D}$ as

$$\mathcal{L}_{\text{MSE}}(\boldsymbol{\theta} \mid \mathcal{D}) := \frac{1}{n} \sum_{i=1}^{n} \left(y_i - \boldsymbol{x}_i^\intercal \boldsymbol{\theta}\right)^2 = \|\boldsymbol{y} - \boldsymbol{X}\boldsymbol{\theta}\|^2. \tag{2.24}$$

To attain the rightmost equation, we stack all inputs $\boldsymbol{x}_i$ vertically next to each other, to form the matrix $\boldsymbol{X} \in \mathbb{R}^{(d-1)\times n}$. The same procedure is applied to y_i, now to be promoted to $\boldsymbol{y}$. The last step allows us to find the set of parameters $\boldsymbol{\theta}$ that minimize

[22] For the sake of simplicity, we consider one-dimensional output. The following derivations, however, can easily be extended to multidimensional output as well.

[23] For one-dimensional input and output, the hyperplane simply is a line.

the MSE. This yields the least-squares estimator (LSE) (for the derivation, see the first half of Appendix B)

$$\theta_{\mathrm{LSE}} = \left(\boldsymbol{X}^{\top}\boldsymbol{X}\right)^{-1}\boldsymbol{X}^{\top}\boldsymbol{y} = \boldsymbol{X}^{+}\boldsymbol{y}, \tag{2.25}$$

where the notation $\boldsymbol{X}^{+}$ denotes the Moore–Penrose inverse [67].

The MSE as the choice of our loss function appears to be self-evident. In fact, we can derive it by maximizing the likelihood of the labeled data given the model parameters $p(\boldsymbol{y} \mid \boldsymbol{X}, \theta)$. To this end, we assume that our targets y are actually sampled from a Gaussian with a mean given by our linear model, that is, $\boldsymbol{x}^{\top}\theta$ with some variance σ that models the noise in the data. We can then write the likelihood of observing the targets $\boldsymbol{y}$ given the locations $\boldsymbol{X}$ and model parameters θ as

$$p(\boldsymbol{y} \mid \boldsymbol{X}, \theta) = \mathcal{N}(\boldsymbol{y} \mid \boldsymbol{X}^{\top}\theta, \sigma^2\mathbb{1}) \tag{2.26}$$

$$= \prod_{i=1}^{n} \mathcal{N}(y_i \mid \boldsymbol{x}_i^{\top}\theta, \sigma^2). \tag{2.27}$$

In the last step, we furthermore assumed a dataset $\mathcal{D}$ of i.i.d. random variables to factorize the multivariate Gaussian. A common assumption is to regard the observed dataset $\mathcal{D}$ as the most probable one of the underlying linear model. Therefore, we seek to maximize the likelihood to finding the set of parameters θ that have led to the most probable data. This is the idea of maximum likelihood estimation (MLE). Its estimator is defined as the argument of the maximum likelihood of Eq. (2.26). We can modify this estimator by including a logarithm and obtain:

$$\theta_{\mathrm{MLE}} := \underset{\theta}{\arg\max}\; p(\boldsymbol{y} \mid \boldsymbol{X}, \theta) \tag{2.28}$$

$$= \underset{\theta}{\arg\max}\; \log p(\boldsymbol{y} \mid \boldsymbol{X}, \theta) \tag{2.29}$$

$$= \underset{\theta}{\arg\max} \left(-\frac{1}{2\sigma^2} \sum_{i=1}^{n} \left(y_i - \boldsymbol{x}_i^{\top}\theta\right)^2 + \text{const.} \right) \tag{2.30}$$

$$= \underset{\theta}{\arg\min} \left(\sum_{i=1}^{n} \left(y_i - \boldsymbol{x}_i^{\top}\theta\right)^2 \right) \equiv \underset{\theta}{\arg\min}\left(\mathcal{L}_{\mathrm{MSE}}\right). \tag{2.31}$$

The constants that appear in Eq. (2.30) can be ignored since they are independent of θ. From the previous results, we hence see that the assumption of i.i.d., together with the concept of MLE, leads to the MSE as the preferred loss function and we conclude that the MLE θ_{MLE} coincides with the LSE θ_{LSE} of Eq. (2.25).

However, the estimator fully ignores the data noise modeled by σ^2, as it was also dropped out in the maximization procedure of the $p(\boldsymbol{y} \mid \boldsymbol{X}, \theta)$. Thus, even if we correctly choose the model, the minimization procedure of the MSE in Eq. (2.24) generally performs well in the provided dataset $\mathcal{D}$ but not on previously unencountered data points. The reason is *overfitting*, which we already introduced as a concept in Section 2.2. This phenomenon occurs when we incorporate the noise on the targets in our model parameters θ_{MLE}. As a way out of this issue, we have introduced the notion of regularization. In our linear model (2.23), we can introduce regularization by means of Bayesian inference. This means that, instead of maximizing only the likelihood of the data in Eq. (2.26), we encode any prior knowledge of the model into the *prior* distribution $p(\theta)$. By virtue of the Bayes theorem from Eq. (2.19), we can calculate the *posterior* distribution[24] $p(\theta \mid \boldsymbol{X}, \boldsymbol{y})$ over the parameters given the dataset and maximize this quantity instead. This yields the maximum a posteriori (MAP) estimator defined as

$$\theta_{\text{MAP}} := \arg\max_{\theta} \; p(\theta \mid \boldsymbol{X}, \boldsymbol{y}) \tag{2.32}$$

$$= \arg\max_{\theta} \; \frac{p(\theta)p(\boldsymbol{y} \mid \boldsymbol{X}, \theta)}{p(\boldsymbol{X}, \boldsymbol{y})}, \tag{2.33}$$

where we have used the Bayes theorem from Eq. (2.19) in the second step. The denominator does not depend on θ and can therefore be ignored. For the likelihood, we keep the assumptions introduced for Eq. (2.26). As the prior, we now draw the parameter values from a Gaussian distribution centered around $\mathbf{0}$ with some variance τ^2, that is,

$$p(\theta) = \mathcal{N}(\theta \mid \mathbf{0}, \tau^2 \mathbb{1}). \tag{2.34}$$

The product of two Gaussian distributions is Gaussian itself, hence allowing us to apply the same trick with the logarithm as before in Eq. (2.29). We arrive at

$$\theta_{\text{MAP}} = \arg\min_{\theta} \left(\mathcal{L}_{\text{MSE}}(\theta \mid \mathcal{D}) + \frac{\sigma^2}{\tau^2} \|\theta\|^2 \right) = \left(\boldsymbol{X}^{\mathsf{T}} \boldsymbol{X} + \frac{\sigma^2}{\tau^2} \mathbb{1} \right)^{-1} \boldsymbol{X}^{\mathsf{T}} \boldsymbol{y}. \tag{2.35}$$

We can picture the parameter $\lambda = \sigma^2/\tau^2$ as a signal-to-noise ratio, which effectively penalizes large magnitudes of parameter values by the additional term in the loss function. Hence, λ is referred to as *regularization strength*. This particular choice of the loss term is called *Tikhonov* regularization. Its corresponding MAP is also called the linear *ridge regression* estimator.

Let us compare the two estimators of Eqs. (2.25) and (2.35). The additional term $\sigma^2/\tau^2 \mathbb{1}$ in the estimator stems from the fact that we take into account both the data noise and a parameter constraint. Both are discarded in the limit of $\tau \to \infty$[25], where we have $\theta_{\text{MAP}} \to \theta_{\text{MLE}}$.

> To consider the underlying noise in the training data, we have to constrain the linear model. Dealing with overfitting in such a way is usually referred to as regularization.

Finally, the choice of the prior in Eq. (2.34) is by no means unique. In fact, there is a plethora of regularization ideas and corresponding penalty terms [13]. An easy variation could, for example, be to replace the ℓ_2-norm with an ℓ_1-norm. This is achieved by choosing a Laplace distribution for the parameters as the prior. The corresponding estimator is the result of least absolute shrinkage and selection operator regression [68]. Because the ℓ_1-norm punishes already small parameter values severely, it favors sparse solutions for the parameters θ instead. This can, for example,

[24] Remember that we call it posterior because it is computed *after* the observation of the dataset $\mathcal{D}$.

[25] This corresponds to a uniform prior of the parameters θ.

be desired to detect the significant features out of a pool of possible candidates in certain tasks [69].

2.4.2 Logistic regression

In Section 2.4.1, we have discussed the linear regression problem. The discussion can be extended to the classification task in a very straightforward way, as we show in the following.

> The basic idea of *logistic regression* is to adapt the linear model, such as to estimate the probability that a given input falls in either one of the possible classes.

Let us consider two classes K_1 and K_2 and an input $\boldsymbol{x}$ to classify. We introduce the class-conditional densities $p(\boldsymbol{x}|K_i)$ and the corresponding baseline class prior probabilities $p(K_i)$. Bayes' theorem of Eq. (2.19) immediately gives us an expression for the posterior probability that the input belongs to class K_1. It reads as

$$\begin{aligned} p(K_1 \mid \boldsymbol{x}) &= \frac{p(\boldsymbol{x} \mid K_1)p(K_1)}{p(\boldsymbol{x} \mid K_1)p(K_1) + p(\boldsymbol{x} \mid K_2)p(K_2)} \\ &= \frac{1}{1+\exp(-\theta)} =: \varsigma(\theta), \\ \text{where } \theta &:= -\log\left(\frac{p(\boldsymbol{x} \mid K_2)p(K_2)}{p(\boldsymbol{x} \mid K_1)p(K_1)}\right) \end{aligned} \tag{2.36}$$

and equips us with the *logistic sigmoid* function ς that maps any real-valued input θ to the interval $[0, 1]$. We can now use the linear model (or any other ML model) to yield a value for θ and map it to the corresponding posterior probability. This additional layer turns the regression model into a classifier.

To extend the situation to more than two classes, we perform a similar reformulation as done in Eq. (2.36). In this case, one obtains the *softmax* function

$$p(K_k \mid \boldsymbol{x}) = \frac{\exp(\theta_k)}{\sum_i \exp(\theta_i)} =: \text{softmax}(\boldsymbol{\theta}) \tag{2.37}$$

that maps the output score vector $\boldsymbol{\theta}$ to a proper probability density over all classes at once. Its name is derived from the fact that in the limiting case of $\theta_i \gg \theta_k \; \forall k \neq i$, the softmax converges to the maximum function, that is, softmax $\to$ max.

In both cases, the model's parameters are trained by parsing the output scores through either Eq. (2.36) or Eq. (2.37) to obtain and subsequently minimize the loss in Eq. (2.2) or Eq. (2.3), respectively. An interesting aspect of any classifier is how it draws a line between data from two different phases, known as the *decision boundary*. In the case of the linear model, the decision boundary is linear, which is a simple consequence of the model choice. Because this boundary is derived from the likelihood of the data due to the particular choice for the loss function, the model is highly prone to outliers. One way to circumvent this issue is to take a geometric approach in finding the decision boundary. This is done in Section 2.4.3.

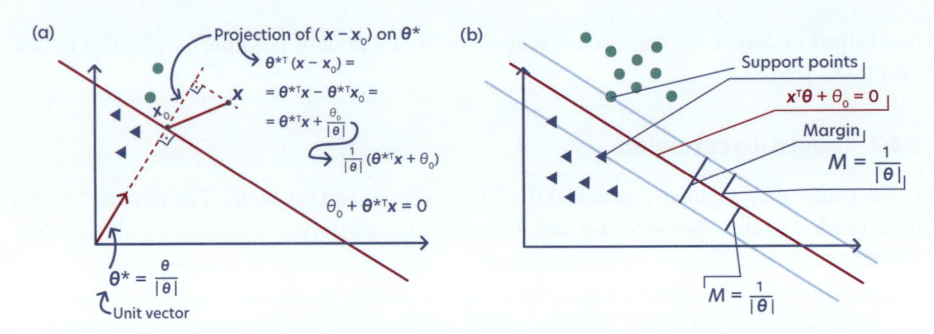

Figure 2.5 Geometric construction of an SVM in a 2D problem. (a) Purple line, described fully by θ, is an exemplary hyperplane separating two classes of data (pink and green points). (b) The optimal hyperplane maximizes the margin, M, between itself and the support points, which are training data points closest to the hyperplane.

2.4.3 Support vector machines

An alternative approach to classification, instead of maximizing a model likelihood, is to analyze the geometrical properties of data. This idea is encapsulated in the framework of SVMs whose origins can be traced back to the 1960s, see [70] and the references therein. For a visualization of the geometry of the data, take a look at the linearly separable problem presented in Fig. 2.5. In panel (a), you see that to classify two types of data, we can draw a line (or, more generally, a hyperplane) that separates the training data. Then, instead of making probabilistic predictions on test data, we can just check on which side of the hyperplane the test points are. Panel (a) also contains a simple geometric analysis that shows that the equation for the hyperplane separating the data is $\boldsymbol{\theta}^{\mathrm{T}}\boldsymbol{x} + \theta_0 = 0$. The unit vector in the direction perpendicular to the hyperplane is $\boldsymbol{\theta}^* = \boldsymbol{\theta}/|\boldsymbol{\theta}|$, and the shortest distance between a point $\boldsymbol{x}$ and the hyperplane is

$$d(\boldsymbol{x}, \boldsymbol{\theta}) = \left(\boldsymbol{\theta}^{\mathrm{T}}\boldsymbol{x} + \theta_0\right) / |\boldsymbol{\theta}| . \tag{2.38}$$

If we do not impose any additional constraints, there are many possible hyperplanes that separate the data into two classes. How do we choose the best? One way is to maximize the distance between the hyperplane and the data points [71]. Therefore, let us formulate the constraint that all data points must be at least the distance M away from the hyperplane. The data points separated from the hyperplane exactly by M, hence the closest to the hyperplane, become the support points presented in Fig. 2.5(b). The classification problem boils down to finding the $\boldsymbol{\theta}$ that maximize the margin. From Eq. (2.38), we can write

$$y_i \left(\boldsymbol{\theta}^{\mathrm{T}}\boldsymbol{x}_i + \theta_0\right) \geq M |\boldsymbol{\theta}| , \tag{2.39}$$

where the elements of the vector of observations y_i are ± 1 that help ensure that this formulation is always positive, regardless of the class to which the data point belongs. Note that if we scale each of the $\boldsymbol{\theta}$ coefficients by the same factor, the above (in)equality still holds. Therefore, we can arbitrarily rescale $\boldsymbol{\theta}$ and θ_0 to have $|\boldsymbol{\theta}| = \frac{1}{M}$, which leads to the following canonical condition for every data point in the dataset:

$$y_i\left(\theta^{\mathrm{T}}\boldsymbol{x}_i + \theta_0\right) \geq 1\,. \tag{2.40}$$

Therefore, to find the optimal hyperplane, we need to minimize $|\theta|$, while ensuring $y_i\left(\theta^{\mathrm{T}}\boldsymbol{x}_i + \theta_0\right) \geq 1$ for every data point. This is the optimization with constraints, and we can use Lagrange multipliers for that! Minimizing $|\theta|$ with constraints boils down to minimizing the following Lagrange function:

$$L = \frac{1}{2}|\theta|^2 - \sum_i^n \alpha_i\left[y_i\left(\theta^{\mathrm{T}}\boldsymbol{x}_i + \theta_0\right) - 1\right], \tag{2.41}$$

where the Lagrange multipliers α_i are chosen such that

$$\alpha_i\left[y_i\left(\theta^{\mathrm{T}}\boldsymbol{x}_i + \theta_0\right) - 1\right] = 0 \text{ for each } i\,. \tag{2.42}$$

Interestingly, the loss function in Eq. (2.41) with the above constraints is a so-called quadratic program as the function itself is quadratic and the constraints are linear with respect to $|\theta|$. It has, therefore, a global minimum found usually via so-called sequential minimal optimization [72] instead of any iterative gradient-based methods.

Also note that the condition put on the Lagrange multipliers in Eq. (2.42) implies the following:

- If $\alpha_i > 0$, then $\left[y_i\left(\theta^{\mathrm{T}}\boldsymbol{x}_i + \theta_0\right) - 1\right] = 0$, which means the point $\boldsymbol{x}_i$ lies on the boundary of the margin slab.
- If $\left[y_i\left(\theta^{\mathrm{T}}\boldsymbol{x}_i + \theta_0\right) - 1\right] > 0$, the points is outside the margin and $\alpha_i = 0$.

Therefore, the final model coefficients are given only in terms of such points $\boldsymbol{x}_i := \boldsymbol{x}_{s,i}$ that lie on the boundary of the slab. These points are the support points and give the SVM its name. The SVM problem relies then on minimizing L[26] numerically to find the coefficients α_i which are nonzero only for support points.

Therefore, classification with SVMs consists of finding the optimal hyperplane separating the data by maximizing the margin between the hyperplane and the support points, which are data points closest to the decision boundary. This optimization problem with constraints is solved with Lagrange multipliers and is convex.

With the found optimal hyperplane $\hat{f}$ we can then make predictions at an arbitrary test point $\boldsymbol{x}^*$:

$$\hat{f}(\boldsymbol{x}^*) = \theta^T\boldsymbol{x}^* + \theta_0 = \sum_i^n \alpha_i y_i \boldsymbol{x}_i^T\boldsymbol{x}^* + \theta_0 = \sum_i \alpha_i y_i \boldsymbol{x}_{s,i}^T\boldsymbol{x}^* + \theta_0, \tag{2.43}$$

where the last summation is only over support points. Finally, to turn this value into a class prediction, we take the sign of $\hat{f}$ as the corresponding class label.

[26] In practice, rather than minimizing L, one maximizes a Lagrange dual, L_D, which provides the lower bound for L. We explain it in more detail in Section 4.2.2.

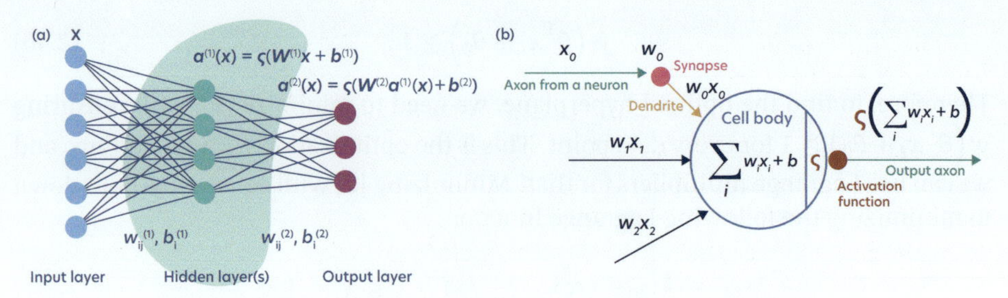

Figure 2.6 Illustration of (a) a typical fully connected (here: two-layer) NN and (b) one of its neurons (simple perceptron) and the computations associated with it.

Until now, we only considered binary classification problems of linearly separable datasets. There are two obvious ways of how to extend the SVM to classification problems that have more than two classes, say K many. The first, known as the one-to-one approach, breaks the multiclass situation down to a binary classification between every combination of two classes, individually. This way, we are required to train $\mathcal{O}(K^2)$ SVMs to make predictions afterward. This numerical overhead is eased in the second approach: one-to-rest classification. Here, we only require a single SVM for each of the K classes that simply predicts whether a test point belongs to the class or not. As a second extension possibility, we can ask about the classification problem that is not linearly separable. We explain this case later in Section 4.2.2.

2.4.4 Neural networks

Artificial neural networks (ANNs), typically referred to as NNs, are a large class of models used to process data in ML tasks. They are parametrized functions that are themselves composed of many simple functions. As the name suggests, ANNs were originally proposed by taking loose inspiration from networks of neurons that constitute our brains. They are typically composed of interconnected layers that sequentially process information, see Fig. 2.6. Each layer contains multiple nodes or units, also called *artificial neurons* or *perceptrons*.[27] Each node i takes as input a vector $\boldsymbol{x} = (x_1, x_2, \ldots, x_m) \in \mathbb{R}^m$, corresponding to the activations of all nodes in the previous layer. It outputs a scalar value $a_i \in \mathbb{R}$ (its *activation*) that is computed as $a_i = \varsigma(\sum_j w_{i,j}x_j + b_i)$, where the parameters $\{w_{i,j}\}_{j=1}^m$ and $b_i \in \mathbb{R}$ are the weights and bias of node i, respectively. The weights of a node control the strength of its connection to the neurons of the previous layer. The function ς is a nonlinear function called *activation function*. Common choices are the rectified linear unit (ReLU)

$$\varsigma(z) = \max(0, z), \tag{2.44}$$

the sigmoid function (Eq. (2.36)), or the tanh function

$$\varsigma(z) = \tanh(z) = \frac{e^z - e^{-z}}{e^z + e^{-z}}. \tag{2.45}$$

[27] Here and in the following, we refer to the modern perceptron introduced by Minsky and Papert [73], which can contain smooth activation functions in contrary to the Heaviside step function utilized in Rosenblatt's original perceptron [74].

The first layer is called the input layer, where the activations of its nodes are set according to the vector $\boldsymbol{x}$ encoding the input data. The last layer is called the output layer, and the activations of its nodes constitute the output of the NN. All intermediate layers are called hidden layers. NNs where each node is by default connected to all nodes in the subsequent layer are referred to as *fully connected*. The number of layers, nodes, and their connections is known as the *architecture* of an NN. NNs are considered deep if they are composed of many hidden layers.[28] ML methods based on deep neural networks (DNNs) as models fall under the name of DL [13].

A central question regarding NNs is what types of functions they can represent (recall our previous discussion on traditional ML vs. DL). First, consider an NN without its nonlinear activation functions. The function realized by such an NN is a simple affine map, that is, consists of multiplying the input by a weight matrix and adding to it an additional bias vector. Thus, the addition of nonlinear activation functions is crucial for NNs to be able to represent a larger class of functions. For example, Kolmogorov and Arnold [75] have shown that any arbitrary continuous high-dimensional function can be expressed as a linear combination of the composition of a set of nonlinear functions

$$f(\boldsymbol{x}) = \sum_{i=0}^{2m} \zeta_i \left(\sum_{j=1}^{m} \varsigma_{i,j}(x_j) \right), \tag{2.46}$$

where $\zeta_i, \varsigma_{i,j}$ are nonlinear functions that act on the individual components of the input $\boldsymbol{x} \in \mathbb{R}^m$. This means that we could represent any function $f(\boldsymbol{x})$ with a polynomial number $O(m^2)$ of one-dimensional nonlinear functions. This strongly resembles the structure of an NN with two hidden layers. Note, however, that the nonlinear functions must be carefully chosen depending on the target function. In NNs, the nonlinearities are typically fixed $\zeta_i = \varsigma_{i,j}\ \forall i, j$. It turns out that fully connected NNs composed of a single hidden layer and nonlinear activation functions are also *universal function approximators*. That is, given that the target function is reasonably well-behaved, it can be approximated to any desired accuracy given that its hidden layer contains enough nodes [13, 76, 77]. Note that this may still require a hidden layer that is exponentially large in the number of nodes. This raises the question of what one can achieve with NNs that have multiple hidden layers.

The universal approximation theorem guarantees that there exists an NN, that is, the choice of NN architecture, as well as weights and biases, which approximates the given target function arbitrarily well. However, it does not guarantee that we are able to find this choice. It turns out that, in practice, DNNs are capable of solving many problems with much fewer nodes, that is, trainable parameters, than shallow NNs. In that sense, choosing a DNN over a shallow NN yields a useful prior over the space of functions that the NN can approximate.

The parameters of an NN are typically optimized by gradient-based methods, such as SGD or Adam, to minimize a given loss function $\mathcal{L}$ (see Section 2.1). Computing the gradient of the loss function with respect to the NN parameter numerically is typically done by means of *backpropagation* [78], which we discuss in more detail in Section 2.5. By contrast, when evaluating an NN with a given input, information flows forward through the networks. As such, this is called *forward propagation*.

[28] There is no clear consensus on the threshold of depth that divides shallow and deep NNs.

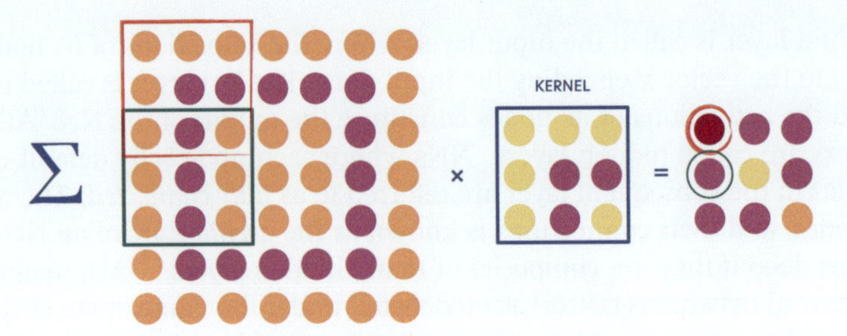

Figure 2.7 Schematic representation of a convolutional layer in two dimensions: A kernel/filter of fixed size (here 3×3) is convolved with a two-dimensional input image. The color intensity corresponds to the magnitude of the neuron activations and kernel (filter) weights.

Convolutional neural networks. Convolutional neural networks (CNNs) are a special class of NNs where, in contrast to fully connected NNs, not every node is connected to all nodes of the subsequent layer. Instead, convolutions replace matrix multiplications in the computation of the activations of subsequent layers. This reduction of the number of parameters per layer allows us to build and train deeper architectures. Moreover, this model architecture makes use of the spatial hierarchy typically present in input data. In image-like data, pixels that are spatially close to each other, generally show better correlation than pixels that are far apart. Replacing the *full* connectivity of standard NN with multiple convolutional layers with *local* connectivity, CNNs make use of this vanishing correlation at large distances.

Figure 2.7 illustrates the working principle behind a CNN – the *convolutions*: A filter (also called kernel) with trainable weights is slid across a given layer. The resulting activation is then obtained by the element-wise multiplication of neuron activations and the filter's weights, followed by an overall sum and the application of a nonlinear activation function. This filtering causes the NN to be only locally connected (as opposed to fully connected). Note that the number of weights, therefore, does not depend on the size of the input but on the size of the filter. The filter size controls the range over which spatial correlations in the input data are registered. One can build one- or two-dimensional CNNs (with filters of corresponding dimension) depending on whether the input data is naturally represented as a vector or a matrix. In a typical CNN, after the application of several such convolutional layers, the activations are flattened to a single feature vector. This corresponds to a lower-dimensional representation of the input data that is further processed using a fully connected architecture. To reduce the dimension of the data representation resulting from the application of convolutional layers, one typically also uses *pooling operations*. These combine the activations resulting from applications of close-by filters, for example, by taking the maximum or mean.

2.4.5 Autoencoders

Autoencoders (AEs) [79, 80] are widely used ML tools for unsupervised learning. Unlabeled data (e.g., images, audio signals, and texts) may often be high dimensional.

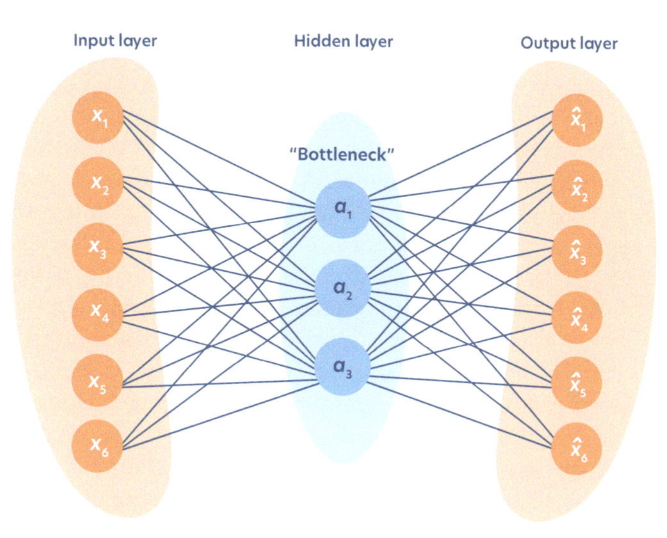

Figure 2.8 Example of the bottleneck architecture of an AE. The input is connected to the bottleneck by an encoder NN on the left, while the decoder NN connects it with the output on the right.

Hence, it is very difficult to analyze and extract any patterns when working in the data domain. However, dimensionality reduction techniques (see, e.g., Section 3.2.1) represent an advantageous approach to extract useful knowledge from such unlabeled data. In a nutshell, the goal of AEs is to precisely *encode* some knowledge, patterns, and attributes of the given input data into some latent variable[29] on a lower dimensional manifold. By means of a so-called bottleneck structure (as shown in Fig. 2.8), the latent representation of the input data is then mapped back into the input space (decoding) by leveraging on the information extracted by the architecture at the time of feature extraction (encoding). This bottleneck architecture is based on two NNs performing the encoding and decoding parts. Such NNs are trained by minimizing the so-called error reconstruction loss, meaning that the optimal setup for such encoder–decoder pair is the one for which the output $\boldsymbol{x}_{\text{rec}}$ is reconstructed as similar as possible to the original input data $\boldsymbol{x}$. These NNs are jointly optimized with an iterative process. In other words, for a given set of possible encoders and decoders, we are looking for the pair that keeps the maximum of information when encoding and, so, has the minimal reconstruction error when decoding. This joint optimization forces the model to maintain only the variations in the data required to reconstruct the input without holding on to redundancies within the input. Henceforth, likewise in PCA, only the most relevant features describing the data are distilled during the learning process. One important remark is that the bottleneck is a key attribute of such a network design; without the presence of an information bottleneck, our network could easily learn to simply memorize the input values by passing these values along through the network. On top of this, by relying on such a pair of NNs,

[29]A latent variable is a random variable that we cannot observe directly. In this case, we call variables latent because we do not observe them in the data.

AEs are inherently more flexible yet expressive compared to standard dimensionality reduction algorithms (e.g., PCA), which rely on a submanifold projection of input data through constrained linear or nonlinear transformations.

There are several types and variations of AEs, all of which share this fundamental bottleneck property as their base structure. A concrete example of a further development of AEs in the context of generative models are variational autoencoders (VAEs). As the name suggests, VAEs [79] have to do with variational inference. What they do in practice is to train the encoding–decoding pair in a slightly more complicated way. The knowledge extracted from the data in the encoding part is nested into a base probability density (e.g., initialized as a Gaussian), which is trained and tuned in such a way that it becomes a good approximation (sampler) of the underlying data distribution. Once the training is done, the latent representation of the input data becomes thus a probability density from which one can sample new, unseen data that resembles the one used for training, as being characterized by the same learned features. As such, the goal here is not only to reconstruct the input data from the extracted knowledge anymore but also to produce new samples as similar as possible to the training set. Further examples of AEs are: sparse AEs [81, 82], denoising AEs [83], and importance weighted AEs [84].

2.4.6 Autoregressive neural networks

To complete this section, let us briefly present ARNNs. These networks were originally inspired by autoregressive models in statistics and economics, which one can employ to predict future values of a time-series (for instance, a financial asset). ARNNs are formalized for the general task of density estimation [85], in which the goal is to estimate a complex, high-dimensional probability density function, see also Section 7.2. They are constructed to satisfy the following property on the outputs of the network, satisfying a conditional structure:

$$f(\boldsymbol{x}) = \prod_{i=1}^{m} f_i(x_i \mid x_{i-1}, \dots, x_1) \tag{2.47}$$

with $\boldsymbol{x} = (x_1, x_2, \dots, x_m)$ the inputs for the model. In the case of time series, the inputs x_i would be values of a variable at times t_i, and the model f tries to predict future values based on past ones. A generic example of such networks is the recurrent neural network (RNN) that was popularized in the context of natural language processing tasks. The main idea behind this class of models is that information *loops back* into the model, introducing correlations between different parts of the network, as opposed to feedforward networks. Broadly speaking, a sentence has a causal order, but the correlations between words are not necessarily highest between words that are close together. Hence, the idea of introducing a back loop, with a memory can be understood somewhat intuitively. The long short-term memory (LSTM) is an extension of this idea with two memory length scales (long- and short-term) and was also found to be successful for such tasks [86]. A sketch of an RNN is presented in Fig. 2.9, with an example use-case from a language processing task. The goal here is to predict the next word in the sentence based on previous words. The parameters of such a

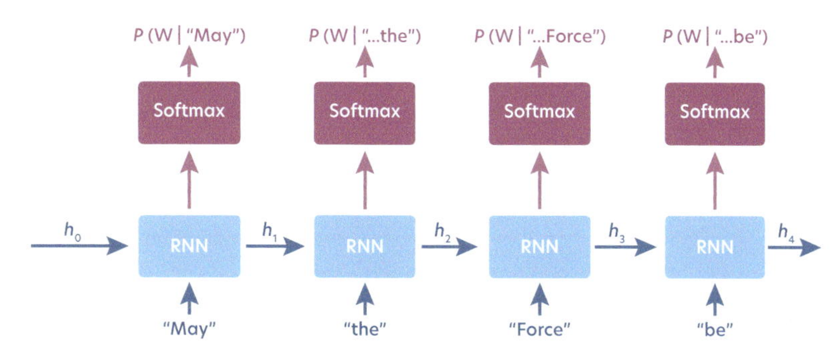

Figure 2.9 Pictorial representation of a recurrent neural network (RNN). One can directly see that the model is autoregressive, as conditional probabilities only depend on the previous input data. Here, the blue box represents a nonlinear transformation as described in the main text. The task here is to be able to generate meaningful sentences. The input data is a sentence, and the output is the probability for the word W to be the next word in the sentence, conditioned on previous words. The $\boldsymbol{h}_i$ are the hidden vectors that take into account memory effects inherited from previous RNN transformations.

network are hidden in the RNN cell and take part in a nonlinear transformation given by:

$$\boldsymbol{h}_i = \varsigma(\boldsymbol{W}_h \boldsymbol{h}_{i-1} + \boldsymbol{W}_x \boldsymbol{x}_i)$$

with $\boldsymbol{W}_h$ and $\boldsymbol{W}_x$ being two weight matrices, $\boldsymbol{h}_i$ the ith hidden vector that represents information coming out of the previous cells, $\boldsymbol{x}_i$ the ith element of the input data, and ς some nonlinear activation function that is applied element-wise.[30] For words, $\boldsymbol{x}_i$ represents the ith word of the sentence that is encoded in some form (e.g., using a one-hot encoding). Note that there exist several variants of this transformation, the most popular being the gated recurrent unit [87]. The autoregressive models have been applied to different problems in physics, such as statistical mechanics [88–90], quantum tomography [91], and ground state search [92]. In Chapter 5 and Section 7.2, we stress the advantages of using such models and present impressive results for quantum physics and chemistry that have been obtained using them.

2.5 Backpropagation

As already mentioned in Section 2.1, NNs are typically trained via gradient-based methods. These approaches require the calculation of the loss function's derivative with respect to each trainable parameter. In principle, given a particular NN architecture, we could derive a closed-form solution for the gradient. However, this computation would need to be performed again given different NN architectures. Such calculations also involve some form of human input, which makes them tedious and prone to errors. Clearly, we would like to automate this gradient calculation

[30] One also has to choose an initialization $\boldsymbol{x}_0, \boldsymbol{h}_0$, which are generally null vectors.

and make it as efficient as possible. The algorithm of choice to train large NNs is *backpropagation* [78]. Backpropagation belongs to a larger class of algorithms known as AD, which allow us to evaluate the derivative of a function represented as a computer program efficiently and in an automated fashion. We describe and compare these methods in detail in Section 7.1.

The basic idea behind backpropagation is to take advantage of the fact that an NN is composed of sequences of many elementary building blocks, such as artificial neurons. Thus, we can compute derivatives through the repeated (reverse) application of the chain rule.

To provide some intuition, we exemplify the use of backpropagation on a simple feedforward network, where nodes in each layer are connected only to nodes in the immediate next layer. However, the main principle carries over to any other architecture, such as the ones introduced in the sections above, like CNNs, AEs, or RNNs, among others. Recall that the activations of the nodes in the lth layer of a feedforward NN are given by

$$\boldsymbol{a}^{(l)} = \varsigma^{(l)}\left(\boldsymbol{z}^{(l)}\right) = \varsigma^{(l)}\left(\boldsymbol{W}^{(l)}\boldsymbol{a}^{(l-1)} + \boldsymbol{b}^{(l)}\right), \tag{2.48}$$

where $\boldsymbol{a}^{(l-1)}$ is a vector that contains the activations of the previous layer, that is, layer $l-1$. The corresponding weight matrix is given by $\boldsymbol{W}^{(l)}$, where $w_{i,j}^{(l)}$ is the weight of the connection from node j in layer $l-1$ to node i in layer l, $\boldsymbol{b}^{(l)}$ is the bias vector of layer l, and $\varsigma^{(l)}$ is the activation function of the lth layer. The function implemented by a feedforward NN with L layers ($L-1$ hidden layers and an output layer) can be obtained by stacking up multiple such layers

$$\mathbf{NN}(\boldsymbol{x}) = \boldsymbol{a}^{(L)}(\boldsymbol{x}) = \varsigma^{(L)}\left(\boldsymbol{W}^{(L)}\boldsymbol{a}^{(L-1)}(\boldsymbol{x}) + \boldsymbol{b}^{(L)}\right), \tag{2.49}$$

where $\boldsymbol{a}^{(0)}(\boldsymbol{x}) = \boldsymbol{x}$ is the input vector. The crucial observation is that the NN output depends on the input $\boldsymbol{x}$ solely through the activations of the previous layer $\boldsymbol{a}^{(L-1)}$, which in turn only depends on the input through $\boldsymbol{a}^{(L-2)}$, and so on (see Eq. (2.48)). This simply arises from the layer-wise processing of information in a feedforward NN.[31]

Eventually, we are interested in computing the derivatives of our loss function $\mathcal{L}$ with respect to all weights $\partial\mathcal{L}/\partial w_{i,j}^{(l)}$ and biases $\partial\mathcal{L}/\partial b_i^{(l)}$. In the following, we focus only on weights. However, the procedure straightforwardly generalizes to biases. For a given training dataset $\mathcal{D} = \{(\boldsymbol{x}_i, \boldsymbol{y}_i)\}_{i=1}^n$, the loss function is typically given as an average

$$\mathcal{L} = \frac{1}{n}\sum_{i=1}^{n} \ell(\mathbf{NN}(\boldsymbol{x}_i), \boldsymbol{y}_i). \tag{2.50}$$

[31] In fact, the concept of a feedforward network can be generalized to any directed acyclic graph. In any case, the information processing occurs in a *forward-directed* manner (from input nodes to output nodes).

Here, ℓ measures the deviation of the prediction $\mathbf{NN}(\boldsymbol{x}_i)$ from the corresponding desired output $\boldsymbol{y}_i$ possibly including an additional regularization term. Thus, we have

$$\frac{\partial \mathcal{L}}{\partial w} = \frac{1}{n}\sum_{i=1}^{n} \frac{\partial \ell(\mathbf{NN}(\boldsymbol{x}_i), \boldsymbol{y}_i)}{\partial w}, \tag{2.51}$$

where w is a single weight of the NN. From Eq. (2.51), we see that the main task boils down to computing derivatives for a fixed input–output pair $(\boldsymbol{x}_i, \boldsymbol{y}_i)$ of the form

$$\begin{aligned}\frac{\partial \ell(\mathbf{NN}(\boldsymbol{x}_i), \boldsymbol{y}_i)}{\partial w} &= \frac{\partial \ell(\mathbf{NN}(\boldsymbol{x}_i), \boldsymbol{y}_i)}{\partial \mathbf{NN}} \cdot \frac{\partial \mathbf{NN}(\boldsymbol{x}_i)}{\partial w}, \\ &= \frac{\partial \ell(\boldsymbol{a}^{(L)}(\boldsymbol{x}_i), \boldsymbol{y}_i)}{\partial \boldsymbol{a}^{(L)}} \cdot \frac{\partial \boldsymbol{a}^{(L)}(\boldsymbol{x}_i)}{\partial w},\end{aligned} \tag{2.52}$$

The first term can be computed manually for a given choice of the loss function (see Section 2.1). For example, for the MSE loss function, we have

$$\ell_{\mathrm{MSE}}(\mathbf{NN}(\boldsymbol{x}_i), \boldsymbol{y}_i) = ||\mathbf{NN}(\boldsymbol{x}_i) - \boldsymbol{y}_i||^2, \tag{2.53}$$

resulting in

$$\frac{\partial \ell_{\mathrm{MSE}}(\mathbf{NN}(\boldsymbol{x}_i), \boldsymbol{y}_i)}{\partial \mathbf{NN}} = 2(\mathbf{NN}(\boldsymbol{x}_i) - \boldsymbol{y}_i). \tag{2.54}$$

Therefore, the central quantity of interest is

$$\frac{\partial \mathbf{NN}(\boldsymbol{x})}{\partial w} = \frac{\partial \boldsymbol{a}^{(L)}(\boldsymbol{x})}{\partial w}, \tag{2.55}$$

which we are going to compute via repeated application of the chain rule. In the following, we drop the explicit dependence on $\boldsymbol{x}$.

The key observation for the backpropagation algorithm is the fact that, due to the layer-wise processing of information in a feedforward NN, the only way a weight in layer l influences the loss is through the next layer $l + 1$. Thus, let us start by looking at the last layer L. Recall that $\boldsymbol{a}^{(l)} = \varsigma^{(l)}(\boldsymbol{z}^{(l)})$ from Eq. (2.48). Using the chain rule, we have

$$\frac{\partial \boldsymbol{a}^{(L)}}{\partial w} = \frac{\partial \boldsymbol{a}^{(L)}}{\partial \boldsymbol{z}^{(L)}}\frac{\partial \boldsymbol{z}^{(L)}}{\partial w} = \frac{\partial \boldsymbol{a}^{(L)}}{\partial \boldsymbol{z}^{(L)}}\left(\frac{\partial \boldsymbol{W}^{(L)}}{\partial w}\boldsymbol{a}^{(L-1)} + \boldsymbol{W}^{(L)}\frac{\partial \boldsymbol{a}^{(L-1)}}{\partial w}\right), \tag{2.56}$$

where $\partial \boldsymbol{a}^{(L)}/\partial \boldsymbol{z}^{(L)} = \boldsymbol{J}_\varsigma^{(L)}$ is the Jacobian matrix of $\varsigma^{(L)}$ containing the derivative of the activation functions $\left(\boldsymbol{J}_\varsigma^{(L)}\right)_{i,j} = \partial \varsigma_i^{(L)}(\boldsymbol{z})/\partial z_j$. Note that $\boldsymbol{J}_\varsigma^{(l)}$ is diagonal for some activation functions, as ReLUs (Eq. (2.44)), but not for others, such as the softmax function (Eq. (2.37)). From Eq. (2.56), if w is a weight of layer L, that is, $w = w_{i,j}^{(L)}$, we have

$$\frac{\partial \boldsymbol{a}^{(L)}}{\partial w} = \boldsymbol{J}_\varsigma^{(L)}\boldsymbol{e}_i^{(L)}a_j^{(L-1)}. \tag{2.57}$$

Here, $\boldsymbol{e}_i^{(L)}$ is an activation vector of layer L, where the activation of all nodes is zero except for the ith node whose activation is one. Otherwise, we have

$$\frac{\partial \boldsymbol{a}^{(L)}}{\partial w} = \boldsymbol{J}_{\varsigma}^{(L)} \boldsymbol{W}^{(L)} \frac{\partial \boldsymbol{a}^{(L-1)}}{\partial w}. \tag{2.58}$$

To evaluate this expression, one needs to go further back in the layers and compute the derivatives $\partial \boldsymbol{a}^{(L-1)}/\partial w$ given by

$$\frac{\partial \boldsymbol{a}^{(L-1)}}{\partial w} = \frac{\partial \boldsymbol{a}^{(L-1)}}{\partial \boldsymbol{z}^{(L-1)}} \frac{\partial \boldsymbol{z}^{(L-1)}}{\partial w} = \boldsymbol{J}_{\varsigma}^{(L-1)} \left(\frac{\partial \boldsymbol{W}^{(L-1)}}{\partial w} \boldsymbol{a}^{(L-2)} + \boldsymbol{W}^{(L-1)} \frac{\partial \boldsymbol{a}^{(L-2)}}{\partial w} \right). \tag{2.59}$$

Again, if w is part of layer $L-1$, that is, $w = w_{i,j}^{(L-1)}$, we have

$$\frac{\partial \boldsymbol{a}^{(L-1)}}{\partial w} = \boldsymbol{J}_{\varsigma}^{(L-1)} \boldsymbol{e}_i^{(L-1)} a_j^{(L-2)}. \tag{2.60}$$

Otherwise, we have

$$\frac{\partial \boldsymbol{a}^{(L-1)}}{\partial w} = \boldsymbol{J}_{\varsigma}^{(L-1)} \boldsymbol{W}^{(L-1)} \frac{\partial \boldsymbol{a}^{(L-2)}}{\partial w}. \tag{2.61}$$

Recognizing the recursive nature of the computation, we have the following relation:

$$\frac{\partial \boldsymbol{a}^{(l)}}{\partial w} = \boldsymbol{J}_{\varsigma}^{(l)} \boldsymbol{W}^{(l)} \boldsymbol{J}_{\varsigma}^{(l-1)} \boldsymbol{W}^{(l-1)} \ldots \boldsymbol{J}_{\varsigma}^{(l'+1)} \boldsymbol{W}^{(l'+1)} \frac{\partial \boldsymbol{a}^{(l')}}{\partial w}, \tag{2.62}$$

given that w is not a weight of the layers l' through l. If w is part of layer l', that is, $w = w_{i,j}^{(l')}$, we instead have

$$\frac{\partial \boldsymbol{a}^{(l)}}{\partial w} = \boldsymbol{J}_{\varsigma}^{(l)} \boldsymbol{W}^{(l)} \boldsymbol{J}_{\varsigma}^{(l-1)} \boldsymbol{W}^{(l-1)} \ldots \boldsymbol{J}_{\varsigma}^{(l'+1)} \boldsymbol{W}^{(l'+1)} \boldsymbol{J}_{\varsigma}^{(l')} \boldsymbol{e}_i^{(l')} a_j^{(l'-1)}. \tag{2.63}$$

Finally, we have all the ingredients to formulate the backpropagation algorithm. Recall that our goal is to compute the derivative in Eq. (2.52) with respect to all tunable weights. To do that efficiently, we first initialize the following *deviation* at the output layer:

$$\boldsymbol{\Delta}^{(L)} = \frac{\partial \ell(\boldsymbol{a}^{(L)}(\boldsymbol{x}_i), \boldsymbol{y}_i)}{\partial \boldsymbol{a}^{(L)}} \odot \boldsymbol{J}_{\varsigma}^{(L)}, \tag{2.64}$$

where $\odot$ denotes an element-wise (Hadamard) product.[32] This intermediate quantity turns out to be useful throughout the computation. Taking the MSE loss as an example, this would correspond to (see Eq. (2.54))

$$\boldsymbol{\Delta}^{(L)} = 2(\boldsymbol{a}^{(L)} - \boldsymbol{y}_i) \odot \boldsymbol{J}_{\varsigma}^{(L)}. \tag{2.65}$$

From Eq. (2.57), it follows that the contributions to the derivative of the loss function with respect to a weight $w_{i,j}^{(L)}$ in layer L are given by

$$2(\boldsymbol{a}^{(L)} - \boldsymbol{y}_i) \odot \boldsymbol{J}_{\varsigma}^{(L)} \boldsymbol{e}_i^{(L)} a_j^{(L-1)} = \boldsymbol{\Delta}^{(L)} \boldsymbol{e}_i^{(L)} a_j^{(L-1)}. \tag{2.66}$$

[32]It simplifies to a regular scalar product given a single output node.

The final derivative $\partial\ell(\boldsymbol{a}^{(L)}(\boldsymbol{x}_i), \boldsymbol{y}_i)/\partial w_{i,j}^{(L)}$ is then obtained by summing up all components of this vector

$$\partial\ell(\boldsymbol{a}^{(L)}(\boldsymbol{x}_i), \boldsymbol{y}_i)/\partial w_{i,j}^{(L)} = \sum_k \left(\boldsymbol{\Delta}^{(L)}\boldsymbol{e}_i^{(L)} a_j^{(L-1)}\right)_k , \tag{2.67}$$

that is, all individual contributions to the inner product given in Eq. (2.52). Having computed the derivative with respect to all weights in layer L, we move one layer backward, hence the name *back*propagation. From Eq. (2.58), it follows that the contributions to the derivative of a weight $w_{i,j}^{(L-1)}$ in layer $L-1$ are given by

$$\boldsymbol{\Delta}^{(L-1)}\boldsymbol{e}_i^{(L-1)} a_j^{(L-2)} , \tag{2.68}$$

where

$$\boldsymbol{\Delta}^{(L-1)} = \boldsymbol{\Delta}^{(L)}\boldsymbol{W}^{(L)}\boldsymbol{J}_\varsigma^{(L-1)} . \tag{2.69}$$

Notice the intimate connection between the above procedure and the expressions in Eqs (2.62) and (2.63). Thus, through recursion we have

$$\boldsymbol{\Delta}^{(l-1)} = \boldsymbol{\Delta}^{(l)}\boldsymbol{W}^{(l)}\boldsymbol{J}_\varsigma^{(l-1)} . \tag{2.70}$$

This process is repeated until one arrives at the first layer. At the end of this reverse pass through the NN, one has computed the desired derivative (Eq. (2.52)) with respect to all tunable weights. During the backpropagation algorithm, the value of all activations $\{\boldsymbol{a}^{(l)}(\boldsymbol{x}_i)\}_{l=0}^L$ and the derivatives of the corresponding activation function evaluated at that activation $\{\boldsymbol{J}_\varsigma^{(l)}(\boldsymbol{x}_i)\}_{l=1}^L$ must be known. To avoid any recomputation, one performs an evaluation of the network for the given input $\boldsymbol{x}_i$, that is, a forward pass, and caches all the required intermediate computation results before executing the backpropagation algorithm, that is, the reverse pass.

To further illustrate how backpropagation works, let us calculate both the forward and reverse passes *explicitly* on the example of a simple two-layer NN with the MSE as the loss function and ReLUs as activation functions, $\varsigma^{(l)}(z) = \varsigma(z) = \text{ReLU}(z) = \max(0, z)$ which act element-wise. The derivative of ReLU is 1 for $z > 0$ and 0 otherwise.[33] We randomly initialize the weights of this NN and ignore biases, see panel (a) of Fig. 2.10. The forward pass for an exemplary input-output pair is presented in Fig. 2.10(b). Importantly, the intermediate computation results are cached. This includes the activations of all nodes, $\boldsymbol{a}^{(1)}$ and $a^{(2)}$, and the derivatives of the corresponding activation functions. Then, the backward pass starts in panel (c) of Fig. 2.10. Here, we only focus on the calculation of the derivative of ℓ with respect to two weights coming from different layers, $w_1^{(2)}$ and $w_{1,2}^{(1)}$. The first step is to compute the deviation $\Delta^{(2)}$ on the last layer from Eq. (2.65), after which $a^{(2)}$ can be erased from memory. Then, using Eq. (2.66), we can calculate the derivatives of ℓ with respect to any weight $\boldsymbol{w}^{(2)}$ in the last layer, and activations $\boldsymbol{a}^{(1)}$ can be discarded. The next step is to compute the deviations $\Delta_1^{(1)}$ and $\Delta_2^{(1)}$ on the second-last layer following Eq. (2.70).

[33] Formally, ReLU is nondifferentiable at $z = 0$. In numerical practice, the derivative at $z = 0$ is usually set to 0.

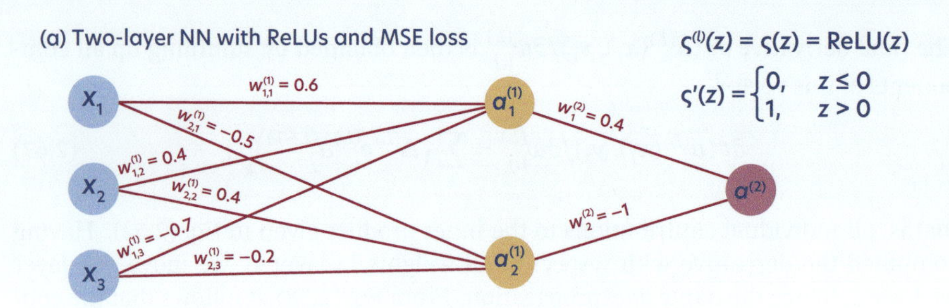

(b) Forward pass for $(\mathbf{x},y) = ([1,1,0], 0)$ caching intermediate results

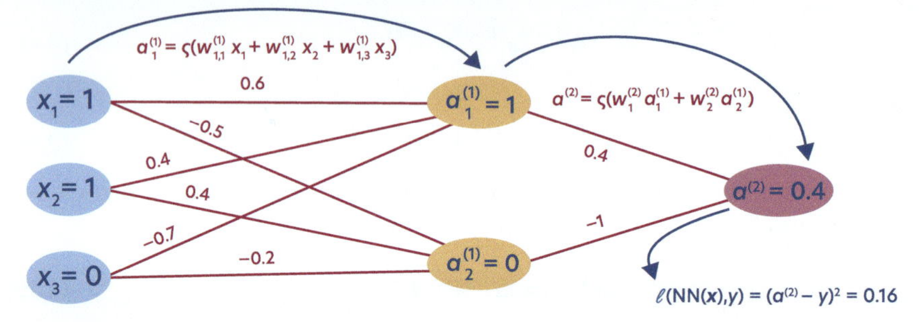

(c) Backward pass for $(\mathbf{x},y) = ([1,1,0], 0)$ to compute gradients taking advantage of cached results from the forward pass

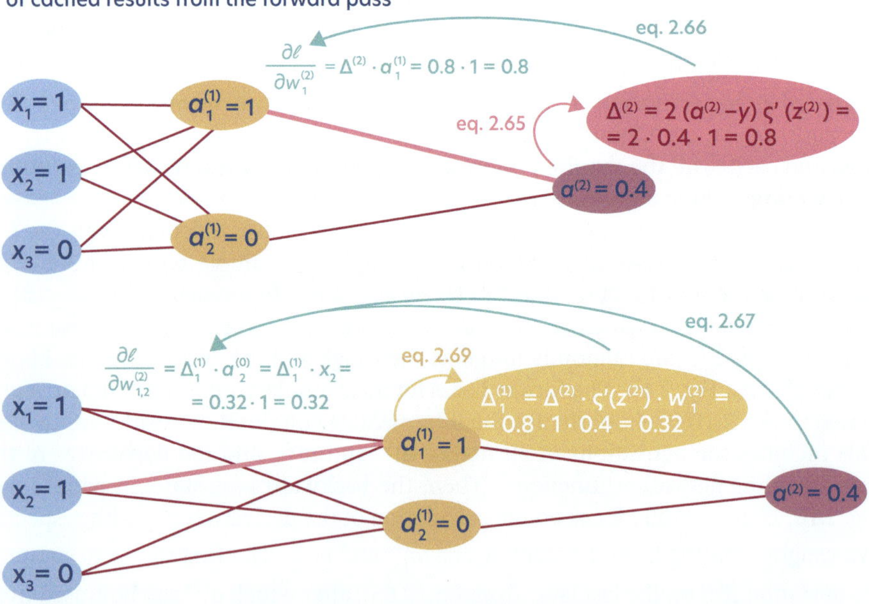

Figure 2.10 Backpropagation on a simple example of a two-layer feedforward NN presented in (a). (b) Forward pass through the network. (c) Reverse pass calculating derivatives.

After this computation, $\Delta^{(2)}$ can be discarded. Using $\Delta_1^{(1)}$ and $a_2^{(0)} = x_2$ and following Eq. (2.68), we can compute $\partial\ell/\partial w_{1,2}^{(1)}$. In the case of a two-layer NN, this concludes the calculation of the gradient. Note that at every step memory can be freed by erasing cached results from the forward pass.

At this point, an interesting question might arise. Why do we compute the derivative in a reverse pass instead of a forward pass? To answer this question, we first have to formulate the corresponding *forward-propagation algorithm*. For each weight $w_{i,j}^{(l)}$ with respect to which one wants to compute a derivative (Eq. (2.52)), one first computes

$$\frac{\partial \boldsymbol{a}^{(l)}}{\partial w_{j,i}^{(l)}} = \boldsymbol{J}_{\varsigma}^{(l)} \boldsymbol{e}_i^{(l)} a_j^{(l-1)}. \tag{2.71}$$

Now, one can directly make use of the relation in Eq. (2.62) to obtain $\partial \boldsymbol{a}^{(L)}/\partial w_{i,j}^{(l)}$, from which the final derivative can be computed via Eq. (2.52). Notice that the difference between the backpropagation and forward-propagation algorithm amounts to evaluating the expression in (2.63) from left-to-right (backward) or right-to-left (forward), respectively.

The forward-propagation algorithm also requires knowledge of the activations and derivatives of the activation functions. In this case, however, one does not need to cache any of the results. Instead, the computation of the derivatives can be carried out in parallel with the forward pass (i.e., the evaluation of the NN). This is because information flows forward from layer-to-layer and no information from the earlier layers is explicitly required at later stages. Besides this difference in memory cost, one can identify two key distinctions between an algorithm based on forward-propagation and the backpropagation algorithm. First, in forward-propagation algorithms, lots of redundant computations are performed, given that the derivatives in later layers have to be computed each time. This redundancy is not present in backpropagation. Second, forward propagation involves unnecessary intermediate calculations where the derivatives of individual nodes are computed. Ultimately, this culminates in the fact that the number of passes through the NN in the forward-propagation algorithm scales with the number of tunable weights, whereas this is not the case in the backpropagation algorithm. This is why backpropagation is generally preferred over forward-propagation-based algorithms for computing gradients in NNs, in particular DNNs featuring a large number of tunable parameters. As such, backpropagation has played a key role in the success of DL and enabled the widespread application of NNs. For a more general in-depth discussion of these concepts, including implementation details, see Section 7.1.

Further reading

1. Bishop, C. M. (2006). *Pattern Recognition and Machine Learning*. Springer "Information Science and Statistics" series. The standard book about standard ML [93].

2. Goodfellow, I., Bengio, Y. & Courville, A. (2016). *Deep Learning*. An MIT Press book. One of the best textbooks on DL with an explanation of all preliminaries [13].
3. Mehta, P. *et al.* (2019). *A high-bias, low-variance introduction to machine learning for physicists*. Phys. Rep. 810, 1–124. For the physicist-friendly introduction to ML [94].
4. Zhang, A. *et al.* (2021). *Dive into Deep Learning*. Interactive DL book with code, math, and discussions. Implemented with NumPy/MXNet, PyTorch, and TensorFlow [95].
5. Recordings of lectures on "ML for physicists" from 2020/21 and "Advanced ML for physics, science, and artificial scientific discovery" from 2021/22 by Florian Marquardt.
6. Introductory ML course developed specifically with STEM students in mind: ML-lectures.org and accompanying content: Neupert, T. *et al.* (2021). *Introduction to machine learning for the sciences*. arXiv:2102.04883 [96].
7. Carrasquilla, J. & Torlai, G. (2021). *How to use neural networks to investigate quantum many-body physics*. PRX Quantum 2, 040201. Tutorial on ML for selected physical problems with code [97].

3 Phase classification

One of the fields in physics where machine learning (ML) and, in particular, neural networks (NNs), could be especially useful is condensed matter physics [5], which revolves around the study of the collective behavior of interacting particles. The difficulties associated with describing such systems arise due to the rapid growth of the number of degrees of freedom as the particle number grows, leading to a larger configuration space. The "standard" approach to circumvent these challenges is to find suitable order parameters – quantities that represent the important "macroscopic" degrees of freedom in a system without keeping track of all the microscopic details. The order as quantified by these order parameters naturally separates matter into different states, that is, phases [98, 99]. For some systems, the order parameter is quite simple: In ferromagnets, for example, the order parameter simply corresponds to the magnetization, which is given by a sum of *local* magnetic moments. In general, however, the identification of order parameters and the classification of matter into distinct phases are difficult tasks. Topological phases of matter, for example, are characterized by topological properties that are intrinsically *nonlocal*. The identification of order parameters represents a crucial first step toward understanding the physics that underlies a many-body system, and identifying an appropriate order parameter for novel phases of matter typically requires lots of physical intuition and educated guessing.

On the other hand, in fields such as computer vision, it has been demonstrated that NNs can be trained to correctly classify intricate sets of labeled data naturally living in high dimensions (see MNIST [31] or CIFAR [33]). This motivates us to explore ML techniques as a novel tool to probe the enormous state space of relevant many-body systems that are currently intractable with other algorithms [6]. Among all potential applications of ML to condensed matter physics, learning phases from (simulated or experimental) data is a particularly intriguing one: It could allow us to discover new phases and new physics without prior human knowledge or supervision. In what follows, we aim to give the reader a first introduction to the field of phase classification using ML.

3.1 Prototypical physical systems for the study of phases of matter

In the following, we briefly describe the two prototypical physical systems for which we demonstrate the task of phase classification in the next sections: the Ising model [100], which exhibits a *symmetry-breaking* phase transition and can be characterized by a simple local order parameter, and the Ising gauge theory (IGT) [101], which shows a *topological* phase without a local order parameter.[1]

[1] In the Landau paradigm of phase transitions [102, 103], changes between phases of matter are fundamentally connected to changes in the underlying symmetries. Interestingly, Landau's symmetry-breaking theory of phase transitions breaks down for topological phases of matter [104].

3.1.1 Ising model

We consider the two-dimensional square-lattice ferromagnetic Ising model, which is one of the simplest classical statistical models to show a phase transition and serves as a simple description of ferromagnetism. Ferromagnetism arises when a collection of spins aligns, yielding a net magnetic moment that is macroscopic in size. In the Ising model, for each lattice site k, there is a discrete (classical) spin variable $\sigma_k \in \{+1, -1\}$ leading to a state space of size 2^N given N lattice sites. The energy of a spin configuration is specified by the following Hamiltonian:

$$H(\boldsymbol{\sigma}) = -J \sum_{\langle i,j \rangle} \sigma_i \sigma_j, \tag{3.1}$$

where the sum runs over nearest-neighboring sites (with periodic boundary conditions), and J is the interaction strength $J > 0$ (ferromagnetic interaction).[2] Let us assume that the system is at equilibrium at an inverse temperature $\beta = 1/k_\mathrm{B}T$, where k_B is the Boltzmann constant and T the temperature. Then, the probability of finding the system in a state with a spin configuration $\boldsymbol{\sigma}$ is described by the Boltzmann distribution

$$P_T(\boldsymbol{\sigma}) = \frac{e^{-\beta H(\boldsymbol{\sigma})}}{Z_T}. \tag{3.2}$$

Here $Z_T = \sum_{\boldsymbol{\sigma}} e^{-\beta H(\boldsymbol{\sigma})}$ is the partition function, where the sum runs over all possible spin configurations. Example spin configurations of the Ising model at various temperatures are shown in Fig. 3.1(a). Using Eq. (3.2), the expectation value of a given observable $O(\boldsymbol{\sigma})$ can be expressed as

$$\langle O(\boldsymbol{\sigma}) \rangle_T = \sum_{\boldsymbol{\sigma}} P_T(\boldsymbol{\sigma}) O(\boldsymbol{\sigma}). \tag{3.3}$$

For example, the observable corresponding to the magnetization per site is given by

$$m(\boldsymbol{\sigma}) = \frac{1}{N} \sum_i \sigma_i. \tag{3.4}$$

In 1944, Onsager [100] obtained the following analytical expression for the critical temperature:

$$T_\mathrm{c} = \frac{2J}{k_\mathrm{B} \ln\left(1 + \sqrt{2}\right)}, \tag{3.5}$$

at which a phase transition between a *high-temperature paramagnetic* (disordered) phase and a *low-temperature ferromagnetic* (ordered) phase occurs, see Fig. 3.1. For temperatures below the critical temperature T_c, spontaneous magnetization occurs, that is, the interaction is sufficiently strong to cause neighboring spins to spontaneously align, leading to a nonzero mean magnetization. At temperatures above T_c, thermal fluctuations completely dominate over any alignment of spins, and a zero magnetization is observed. As such, the magnetization serves as an order parameter, which is zero within the disordered (paramagnetic phase) and approaches one in the ordered (ferromagnetic phase), see Fig. 3.1(b).

[2] The case $J < 0$ corresponds to the two-dimensional square-lattice *antiferromagnetic* Ising model which exhibits a phase transition at the same critical temperature.

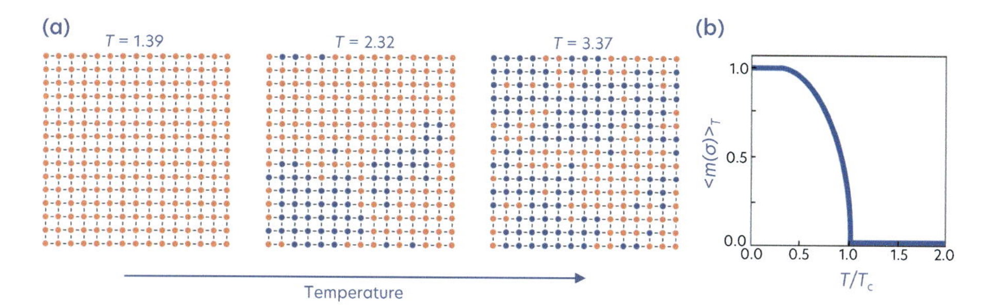

Figure 3.1 (a) Example spin configuration samples of the Ising model with $J = 1$ and $k_B = 1$ at various temperatures, where $T_c \approx 2.27$ (see Eq. (3.5)). Here, the blue (orange) colored dots on each lattice site denote the value of the spin variable at that site $\sigma_k = 1$ ($\sigma_k = -1$). Panel reproduced from [2, Notebook A1]. (b) Mean magnetization per site $\langle m(\boldsymbol{\sigma})\rangle_T$ of the Ising model as a function of the temperature T.

3.1.2 Ising gauge theory

One of the most exciting research areas is the classification of phases that do not have a local order parameter but rather a global one. Examples of systems that exhibit such phases are band topological insulators and topological superconductors [105]. Detecting topological phases is a challenging task from the experimental point of view because, in general, experimentalists have access only to local observables. In this context, ML techniques can be of great help [106–112].

The IGT [101] is the prototypical example of a system that exhibits a topological phase of matter. Like the Ising model, the IGT is also a classical spin model ($\sigma_k \in \{+1, -1\}$) defined on a square lattice (with periodic boundary conditions). Here, however, the spins are placed on the lattice bonds. It is described by the following Hamiltonian:

$$H(\boldsymbol{\sigma}) = -J \sum_{p} \prod_{i \in p} \sigma_i, \tag{3.6}$$

where p refers to plaquettes on the lattice; see Fig. 3.2. The ground state of this Hamiltonian is a highly degenerate manifold spanned by all states that meet the local constraint that the product of spins along each plaquette is $\prod_{i \in p} \sigma_i = 1$. As such, this ground state corresponds to a topological phase of matter. In systems of finite size, the violations of the local constraints are strongly suppressed, and the system exhibits a slow transition from the low-temperature topological phase to the high-temperature phase with violated constraints. This allows for the definition of a crossover temperature T_c defined by the first appearance of a violated local constraint.[3]

There exists an interesting representation that highlights the topological character of the ground state of the IGT: connect the edges of the lattice that contain spins with the same orientation and form loops. The ground-state phase is then characterized by the property that all these loops are closed; the violation of a constraint results

[3]Note that as we increase the system size $T_c \to 0$, that is, the crossover temperature vanishes in the thermodynamic limit. As such, the IGT does not exhibit a phase transition at a nonzero temperature.

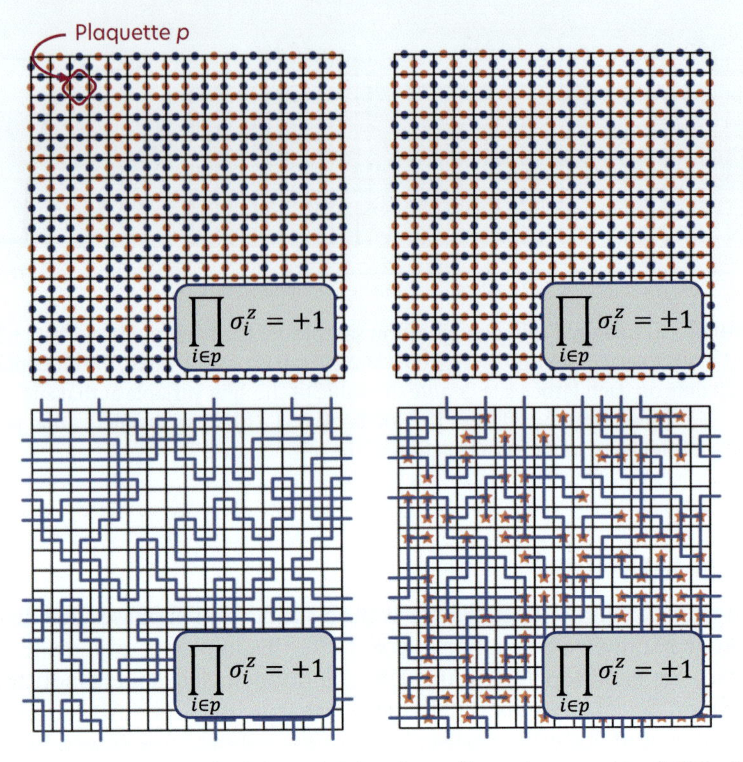

Figure 3.2 The upper panels show example spin configuration samples of IGT at $T = 0$ (left panel) and $T \to \infty$ (right panel) where local constraints are satisfied for all (some) plaquettes, respectively. An exemplary plaquette is marked in red. The lower panels show the corresponding dual representation, where the stars highlight loop breakage. Reproduced with [2, Notebook A1].

in the appearance of an open loop, see Fig. 3.2. Looking at typical spin configuration samples of the IGT makes clear that its phases are hard to distinguish visually without prior knowledge of the local constraints or the corresponding dual representation. As such, IGT and other systems characterized by nonlocal and long-range correlations pose a hard problem for any phase classification algorithm.

3.2 Unsupervised phase classification without neural networks

Having introduced the Ising model and the IGT, let us discuss how we can classify their respective phases of matter. In particular, we are concerned with unsupervised ML algorithms. They work with training data that do not need to be labeled (see Section 1.5). Unsupervised learning algorithms must *by itself* discover the relevant patterns in a training dataset. As such, these algorithms represent a primary candidate for the autonomous discovery of new phases as they do not require prior labeling of the samples by the phase they belong to.

In particular, we discuss algorithms that perform a dimensionality reduction. In dimensionality reduction, we are concerned with projecting the input data into a lower-dimensional space. While any dimensionality reduction necessarily leads to an information loss, one aims to discard only information in the input data that is less relevant to the problem at hand. In particular, it is believed that real-world data often resides on a low-dimensional manifold within the original space [113]. For example, one expects that the set of images one would like to classify constitutes a small subspace of all possible images. In this case, the data can be effectively described by fewer degrees of freedom. Clearly, such an approach lends itself naturally to distinguishing between different phases of matter and detecting phase transitions in condensed matter systems: We want to discard the information-rich but complicated microscopic description of the system for the sake of a simpler macroscopic description, for example, in the form of an order parameter.

Once we have performed the dimensionality reduction, we may already learn a lot about the given problem by visualizing the data within the low-dimensional representation space. We tend to think that samples from the same phase of matter should be more similar to each other than to samples from another phase. If the dimension reduction technique preserves some of this similarity, we expect this to reveal itself in the data visualization. However, this is not guaranteed to work in general. We see an example of such a failure in Section 3.2.1.

Going beyond visualization, we can process the data further, for example, using *clustering* methods. Clustering is one of the most fundamental unsupervised learning methods used to group unlabeled data into clusters of similar data points, where the similarity is assessed by a distance measure. In our context, the clusters would ideally correspond to the different phases of matter present in the data. There exist many different clustering algorithms suited for different types of data, with k-means clustering being one of the simplest (see [94] for further details).

Clustering can, in principle, be performed without dimensionality reduction as a preprocessing step. However, dimensionality reduction may help in several aspects [94, 114]. First, clustering typically relies on the Euclidean distance being a good measure of similarity.[4] The distance between two data points in the original high-dimensional representation may, however, not be particularly relevant as it is believed to often reside on a non-Euclidean manifold. Dimensionality reduction techniques can allow for the identification of a low-dimensional Euclidean representation of the data. The Euclidean distance between data points within this representation is often physically more meaningful, resulting in a better clustering. Secondly, performing dimensionality reduction as a preprocessing step helps to alleviate the problems of the curse of dimensionality experienced when clustering data in high-dimensional spaces. Finally, identifying a low-dimensional representation also helps to better visualize and understand the clustering that is eventually obtained.

3.2.1 Principal component analysis

As an example, we consider principal component analysis (PCA), which is a common method to perform dimensionality reduction. PCA identifies mutually orthogonal

[4] In general, whether clustering succeeds or not depends on whether the choice of distance measure (be it Euclidean or not) is a good measure of similarity.

directions, called principal components (PCs), in the data space along which the linear correlation in the data vanishes. We rank each PC based on the variance of the data along the corresponding direction. To reduce the dimensionality of our space, we discard the PCs along which the data shows the least variance. As such, in PCA directions along which the data exhibits a *large variance* are considered to contain the most *important information*. In our case, ideally, the data (raw spin configuration samples) naturally splits into different clusters corresponding to the individual phases of the system when displayed in their low-dimensional representation.

To be more precise, we consider the case where we are given n data points $\{\boldsymbol{x}_1, \boldsymbol{x}_2, ..., \boldsymbol{x}_n\}$ each living in a m-dimensional feature space $\mathbb{R}^m$ with zero mean $\bar{\boldsymbol{x}} = \frac{1}{n}\sum_{i=1}^{n} \boldsymbol{x}_i = 0$. Note that real-life data typically does not have zero mean. In this case, the data first needs to be transformed by subtracting the mean element-wise. We define the $n \times m$ design matrix $\boldsymbol{X} = [\boldsymbol{x}_1, \boldsymbol{x}_2, ..., \boldsymbol{x}_n]^{\mathrm{T}}$. The symmetric $m \times m$ empirical covariance matrix is then given as $\boldsymbol{\Sigma} = \frac{1}{n}\boldsymbol{X}^{\mathsf{T}}\boldsymbol{X}$. Here, the ith diagonal entry of the covariance matrix $\boldsymbol{\Sigma}_{ii}$ corresponds to the variance of the ith feature over the entire data and the off-diagonal entries $\boldsymbol{\Sigma}_{ij}$ correspond to the covariance between feature i and feature j. The basis in which the correlations between features vanish corresponds to the eigenbasis of $\boldsymbol{\Sigma}$ in which $\boldsymbol{\Sigma}$ appears diagonal. Consequently, the problem of finding directions along which the linear correlation in the data vanishes reduces to diagonalizing $\boldsymbol{\Sigma}$, that is, finding its eigenvectors (or PCs) $\{\boldsymbol{v}_1, \boldsymbol{v}_2, ..., \boldsymbol{v}_m\}$ and eigenvalues $\{\lambda_1, \lambda_2, ..., \lambda_m\}$. Here, the eigenvalue λ_i corresponds to the variance of the data along the direction given by $\boldsymbol{v}_i$. We denote $\tilde{\lambda}_j = \lambda_j / \sum_{i=1}^{m} \lambda_i$ as the ratio of explained variance contained in the jth PC. We refer to the appendix for a mathematical derivation of the procedure. Dimensionality reduction is then performed by selecting the first k PCs with the largest ratios of explained variance $\tilde{\lambda}$ and projecting the data into this space of reduced dimensionality. The projection is performed by the linear transformation $\tilde{\boldsymbol{X}} = \boldsymbol{X}\tilde{\boldsymbol{V}}$, where $\tilde{\boldsymbol{V}} = [\boldsymbol{v}_1, \boldsymbol{v}_2, ..., \boldsymbol{v}_k]$ and $\tilde{\boldsymbol{X}}$ is the projected design matrix. Note that one has to choose k, the number of PCs to keep. This can be done in an ad hoc fashion that may be problem-specific or, for example, by choosing the minimal number of PCs such that $\sum_{i=1}^{k} \tilde{\lambda}_i \geq \tilde{\lambda}_{\text{thresh}}$, where $\tilde{\lambda}_{\text{thresh}}$ is the desired threshold explained variance ratio. The procedure is summarized in Algorithm 2. For an intuitive understanding of the procedure, we refer to Fig. 3.3(a) and (b): In this example, the data resides in a two-dimensional feature space. After subtracting the data mean, PCA identifies the first PC that contains the largest proportion of the data variance. PCA can not only be understood as variance maximization but also as a minimization

---

Algorithm 2 Principal component analysis (PCA)

Require: Hyperparameter k (dimensionality of the projected data)
Require: Design matrix $\boldsymbol{X} \in \mathbb{R}^{n \times m}$

$\boldsymbol{X} \leftarrow \boldsymbol{X} - \text{mean}(\boldsymbol{X})$ ▷ Remove mean element-wise
$\boldsymbol{\Sigma} \leftarrow \boldsymbol{X}^{\mathsf{T}}\boldsymbol{X}/n$ ▷ Construct empirical covariance matrix
$\boldsymbol{V} \leftarrow \text{Eigenvectors}(\boldsymbol{\Sigma})$ ▷ Find eigenvectors and orer them by descending eigenvalue
$\tilde{\boldsymbol{V}} \leftarrow \boldsymbol{V}[:, : k]$ ▷ Keep only first k eigenvectors **return** $\tilde{\boldsymbol{X}} \leftarrow \boldsymbol{X}\tilde{\boldsymbol{V}} \in \mathbb{R}^{n \times k}$

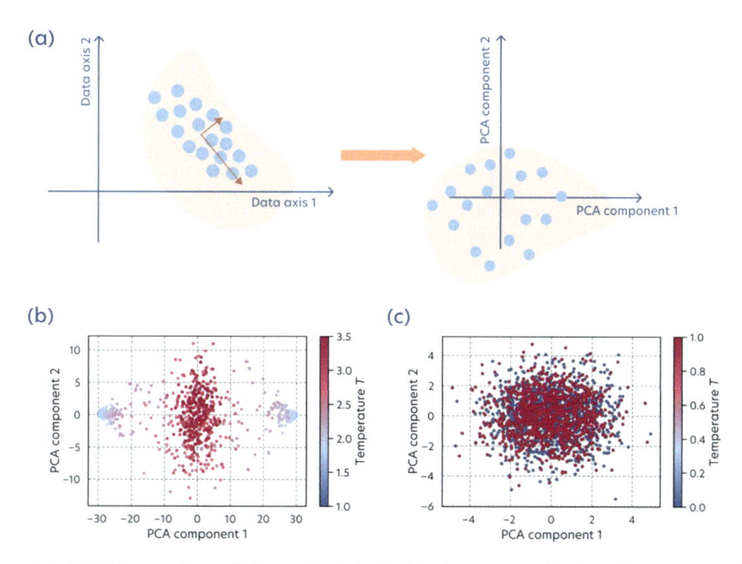

Figure 3.3 (a) Illustration of the principle behind PCA applied to data points (blue) living in a two-dimensional feature space. Orange vectors denote the first two principal components on which we project the data (right panel). PCA applied to the spin configuration samples of (b) the Ising model and (c) IGT with $k = 2$. For the Ising model, the data consists of 50 spin configurations (linear lattice size $L = 30$) sampled using Monte Carlo methods at temperatures T ranging from $T_1 = 1$ to $T_{20} = 3.5$ in equidistant steps. For the IGT, the data consists of 1 000 spin configurations (linear lattice size $L = 16$) drawn within the topological phase and the disordered phase at high temperatures. Panels (b) and (c) are reproduced from [2, Notebook A1].

of a reconstruction error of a linear transformation. The proof of this equivalence can be found in Appendix A. For further details, see, for example, [94].

Now, we can readily apply PCA to our spin configuration samples. Figure 3.3(b, c) shows the results of PCA applied to spin configuration samples of the Ising model and IGT, respectively. For the Ising model, PCA separates the data into three clusters – a high-temperature cluster corresponding to the disordered phase, as well as two low-temperature clusters corresponding to the ordered phase with either positive or negative magnetization. Further analysis shows that the first PC corresponds to the magnetization [114]. By drawing a vertical decision boundary (perpendicular to PC1), which separates the high-temperature cluster and a low-temperature cluster, a rough estimate of the critical transition temperature can be obtained as $T_{\mathrm{c,PCA}} \approx 2.3$ which is in agreement with the Onsager solution. In the case of the IGT, PCA fails to cluster the data into the two prevalent phases [see Fig. 3.3(c)]. This is because PCA is restricted to linear transformations of the input data. While this is sufficient to encode simple local order parameters [114–116], such as the magnetization in the case of the Ising model, linear transformations are not sufficient to compute topological features, that is, nonlocal correlations in the data [116].

As illustrated by the failure of PCA in the case of the IGT, the restriction of PCA to linear transformations of the input space severely limits its performance. That is, one may not be able to find the optimal set of directions to perform dimensionality reduction using PCA. In particular, the low-dimensional manifold on which the data resides within the original space may not necessarily be parametrized by linear transformations of the original coordinates. In such cases, a dimensionality reduction using PCA does not preserve the relative pairwise distance, or similarity, between data points with respect to the manifold. However, this is a desired property for any algorithm that aims at performing dimensionality reduction. This problem is tackled by nonlinear dimensionality reduction techniques, such as the kernel PCA (kPCA) [117] (see Chapter 4 on the kernel trick), the t-distributed stochastic neighbor embedding (t-SNE) [118], or uniform manifold approximation and projection [119]. In Section 3.2.2, we briefly describe t-SNE.

3.2.2 t-Distributed stochastic neighbor embedding

Stochastic neighbor embedding [120] and its variant called t-SNE [118] are techniques for nonlinear dimensionality reduction, which aim to preserve the local structure of the original data. That is, points that are close in the high-dimensional dataset tend to be close to one another in the low-dimensional representation.

Let us consider an initial m-dimensional space with n points, that is, $\boldsymbol{x}_i \in \mathbb{R}^m$. We define the conditional probability $p_{i|j}$ that two points $\boldsymbol{x}_i$ and $\boldsymbol{x}_j$ are similar (i.e., close to one another) as

$$p_{i|j} = \frac{e^{-||\boldsymbol{x}_i - \boldsymbol{x}_j||^2/2\sigma_i^2}}{\sum_{k \neq l} e^{-||\boldsymbol{x}_k - \boldsymbol{x}_l||^2/2\sigma_i^2}}, \tag{3.7}$$

where $||\boldsymbol{x}_i - \boldsymbol{x}_j||$ is the Euclidean distance between the two points. The fact that Gaussian likelihoods are used in $p_{i|j}$ implies that only points near $\boldsymbol{x}_i$ contribute significantly to its probability. The variance σ_i^2 depends on the *perplexity* defined as

$$P_i = 2^{-\sum_{j=1}^{n} p_{j|i} \log_2 p_{j|i}}, \tag{3.8}$$

which is a measure based on Shannon entropy. In the first step of the t-SNE algorithm, the variances σ_i^2 are optimized for each point $\boldsymbol{x}_i$ to have a fixed perplexity value $P_i =$ const. Points in regions of high density have a smaller variance, while regions of low density have a larger variance. In practice, the perplexity is usually set between 5 and 50. Note that $p_{i|j} \neq p_{j|i}$ due to the dependence on σ_i^2. To recover a symmetric relation $p_{i|j} = p_{j|i}$, we define the joint probability distribution as

$$p_{ij} = \frac{p_{i|j} + p_{j|i}}{2n}. \tag{3.9}$$

The objective of the t-SNE algorithm is to find another set of points in lower dimensional representation $\boldsymbol{y}_i \in \mathbb{R}^{n \times d_{\text{red}}}, d_{\text{red}} < m$ and corresponding probability distribution q_{ij} in a new representation for which the KL divergence

$$D_{KL}(p||q) = \sum_{i,j} p_{ij} \log \frac{p_{ij}}{q_{ij}} \tag{3.10}$$

is minimal.

The procedure starts with randomly sampling n points $\boldsymbol{y}_i(z_1, z_2, \ldots, z_{d_{\text{red}}})$ in a d_{red}-dimensional space. For each point, we define the probability distribution q_{ij} in a similar way as in the high-dimensional space but using the t-Student probability distribution instead of Gaussian distributions:

$$q_{ij} = \frac{(1 + ||\boldsymbol{y}_i - \boldsymbol{y}_j||^2)^{-1}}{\sum_{k \neq l}(1 + ||\boldsymbol{y}_k - \boldsymbol{y}_l||^2)^{-1}}. \tag{3.11}$$

In the last step, we minimize the Kullback–Leibler (KL) divergence from Eq. (3.10) (see Section 2.3) by optimizing the position $(z_1, z_2, ..., z_{d_{\text{red}}})$ of each point $\boldsymbol{y}_i = \boldsymbol{y}_i(z_1, z_2, ..., z_{d_{\text{red}}})$ in the d_{red}-dimensional space which eventually yields a low-dimensional data representation. The t-SNE algorithm is summarized in Algorithm 3.

Algorithm 3 t-distributed stochastic neighbor embedding (t-SNE)

Require: Hyperparameters: d (dimensionality of the projected data), perplexity P, learning rate η
Require: Original data set of n points in m dimensional space $\boldsymbol{X} \in \mathbb{R}^{n \times m}$
Require: Random set of n points in lower dimensional $d_{\text{red}} < m$ representation $\boldsymbol{Y}^{(0)} \in \mathbb{R}^{n \times d_{\text{red}}}$
  for each $\boldsymbol{x}_i$ **do**
    for each $\boldsymbol{x}_j$ **do**
      Calculate pairwise conditional probability distribution $p_{i|j}$ with fixed perplexity P
    end for
  end for
  Calculate probability distribution p_{ij}
  for $t = 1$ to T **do**
    for each $\boldsymbol{y}_i^{(t-1)} \in \boldsymbol{Y}^{(t-1)}$ **do**
      for each $\boldsymbol{y}_j^{(t-1)} \in \boldsymbol{Y}^{(t-1)}$ **do**
        Calculate probability distribution q_{ij}
      end for
      Calculate gradients of the KL divergence with respect to coordinates of each $\boldsymbol{y}_i^{(t-1)}(z_1^{(i)}, z_2^{(i)}, \dots, z_{d_{\text{red}}}^{(i)}) \in \boldsymbol{Y}^{(t-1)}$, i.e., $\frac{\partial D_{KL}(p||q)}{\partial \boldsymbol{z}^{(i)}}$
    end for
    $\boldsymbol{Y}^{(t)} \leftarrow \boldsymbol{Y}^{(t-1)} - \eta \frac{\partial D_{KL}(p||q)}{\partial \boldsymbol{z}}$ ▷ Update the coordinates of each point
  end for

The low-dimensional data representation preserves the local structure of the original dataset, that is, similar points in the original dataset are now clustered in the d_{red}-dimensional representation space. However, the distance between the resulting clusters loses its meaning in representation space.

In general, clustering in combination with dimensionality reduction works elegantly for simple problems, such as the Ising model. However, such approaches typically do not perform well when applied to more difficult phase classification tasks, for example, in the presence of topological phases such as in the IGT [121], or when a large number of phases is present [122].

3.3 Supervised phase classification with neural networks

One may wonder whether the issues encountered by clustering methods introduced in Section 3.2 can be tackled by making use of the powerful machinery of NNs introduced in Section 2.4.4. The idea is the following [123]. We train an NN to take spin configuration samples as input and correctly label them by the phase they belong to, see Fig. 3.4(a). Typically, the label is encoded as a binary bit string in a *one-hot encoding*. In the case of the Ising model, this would correspond to the label 1 for all samples drawn within the ordered phase ($T < T_{\mathrm{c}}$) or the label 0 for all samples drawn within the disordered phase ($T > T_{\mathrm{c}}$).[5] To ensure that the output of the NN can be used to predict a binary label, we choose the output layer to be composed of two nodes to which we apply the softmax activation function introduced in Eq. (2.37) over the activations $\boldsymbol{x}_j$ of all nodes within the output layer. This ensures that the output layer encodes a valid probability distribution over the classes. The predicted label is then typically chosen based on the node that yields the maximum probability. For training, one typically employs the binary cross-entropy (see Eq. (2.2)) which, for a fixed input $\boldsymbol{x}$ is given as

$$\mathcal{L} = -\sum_j p_j(\boldsymbol{x}) \log(\mathrm{NN}(\boldsymbol{x})_j). \tag{3.12}$$

Here, $\mathrm{NN}(\boldsymbol{x})$ denotes the output of an NN, which contains a softmax activation function in its last layer, applied to the input $\boldsymbol{x}$. The sum runs over all output nodes, that is, the number of distinct classes. $p_j(\boldsymbol{x})$ is the true label of the input $\boldsymbol{x}$ as specified by the one-hot encoding. For example, given two classes and an input whose true label is 0, we have $p_0(\boldsymbol{x}) = 1$ and $p_1(\boldsymbol{x}) = 0$ such that $\sum_j p_j(\boldsymbol{x}) = 1$. In Eq. (3.12), this is compared to $\mathrm{NN}(\boldsymbol{x})_j$ which is the activation of the jth output node and corresponds to the predicted probability of the input $\boldsymbol{x}$ to belong to class j.

In our example, the training set consists of labeled spin configuration samples for a wide range of temperatures far above and below T_c, whereas the test set is chosen over the entire temperature range. After training the NN (see Section 2.4.4) on the training set, it is evaluated on the test set. In particular, we average the activation of the two nodes in the output layer, which encode the probability of the input sample belonging to phase 0 or 1, respectively, over the test set. Remarkably, Fig. 3.4(b) shows that these activations cross over precisely at T_{c} enabling us to extract the correct critical temperature. Similarly, this method is capable of correctly identifying the crossover temperature in the IGT [123]. The fact that NNs can generalize to unseen input data can, for example, be exploited as follows. An NN trained on configurations for the square-lattice ferromagnetic Ising model can also highlight the critical temperature of the Ising model with a different lattice geometry, such as a triangular lattice [123]. Note that the ferromagnetic Ising model on a triangular lattice is a typical example of a frustrated system.

[5] Of course, the opposite choice for labeling the two phases with label 0 for the ordered phase and 1 for the disordered phase is equally good.

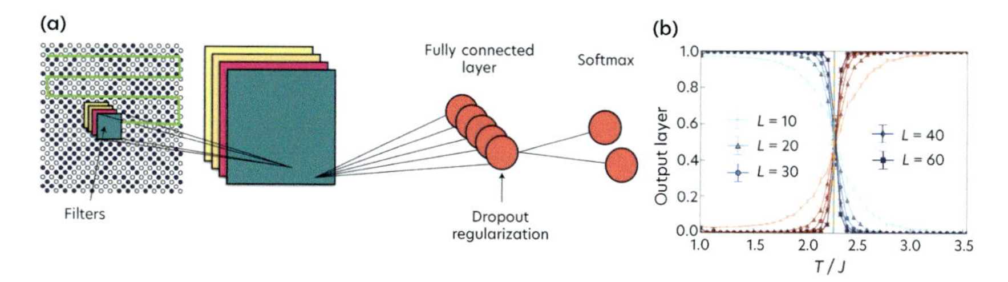

Figure 3.4 Supervised phase classification with an NN performed on the Ising model of varying linear lattice size L ($N = L^2$) (see [123] for further details). (a) (Convolutional) NN applied to a configuration sample of the Ising model. (b) The average activations of the two nodes in the output layer is given by blue and red curves, respectively. The predicted critical temperature is marked by their crossover and is in good agreement with the Onsager solution (Eq. (3.5)) depicted as an orange vertical line. Adapted from [123] with permission from Springer Nature.

3.4 Unsupervised phase classification with neural networks

In Section 3.3, we showed that NNs can perform supervised phase classification. Due to its supervised nature, this approach requires partial knowledge of the phase diagram of the system. One can determine the critical temperature (through "interpolation") if one knows the labels of samples deep within two neighboring phases. Ideally, to discover new phases of matter, a phase classification algorithm should not rely on such a priori knowledge about the phases, that is, it should be unsupervised in that regard. While clustering is unsupervised, we have seen that its power can be limited. In Sections 3.4.1–3.4.3, we discuss three methods that use NNs to perform unsupervised phase classification.

3.4.1 Learning with autoencoders

A natural NN-based unsupervised method is based on the analysis of the latent data representation given by an AE. As we have briefly explained in Section 2.4.5, AEs are NNs with a bottleneck in their center, which are trained to reconstruct the input at the output. The architecture of a typical AE is depicted in Fig. 3.5. Due to the bottleneck, the information passing through the network needs to get compressed at the bottleneck and then decompressed to recover the input. As a consequence of the compression, some information may be lost.[6] However, the retained information in the bottleneck should ideally contain everything relevant for the reconstruction of the input. Therefore, the bottleneck forms a *latent space* that contains a compressed representation of the input data. This is akin to the dimensionality reduction schemes we discussed previously (see Section 3.2), which preserve the most important features for the reconstruction. As such, we can analyze the latent representation of the

[6]In general, it is possible that NNs could compress more dimensions into a single neuron. However, in practice NNs tend to learn smooth functions, which penalizes this behavior.

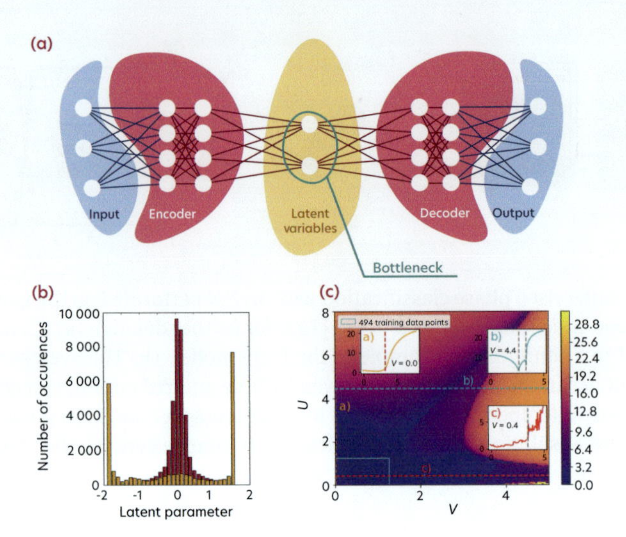

Figure 3.5 (a) Illustration of a natural bottleneck (here two neurons) in an AE architecture. (b) Analysis of bottleneck neurons of an AE trained to reconstruct spin configurations of a two-dimensional Ising model. Latent representation of Ising configurations clusters into two phases visible as a histogram. (c) Anomaly detection scheme allows for the recovery of the phase diagram from the reconstruction loss of an AE trained on one phase (blue box in the bottom left). Panel (b) is taken from [115] and panel (c) is from [126].

input data in a similar way as the lower-dimensional representation obtained by PCA in Section 3.2.1.[7]

Let us apply an AE to reconstruct Monte Carlo samples of the two-dimensional Ising model [115]. Clearly, this represents an unsupervised phase classification scheme because we do not provide any labels. The relevant loss function to be minimized is given by the reconstruction error between the input and output spin configurations (e.g., mean-squared error (MSE)). If we look at how the latent representation of the spin configurations in the trained AE change with the temperature (see Fig. 3.5(b)), we can immediately observe a clustering of the latent parameters. The clusters correspond to the two phases of the Ising model.[8] Red points correspond to the high-temperature paramagnetic phase, while yellow points correspond to the low-temperature ferromagnetic phase. Note the two large yellow bins at the edges of the histogram in Fig. 3.5(b). These are formed due to the degeneracy of the ground state, which has either all spins pointing up, or all spins pointing down.

Analysis of the AE latent representation of the input data is not the only way of an AE-based unsupervised phase classification. Another successful and robust

[7]The quantum versions of AEs are also being developed and applied to phase classification [124] and clustering of subspaces of the Hilbert space [125]. For more details, see Section 8.2.7.

[8]Beware, clustering of data in the latent space according to the phases present in the system is not a general property of AEs. The clustering occurs when input data causes distinctive activations in the bottleneck, which *often* corresponds to different phases.

scheme based on anomaly detection[9] was presented in [126]. The basic idea is as follows. Imagine training an AE to reconstruct states coming from one phase. Then, the AE is used to reconstruct states coming from the rest of the phase diagram. Such a task is difficult because the training data is limited only to one phase, and the AE is bound to make reconstruction errors in other phases. Moreover, we expect that the error is lower for phases that are similar to the "training" phase and higher for phases that contain states that look very different. Finally, the quantum states from the transition regimes are usually distinctive and the most unique from the rest of the phase diagram. Altogether, the reconstruction error across the phase diagram, made by an AE trained to reproduce states from one phase, is expected to vary according to the phases and the phase boundaries in the system. This scheme enables the discovery of phases in a fully unsupervised way. The authors of [126] used this scheme based on anomaly detection to recover a full phase diagram of the extended Bose–Hubbard model in one dimension at exact integer filling. This result is presented in panel (c) of Fig. 3.5. Interestingly, their work also revealed within the phase diagram a phase-separated region[10] with unexpected properties, which may be one of the first fully unsupervised discoveries in the ML-guided phase classification.

3.4.2 Learning by confusion

Learning by confusion [128] is another NN-based unsupervised method that works as follows. We start by partitioning the temperature range into two regions with distinct labels. Based on these labels, we perform supervised learning over the entire temperature range as described in Section 3.3 and keep track of the final overall classification accuracy of the model. This classification accuracy is associated with the guess for the critical temperature located at the boundary of the two regions. We repeat this procedure systematically for multiple bipartitions of the temperature range, that is, guesses for the critical temperature. Finally, we plot the classification accuracy against the guessed critical temperature. This procedure is summarized in Algorithm 4. Note that each partitioning requires the training of a separate NN.[11] The results of this algorithm applied to the Ising model are depicted in Fig. 3.6. We observe that the classification accuracy is *W-shaped*. The high classification accuracy at the extremes of the temperature range arises due to the fact that, in these cases, almost all samples are assigned the same label. In particular, in the extreme case where all samples are assigned the same label a classification accuracy of 1 can be achieved trivially because the NN simply needs to learn to output the same label independent of the input. The middle peak, however, is nontrivial and corresponds to the predicted critical temperature of the method. Here, the predicted critical temperature is in good agreement with the Onsager solution. The presence of this middle peak can be explained as fol-

[9]The AE-based anomaly detection scheme was also successfully applied to quantum dynamics problems [127].

[10]This phase-separated region is located between supersolid and superfluid phases, for more details see [126].

[11]Retraining a model for each choice of a bipartition can become computationally expensive, in particular when increasing the resolution of the method. There has been an extension of the learning-by-confusion scheme that uses two NNs [129] to try to circumvent this issue by choosing bipartition points one at a time in a guided manner.

Algorithm 4 Learning by confusion

Require: Data set of (spin configuration) samples $\mathcal{D}_0 = \{\boldsymbol{x}\}$, guesses for critical temperature $\mathcal{T} = \{T_1, ..., T_{\max}\}$
for $T_c^* \in \mathcal{T}$ **do**
Partition data set $\mathcal{D}_0$ into two regions with $T \leq T_c^*$ and $T > T_c^*$
Set label $\boldsymbol{y}$ of all samples in region with $T \leq T_c^*$ as 0 and $T \leq T_c^*$ as 1
Split resulting data set into training and test set
Perform supervised learning on the training set, i.e., train an NN to minimize loss in Eq. (3.12)
Evaluate classification accuracy on the test set
end for
Plot accuracy vs. T_c^* $\forall T_c^* \in \mathcal{T}$ (see Fig. 3.6) ▷ Critical temperature T_c^* at which the accuracy peaks corresponds to the best guess for the location of the phase transition

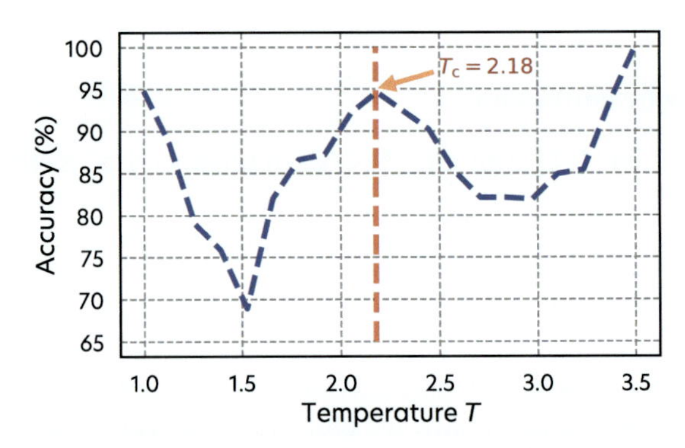

Figure 3.6 Result of the learning by confusion scheme applied to the Ising model. The data consists of 100 spin configurations (linear lattice size $L = 30$) sampled using Monte Carlo methods at temperatures T ranging from $T_1 = 1$ to $T_{20} = 3.5$ in equidistant steps. The dataset is split into equally sized training and test sets (such that 50 spin configurations are present at each sampled temperature). The blue curve shows the classification accuracy on the test set for various choices of bipartitions. It has a characteristic W-shape whose middle peak is at $T \approx 2.3$, which is in good agreement with the Onsager solution. Reproduced from [2, Notebook A3].

lows. Let us assume that the data can naturally be classified into two distinct groups realized by a particular choice for the bipartition of the temperature range. Then, the closer our choice of bipartition matches the "correct" bipartition underlying the data, the larger the classification accuracy of our algorithm.

Here, we have discussed the case where there are precisely two distinct phases present in the parameter range under consideration. In this case, the accuracy ideally displays a characteristic W-shape; see Fig. 3.6. If multiple phases are present, this characteristic W-shape is modified. The shape of the signal (in particular, the number of obtained peaks) could then be used to identify the number of different phases present in the data [128, 130, 131].

3.4.3 Prediction-based method

The learning by confusion scheme is difficult to efficiently extend to high-dimensional parameter spaces, which may feature several distinct phases.[12] Moreover, it has been shown that the learning by confusion scheme has difficulties in correctly identifying the crossover in the IGT [121]. These limitations can be circumvented through the so-called prediction-based method [121, 122, 132], which works as follows.

We train an NN to predict the tuning parameter (here, the temperature) for each configuration sample. The value of the tuning parameter at which a given configuration sample has been generated is readily available in both experiment and simulation. If the system does not undergo any phase transition, the predicted tuning parameter is *linearly* dependent on the true tuning parameter as shown in Fig. 3.7(a). Consequently, the derivative of the predicted tuning parameter with respect to the true tuning parameter is constant; see Fig. 3.7(b). For systems that exhibit a phase transition, the situation is different. In this case, the tuning parameter cannot be predicted with perfect accuracy resulting in a *nonlinear* relationship between the predicted and the true value of the tuning parameter. Figure 3.7(c) illustrates this in the case of the

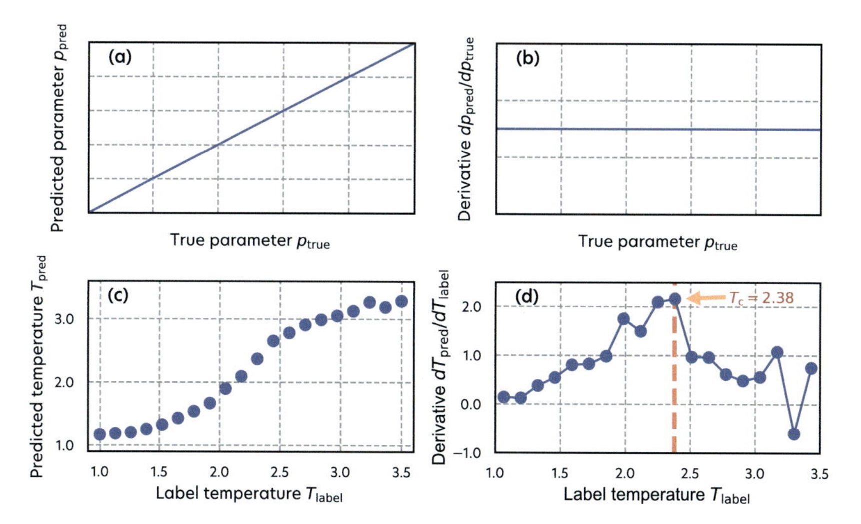

Figure 3.7 (a,b) Illustration of the output of the prediction-based method if the system does not undergo any phase transition. (c,d) Result of the prediction-based method applied to the Ising model. The data consists of 100 spin configurations (linear lattice size $L = 30$) sampled using Monte Carlo methods at temperatures T ranging from $T_1 = 1$ to $T_{20} = 3.5$ in equidistant steps. The dataset is split into equally sized training and test sets (such that 50 spin configurations are present at each sampled temperature). (c) Average predicted temperature for the test data as a function of the true underlying temperature. (d) Derivative of the average predicted temperature for the test data as a function of the true underlying temperature that peaks at the critical temperature of the Ising model. Panels (b) and (c) reproduced from [2, Notebook A3].

[12] In [129], an approach to extend the scheme to two-dimensional parameter spaces featuring two distinct phases is presented.

Ising model. Consequently, the derivative is not constant, and at the critical tuning parameter, the largest rate of change occurs; see Fig. 3.7(c) and (d). In other words, the parameter value for which the NN predictions are most susceptible identifies the position of the phase transition.

Let us elaborate on this point: While the tuning parameter only changes marginally in the vicinity of the phase transition, the system's state and its corresponding order parameter change dramatically as the tuning parameter crosses its critical value. As a result, the NN is the best at distinguishing samples originating from two different phases, whereas it has difficulties in distinguishing samples from within the same phase. That is, its predictions change the most as the tuning parameter is swept across its critical value. Figure 3.7(c) shows that the predictions start to saturate deep within each phase, whereas they vary most strongly with the tuning parameter around the transition point.

So far, we have seen that phase classification methods based on NNs are capable of locating the phase transition of the two-dimensional Ising model. To date, these methods have successfully revealed a plethora of other phase transitions in various physical systems.[13] This fact highlights that these methods are *generic* and have been formulated in a system-agnostic fashion. There may exist various physical observables (such as order parameters) that can be used to identify a given phase transition. However, finding these quantities is typically a hard task and requires a deep understanding of the physical system at hand. Remarkably, the NN-based methods we showcased here can successfully classify different phases of matter in an automated fashion without a priori knowledge of the underlying physics. Note that there exist similar system-agnostic tools that do not rely on ML, such as the specific heat for thermal phase transitions or the fidelity susceptibility [139] for quantum phase transitions.[14] However, these tools can still fail for a given system and can be expensive to compute or difficult to measure in an experiment. For example, the specific heat fails to locate the crossover temperature in the IGT. In the case of the fidelity susceptibility, one investigates the change in the overlap $\langle\Psi_0(p)|\Psi_0(p+\epsilon)\rangle$, where $|\Psi_0(p)\rangle$ is the ground state of the Hamiltonian $H(p)$, p is the tuning parameter, and ϵ is an infinitesimal perturbation. Because one typically does not have access to the full wave function, the fidelity susceptibility typically remains difficult to evaluate. The NN-based methods we discussed constitute alternative tools. In particular, they can *in principle* be applied using various properties of the system's state at different values of the tuning parameter as input. This allows them to identify phase transitions based on experimentally accessible measurement data [106, 140, 141].

While the phase classification methods we have discussed up to now are capable of locating phase transitions, we have not yet gained any insights into the specific type of phase transition that the system undergoes. The crucial question is whether one

[13] There exist various other ML methods for detecting phase transitions and classifying phases of matter [122, 133–138]. For example, in [135], phase transitions can be inferred by training an ML model to fit the properties within one phase and extrapolating toward other regions in parameter space. Here, the model is based on a Gaussian process (GP) utilizing kernels. We discuss kernel methods, including GPs, in detail in Chapter 4.

[14] A quantum phase transition [98] corresponds to nonanalytic behavior of the ground-state properties at the critical value of the tuning parameter p_c, where the system Hamiltonian is $H(p)$. It emerges due to the competition of individual terms in the Hamiltonian, which depends on the tuning parameter.

can extract physical insights from the NNs concerning the underlying phase classification tasks. In particular, one can ask whether it is possible to extract novel order parameters from such NNs, which is an ultimate goal of the interpretable ML applied to phase classification problems.

3.5 Interpretability of machine learning models

As seen in Sections 3.3 and 3.4, NNs are powerful tools to identify phases in physical data. Now imagine applying these methods to a novel physical system whose phases and corresponding order parameters are not yet known. The natural questions that arise in this scenario are: Can we trust the NN predictions? In particular, how can we know that the model correctly located a phase transition in the parameter space? Moreover, assuming that the methods correctly classified the data into different phases of matter, how can we gain physical insights into the problem at hand? For instance, can we analyze the trained NNs to determine what types of phase transitions the system undergoes? Or would it even be possible to extract novel order parameters from them? When using ML (and especially DL) models, answers to these questions are not easy to find. Such challenges are being addressed by the research on the ML reliability and interpretability.[15]

> Reliability is about trusting our ML model predictions. Our trust in the model is increased, for example, when we have access to the uncertainty of model predictions. Interpretability is about understanding what an ML model learns and how it makes its predictions. As such, these two ideas are closely intertwined.

Both concepts are particularly important on our way toward scientific discovery using ML. If we are not able to understand what an NN learns when given a problem, *our* understanding of the problem remains limited![16]

We have already mentioned that a priori, DL models are usually neither reliable nor interpretable. As such, they largely serve as black-box models that provide us with suitable predictions (from which we, e.g., can locate phase transitions in the underlying input data). There are several reasons for that: First, their learning dynamics are largely opaque and not well understood.[17] Second, the direct analysis of trained NNs is challenging, as we explain in Section 3.5.1. In particular, the "reasoning" of NNs does not necessarily have to be based on the same observations on which a human would base its decisions. Tackling these challenges is important for all ML applica-

[15]Note that the formal definitions of these terms are not agreed upon in either the physical or computer science community [142]. To circumvent the problem, here we provide intuitions about the meanings of these terms.

[16]We can imagine a noninterpretable black-box NN that after training can give insights to the problem, for example, Ref [143]. However, the model still needs to be reliable so we can trust the new insights, and we need to have previous deep insights into the problem.

[17]We show you how some of these questions can be answered with tools from statistical physics in Section 8.1.

tions but especially crucial, for example, for medical diagnosis or insurance and hiring decisions.

3.5.1 Difficulty of interpreting parameters of a model

When looking at a DNN with possibly billions of trainable parameters, it is hard for us humans to decipher what the NN is really doing under the hood. It may be that an NN actually computes a simple, physically relevant function, such as an order parameter, to make its predictions. Recognizing whether that is the case is hard because the computation and relevant information are spread over the multiple layers containing a large number of neurons each. However, if an NN is sufficiently small, a direct interpretation by looking at its trainable parameters may be possible. Consider the limiting case of a single-layer NN without any nonlinear activation function. This corresponds to a simple linear regression model, described in Section 2.4.1:

$$\hat{\boldsymbol{y}} = \boldsymbol{w}\boldsymbol{x} + \boldsymbol{b}, \tag{3.13}$$

where $\boldsymbol{w}$ is a vector of weights and $\boldsymbol{b}$ is a vector of biases. Evidently, such a linear model allows for a direct interpretation in terms of its weights: The larger the magnitude of a given weight (connection), the more important the corresponding normalized feature for solving the problem at hand. For an example of weight interpretation in the context of phase classification, see [144].

However, a reduction in depth and loss of nonlinearity come at the cost of expressivity. For such a model to be accurate, it generally requires highly pre-processed inputs $\boldsymbol{x}$ whose processing takes care of the necessary nonlinearities. Moreover, the importance of a given feature has more meaning if it is already present in a compact, physically relevant form. This largely limits the domain of applicability of small predictive models to problems of which we (at least) have partial knowledge.

Reducing the number of effective parameters via regularization. One way to obtain NNs with a reduced number of effective parameters is regularization – in particular, the addition of a ℓ_1 regularization term in the loss function given by $\ell_1 = \lambda \sum_i |w_i|$, where λ parametrizes the regularization strength, and the sum runs over all (trainable) weights within the NN. This term forces the weights to vanish, that is, for connections to be cut. Ideally, this results in a sparser, and thus effectively smaller NN, which improves interpretability. In [145], for example, the authors could extract analytical expressions for force laws and dark matter distributions from graph NNs trained to predict planetary and dark matter dynamics. This was achieved by performing symbolic regression on the corresponding sparse networks.[18] Regularization is also important when interpreting linear models as in Eq. (3.13). Often, learning problems do not have a unique solution. This means that the weights can vary given the same data and optimization procedure, which would result in different "interpretations" of the NN's inner workings. Regularization terms help to remove the remaining degrees of freedom of the weights and enforce Occam's razor.

[18] A graph NN is similar to a CNN in the sense that the spatial location of the input is crucial for the meaning of the input. While in CNNs, neighbors are determined by their position of the input data grid, in graph NNs, neighbors can be defined much more broadly through custom connections between graph nodes. NN layers can act on each node or through message-passing between nodes.

Extracting order parameters with support vector machines (SVMs). A large class of ML algorithms that allow for a direct interpretation in terms of model parameters are SVMs, see Sections 2.4.3 and 4.2.2. SVMs were first proposed for solving phase classification tasks in [146] and were later expanded and applied to higher-order spin systems in [147, 148]. As discussed above, while these algorithms might not be as powerful as NNs, a major advantage is the possibility of having an interpretable decision function from which order parameters can be inferred.[19]

3.5.2 Interpretability via bottlenecks

As we have explained in Section 3.5.1, interpretability is an inherent characteristic of small models. Fortunately, there are alternative approaches to interpretability that are not limited to simple small models. What we can do in large architectures is to identify bottlenecks in the information flow and focus our attention there. A bottleneck in an NN is just a layer with fewer neurons than the layer before and after it. An example of an NN with a natural bottleneck has already appeared in Section 3.4.1 and in Fig. 3.5(a), namely, an AE. Its bottleneck forces the NN to distill the relevant information within the inputs such that it can flow through this constriction. As such, the NN performs a dimensionality reduction and finds a suitable low-dimensional feature representation. While the entire NN architecture can be large and have many trainable parameters, the bottleneck itself can have as little as a few parameters. Because all the relevant information for the predictions of the NN must eventually flow through the bottleneck, we can limit our analysis to the small number of trainable parameters of the bottleneck as opposed to the entire NN.

In particular, we can perform a regression on the output of such bottleneck neurons and extract the mapping between the input features and the activations of the bottleneck neurons. There is a natural bottleneck in almost every NN – its output neuron. However, performing a regression on it without imposing any additional bottlenecks is challenging because you need to take into account all input features, which can grow quickly in number. Apart from the output neuron, other types of bottlenecks can appear naturally in NNs architectures, such as in AEs (see Section 2.4.5) [115, 149] and convolutional neural networks (CNNs) (see Section 2.4.4) [150]. However, we can also introduce bottlenecks into our architecture on purpose to have more interpretable ML models. This idea gave birth to, for example, Siamese NNs [151]. We look at these approaches in more detail in the next paragraphs and see what information on physical systems we can extract using them.

Interpretability with autoencoders. As we have explained in Sections 2.4.5 and 3.4.1, AEs are NNs with a bottleneck in the middle that are trained to reconstruct the input at the output. We have already shown in Section 3.4.1 and in Fig. 3.5(b) that we can obtain clustering in the latent representation corresponding to phases present in the input data, achieving the unsupervised phase classification. Moreover, thanks to the bottleneck, which ideally should contain all information that is relevant for the reconstruction, we can also extract additional information and interpret what property of the input data is preserved by an AE. In particular, if we plot the

[19] The decision function determines the distance of a given sample $\boldsymbol{x}$ from the hyperplane.

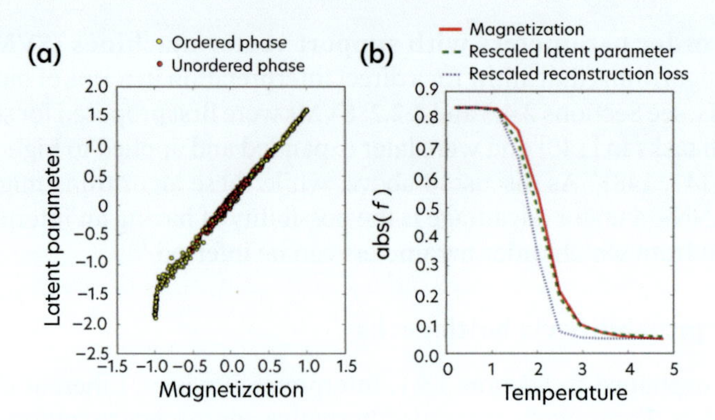

Figure 3.8 Analysis of bottleneck neurons of an AE trained to reconstruct spin configurations of a two-dimensional Ising model. (a) Dependence of latent space parameter on the magnetization. The red (yellow) color corresponds to samples from the low(high)-temperature regime. (b) Absolute magnetization, absolute rescaled values of the latent parameter, and reconstruction loss averaged for fixed temperature. Adapted from [115].

latent parameter against the magnetization of the respective two-dimensional Ising spin configuration, as in Fig. 3.8(a), we see a linear dependence. This suggests that the compressed representation learned by the AE is connected to the magnetization. To be more precise, the behavior deviates from a strict linear dependence at values of the magnetization close to -1. However, we still can make the statement that the AE learned a property related to the magnetization, given that the mapping between the latent parameter and the magnetization is bijective. For example, such a statement would hold even if the latent parameter as a function of the magnetization would vary according to a sigmoid function.

Another example of analysis of latent space of AEs is work by Iten *et al.* [149]. They used a special AE architecture with a *question neuron*, that is, an additional neuron connected to the first decoding layer after the bottleneck. The input of this question neuron is provided by a user. You can think of it as an alternative way of providing data to the network. The authors showed that you can train such a special AE in a way that a user can ask a question via the question neuron, and the answer is encoded in the latent space.

As you see, AEs are to some degree inherently interpretable by virtue of their low-dimensional latent space. However, the analysis of latent space does not give us any hint about the order parameter or important features. We can only compare it against the quantities or features we suspect to be important. If you look for a more automated way of detecting order parameters, we can turn to very special CNNs.

Extracting order parameters with convolutional neural networks. We have already mentioned that CNNs have natural bottlenecks in their architectures. These bottlenecks are their filters or kernels, that is, the structures with which they "scan" the data. Their size can be thought of as a receptive field size and tells us how many neighboring features (e.g., pixels) the network can analyze at the same time. Of

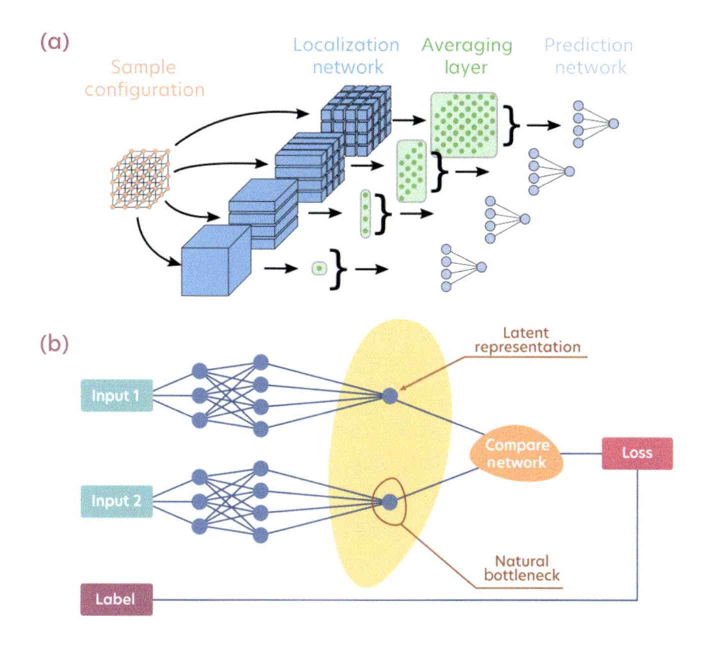

Figure 3.9 (a) Interpretation Net where the size of the receptive field is systematically reduced in each training step. The localization network consists of one or more subsequent convolutional layers. The averaging layer collapses convolved features to a single number. (b) Scheme of a Siamese NN, whose two subnets are identical and share the same tunable parameters. The input is a pair of data points, and the resulting label is either "same" or "different." Panel (a) is adapted from [150].

course, if you have multiple convolutional layers with multiple kernels of different sizes intertwined with pooling layers, their analysis is still challenging. But if you consider a simple CNN with only one or few subsequent convolutional layers with kernels of a fixed size and only one averaging layer at the end of the architecture, such a regression becomes tractable.[20] The mentioned architecture was proposed by Wetzel *et al.* (2017) [150] and is called *Interpretation Net* or *Correlation-Probing NN*, see Fig. 3.9(a). Such an architecture allows us to perform a regression on the output neuron with features extracted by kernels. Eventually, we obtain an analytical expression for the CNN decision function. If applied to a phase classification problem, such a decision function could unravel the order parameter. It seems, however, that such a decision function, and therefore the order parameter that may potentially be discovered through the CNN, depends on the choice of kernel size. What is the appropriate choice of kernel size and, thus, decision function? Occam's razor tells us we should be interested in the simplest decision function. That is, one should aim to take into account only a small number of input features. Crucially, this also makes the task of symbolic regression easier.

[20] It remains nontrivial and involves careful zeroing of weights, Fourier series, and other tricks. If you are interested, see [150].

Therefore, the idea of the Interpretation Net is to systematically reduce the size of the kernel (and thus the input dimension for the symbolic regression task) by cutting connections until there is a significant drop in the CNN performance. This drop corresponds to the CNN becoming "blind" to the correlations that are crucial for detecting and distinguishing different phases.

Imagine starting from a large kernel whose size corresponds to the size of the entire input image, for example, 28×28. We train our Interpretation Net with such 28×28 kernels and see that it yields good results. Now, we reduce the receptive field size, for example, to 20×20, retrain, and observe the performance. We repeat this process of reducing the kernel size and retraining until we see a significant drop in the CNN performance. Such a drop occurs as soon as the CNN gets blind to correlations in the system that are crucial for the phase classification, for example, next-nearest-neighbor correlations. Finally, we can perform a regression on the output neuron of the CNN with the smallest kernel size that still yields good performance. The decision function we recover in the process is ideally connected to the underlying order parameter. With this approach, the Interpretation Net is capable of successfully classifying the phases of the two-dimensional Ising model[21] or $SU(2)$ lattice gauge theory and allows for extracting the corresponding decision functions. We stress that *the learned decision function can strongly depend on the choice of the network architecture and training procedure as well as the available data*. Therefore, various networks can detect different order parameters, for example, in the Ising model, they can detect the expected energy per site, magnetization, or a scaled combination of those.

A related approach was used by Miles *et al.* (2021) [152] when designing a so-called *Correlator CNN* depicted in Fig. 3.10. This CNN performs automatic feature engineering by probing for one-, two-, and higher-body correlations in the first few layers. The subsequent layers are designed to check which correlations are the most important for the classification. Again, the network is rendered blind to certain features by tuning what the NN can learn. In contrast to the previous unnumbered section, the authors penalize the learning of certain filters through regularization. By increasing the regularization strength, it is possible to successively disable certain features depending on their importance for the prediction accuracy, thus leading to a hierarchy of important correlations corresponding to the underlying physics.

The Correlator CNN was used to detect the key many-body correlators differentiating between two theoretical quantum models serving as two candidate theories approximating the doped Fermi–Hubbard model [152]. As such, it explained the results of another work [143], where a CNN was trained to differentiate between numerically generated snapshots of the quantum system following two candidate theories. Then the trained CNN was tested on experimental snapshots and indicated which of the two theoretical quantum models described those snapshots better. This represents one of the first examples of scientific discovery with NNs.

[21] Interestingly, in their paper, the quantity that leads to the better CNN performance in the case of the Ising model is the expected energy per site ($\frac{-J}{N}\sum_{\langle i,j\rangle}\sigma_i\sigma_j$), which can still be detected with a 2×1 kernel, not the magnetization ($\frac{1}{N}\sum_i\sigma_i$), which can still be detected with a 1×1 kernel.

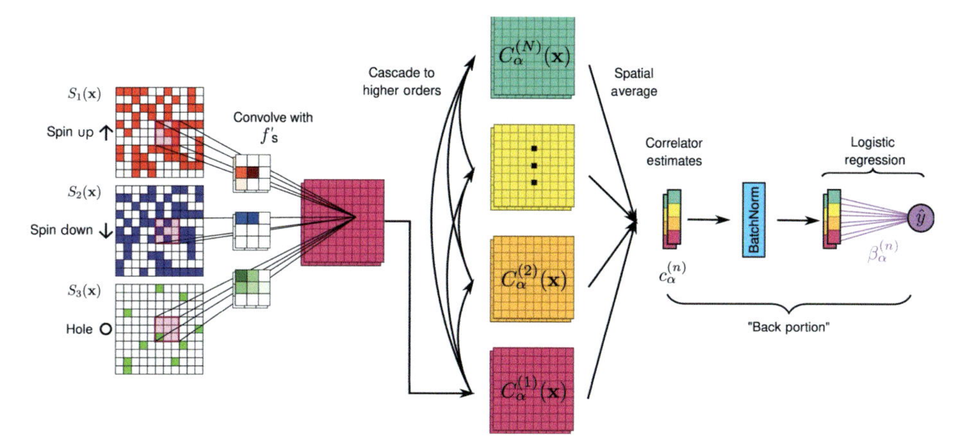

Figure 3.10 The Correlator CNN has learnable filters that can be activated and deactivated through regularization. The input is first convolved with learned filters to produce a set of convolutional maps from which information about higher-order local correlations can then be recursively constructed. Taken from [152] with permission from Springer Nature.

Interpretability with Siamese neural networks. One can consider even more complex architectures with artificial bottlenecks allowing for an interpretation via symbolic regression. An interesting example of this is the Siamese NN [151] (sometimes also called twin NN), presented in Fig. 3.9(b). It takes two input data points at the same time and is composed of two twin subnetworks with the same parameters and architecture. Their output neurons form two bottlenecks, which, in turn, are inputs for the third subnetwork, the aim of which is to connect and compare the twin outputs. The task of the network is to determine whether two input data points are similar or not.[22] This means that Siamese NNs are able to perform a multiple-class classification without fixing the number of classes a priori and with relatively little training data per class. Moreover, by analyzing the bottlenecks, we can extract what the NN learns in a problem.

These networks provide a powerful tool to discover phase transitions in an unsupervised manner as outlined in [154, 155]. Learning phase transitions with Siamese NNs is very similar to using CNNs. However, due to the unsupervised nature, one is not able to supply phase labels. Hence, the labels are initialized such that all data pairs sampled from the same point in the phase diagram get the *same* label, while pairs from different points in the phase diagram obtain the *different* label. After training, the Siamese NN can be used to detect phase boundaries by sampling pairs from adjacent points in the phase diagram and predicting whether they are *similar* or *different*. A spike in dissimilarity then marks the phase transition.

Before we conclude, let us look at a few other applications beyond classifying phases of matter where the interpretation of NNs via bottlenecks comes in handy. The authors of [151] applied the Siamese NN, for example, to the motion of a particle in a central potential. The task was to learn whether two observations of a particle

[22] A quantum version of Siamese NNs was developed in [153], and, interestingly, it goes beyond distance-based notions of similarity.

correspond to the same particle trajectory (or two distinct trajectories). After successful training, we can perform a polynomial regression on the bottleneck with respect to the input data features. In this case, the features are comprised of the position of the particle in two-dimensional space and its two-dimensional velocity vector. By analyzing the dominant regression terms, they observed that the result of the regression is proportional to the angular momentum of the particle. Such an analysis of bottlenecks of a successfully trained Siamese NN indicates that the network learns conserved quantities and invariants.[23] Similar results can be obtained for problems in special relativity and electromagnetism [151].[24]

So far, we have learned that we can interpret ML models by analyzing bottlenecks in their architecture. These bottlenecks can either appear naturally, such as in AEs, or be imposed explicitly, such as in CNNs where the kernel size is systematically reduced or in Siamese NNs. Another approach toward interpretability is based on the analysis of the minimum of the training loss function reached by a model during the optimization. Because such an analysis is based on the minimum, it is generic and can be applied independent of the particular choice of ML model architecture or learning procedure.

3.5.3 Hessian-based interpretability

As described in Section 2.1, ML models learn by minimizing a training loss function $\mathcal{L}$ describing the problem through the variation of their parameters $\boldsymbol{\theta}$. The training loss landscape of deep NNs is, however, highly nonconvex. This renders the optimization problem difficult, for example, due to the presence of many local minima [see Fig. 3.11(a)]. Moreover, these minima may not have equally good generalization abilities, and it seems these abilities are connected to the curvature around a minimum.[25] The connection is an instance where the shape of the reached minimum can tell us something useful about trained ML models. The shape or curvature around the minimum $\boldsymbol{\theta} = \boldsymbol{\theta}^*$ is described by a Hessian matrix calculated at the minimum, that is:

$$\boldsymbol{H}_{\boldsymbol{\theta}^*,ij} = \frac{\partial^2}{\partial\theta_i\theta_j}\mathcal{L}_{\text{train}}|_{\boldsymbol{\theta}=\boldsymbol{\theta}^*}\,. \tag{3.14}$$

The knowledge of the curvature around the minimum also allows us to approximate how our ML model (and, as a result, its predictions) would change upon some action. Possible actions could be the removal of a single training point or a slight modification of the model parameters $\boldsymbol{\theta}^* \to \tilde{\boldsymbol{\theta}}$ toward an adjacent minimum with identical training error. The study of how a model reacts to such actions is at heart of the Hessian-based toolbox summarized in Fig. 3.11(b), which contains influence functions [165], the resampling uncertainty estimation (RUE) [166], and local ensembles (LEs) [167] whose conceptual ideas we introduce in the following.

---

[23]It is much easier to detect invariants that can be represented as polynomial functions of input features.

[24]The process of finding symmetry invariants and conserved quantities with ML has emerged as its own subfield, and many improved methods to detect these have been devised in [156–158].

[25]There is a general consensus that wide, flat minima generalize better than sharp minima [159–162]. Keep in mind that flatness is not a well-developed concept in nonconvex landscapes of deep models [163].

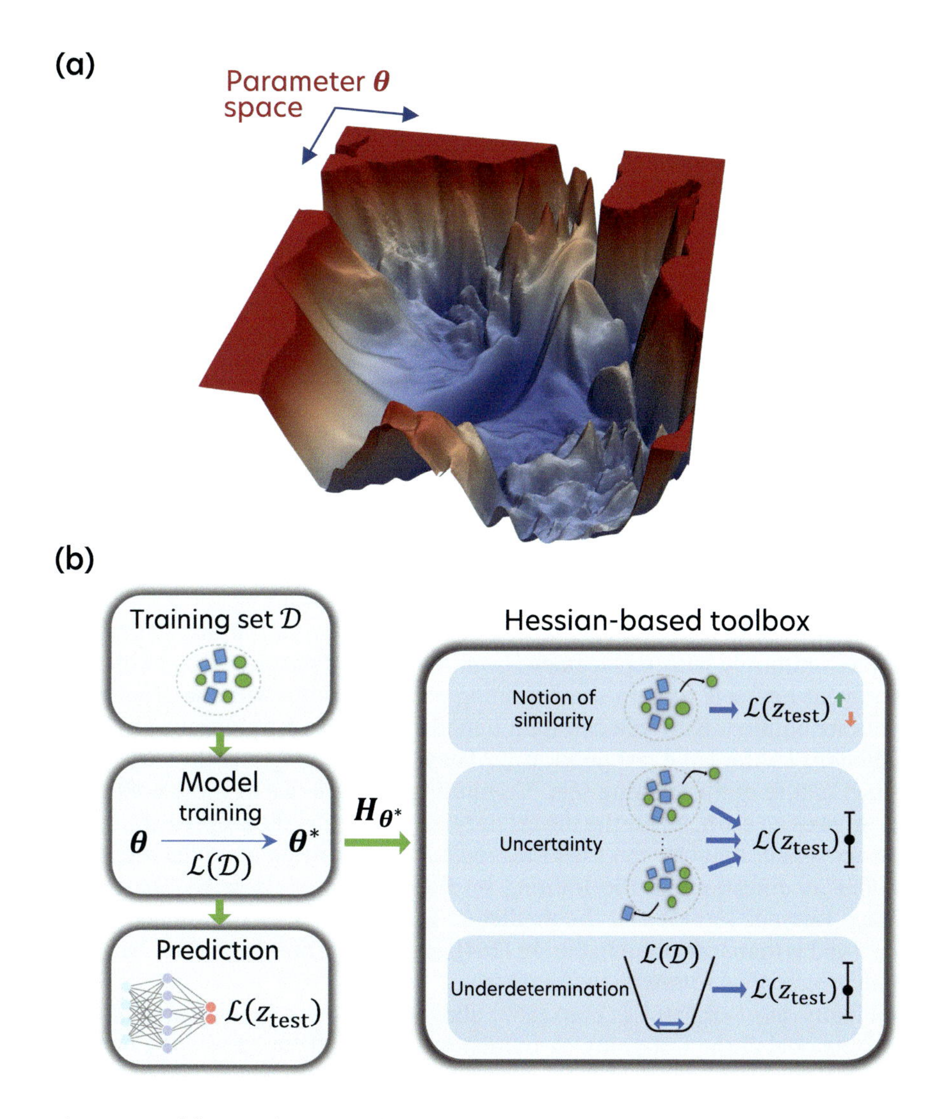

Figure 3.11 (a) Low-dimensional visualization of a nonconvex loss landscape of a DNN called VGG-56 trained on CIFAR-10. Taken from [44] under the MIT license. (b) Hessian-based toolbox to increase the interpretability and the reliability of a trained ML model. It is based on the Hessian of the training loss at the minimum (or an approximation thereof). Adapted from [164] under the CC BY 4.0 DEED license.

Influence functions are an approximation of the procedure known as leave-one-out training[26] and estimate how the model prediction on a test point $\boldsymbol{x}_{\text{test}}$ change if

[26]Leave-one-out training for DL models with nonconvex loss landscapes is tricky because if we land in a different local minimum, we cannot make any claims on the perturbation caused by the removal of a single training point. This is why we usually retrain carefully, starting from the minimum reached by the original model.

a certain training point $\boldsymbol{x}_{\mathrm{R}}$ is removed from the training set. You can imagine three outcomes of such a removal: (1) the prediction stays the same because the removed training point has no impact on the model prediction, (2) the prediction gets better (i.e., it leads to a lower test loss for $\boldsymbol{x}_{\mathrm{test}}$), so $\boldsymbol{x}_{\mathrm{R}}$ is a "harmful" training point for making a prediction on z_{test}, (3) the prediction gets worse (i.e., it leads to a larger test loss for $\boldsymbol{x}_{\mathrm{test}}$), so $\boldsymbol{x}_{\mathrm{R}}$ is a "helpful" training point for making a prediction on z_{test}, and its removal made the task of constructing an accurate model harder. With such an analysis we can determine how influential training data points are to predictions at test points, which may give us a hint at how the model reasons. We can even go a step further and say that if two data points strongly influence each other, it is because they are very similar from the model's perspective.[27] This concept of similarity learned by an ML model can be understood as a distance between data points in the internal model representation and is a powerful tool for *detecting additional phases in mislabeled data* [106, 168], *detecting influential features* [106], and *detecting anomalies* [164].[28]

Another tool in the Hessian-based toolbox is the RUE [166]. It estimates the uncertainty of model predictions. It is an approximation of the classical procedure known as bootstrapping. You start with your original training set containing each training data point once. Imagine now that you create b new training sets by drawing samples uniformly with replacement from the original dataset. Due to the replacement, your new sets contain some training points in more than one copy, and some points are omitted. Now you can train b models on these b training sets and make b predictions on the same test point, z_{test}. These predictions generally vary due to the distinct nature of the training sets. Computing the variance of these predictions on z_{test} gives us an estimate for the uncertainty of the original model prediction. A small variance signals that one can trust the prediction of the original model because small random modifications to the training set do not change its prediction too much. A large variance signals that the prediction is based on a small number of training points and is therefore not reliable. In [164], you can see how such error bars indicate the *sharpness of quantum phase transitions*.

Finally, *local ensembles* (LEs) [167] allow us to detect the *underspecification* of a given model at the test point. A trained model is underspecified at a test input if many different predictions at that input data are all equally consistent with the constraints posed by the training data and the learning problem specification (i.e., the model architecture and the loss function). As described in Section 2.1, the minimum reached within the optimization is usually surrounded by a mostly flat landscape. This means that if the model had ended up in one of these flat neighboring points, the training error would have stayed exactly the same. Thus, such changes should not impact the predictions – unless a prediction is *underdetermined*, that is, unstable and not well explained by the training data. Therefore, we can again create multiple models by shifting the parameters of the original model by small amounts. As such, these new models explore the flat landscape around the original minimum.

[27] This argument is well based on the geometric interpretation of influence functions; for details, see section 2.3.3 in [164].

[28] We stress that *similarity*, which is arguably the central concept behind classification tasks, has various meanings when it comes to Siamese NNs, influence functions, and kernel methods (which will be covered in Chapter 4).

Eventually, we make predictions with these new models. If a prediction on a test point z_{test} changed due to such modifications, this point may be an out-of-distribution point, that is, a point coming from a distribution that is significantly different from the distribution underlying the training data. LEs allow for the detection of such out-of-distribution test points, which increases the reliability of the ML model. Moreover, the authors of [167] successfully used LEs for active learning, that is, they built a much smaller, yet similarly informative training dataset by iteratively adding to it test points with the largest underspecification score detected by LEs.

Therefore, we have answered our initial questions regarding interpretability: It is indeed possible to look inside the black box of ML models. If you focus your attention on the bottlenecks present in NN architectures, you can determine which quantities dominate in the NN prediction using regression methods. Further, it is possible to interpret the filters in the early layers of NNs. If these quantities are physically relevant, we can argue that the NN indeed bases its predictions on physically relevant quantities. Additionally, you can turn your attention to the curvature around the minimum of the training loss. It contains information on the similarity learned by a model and allows for estimating the uncertainty.

3.5.4 A probabilistic view on phase classification

In Section 2.3, we have introduced a probabilistic view on ML. This viewpoint turns out to be particularly useful when applying ML to the task of classifying phases of matter. In the following, we focus on the simple case of supervised learning (Section 3.3). However, the idea also generalizes to other NN-based phase classification methods, such as learning by confusion (Section 3.4.2) or the prediction-based method (Section 3.4.3), see [169, 170]. Let us first consider the scenario where we would like to distinguish between images of cats and dogs in a supervised setting. Recall from Section 2.3 that a Bayes classifier outputs

$$y_{\text{Bayes}}(\boldsymbol{x}) = \arg\max \{p(\text{cat} \mid \boldsymbol{x}),\ p(\text{dog} \mid \boldsymbol{x})\}, \tag{3.15}$$

where $p(\text{cat} \mid \boldsymbol{x})$ and $p(\text{dog} \mid \boldsymbol{x})$ are the probabilities that the given sample $\boldsymbol{x}$ is a cat or a dog, respectively. The *Bayes classifier* is optimal as it outperforms any other classifier in the classification task at hand, that is, achieves the lowest possible misclassification probability. However, for most real-world datasets, such as images of cats and dogs, the ground-truth class-conditional probabilities are inaccessible, abstract quantities, and we do not know them (or their form) a priori. One way to tackle the classification task is thus to parametrize these conditional probabilities by an NN and train it to minimize the misclassification probability (i.e., the corresponding loss function). Because the NN is a universal function approximator, its predictions are expected to approach those of a Bayes classifier as we make the NN more expressive, train it better, and increase the size of our dataset. Similarly, the misclassification probability of our NN is expected to approach the Bayes error from above.

As discussed, in realistic scenarios, it is typically impossible to construct a Bayes classifier. Interestingly, this can change when we move to the realm of physics, particularly statistical physics and quantum physics, where the data underlying the task of classifying different phases of matter resides. To illustrate this, let us consider the case of supervised learning, where we want to distinguish between two phases: phase A and phase B. The optimal outputs of a Bayes classifier are then given as

$$y_{\mathrm{Bayes}}(\boldsymbol{x}) = \arg\max \{p(\text{phase A} \mid \boldsymbol{x}),\ p(\text{phase B} \mid \boldsymbol{x})\}. \tag{3.16}$$

Using Bayes' rule, Eq. (2.19), we have

$$p(\text{phase A} \mid \boldsymbol{x}) = \frac{p(\boldsymbol{x} \mid \text{phase A})\, p(\text{phase A})}{p(\boldsymbol{x})}, \tag{3.17}$$

and similarly for $p(\text{phase B} \mid \boldsymbol{x})$. Now assuming both phases are represented equally in terms of their labels in our dataset, we have $p(\text{phase A}) = p(\text{phase B}) = 1/2$. Moreover, we have

$$p(\boldsymbol{x}) = p(\text{phase A})p(\boldsymbol{x} \mid \text{phase A}) + p(\text{phase B})p(\boldsymbol{x} \mid \text{phase B}), \tag{3.18}$$

which, inserting in Bayes' rule, yields

$$p(\text{phase A} \mid \boldsymbol{x}) = \frac{p(\boldsymbol{x} \mid \text{phase A})}{p(\boldsymbol{x} \mid \text{phase A}) + p(\boldsymbol{x} \mid \text{phase B})}. \tag{3.19}$$

Hence, we can compute the class-conditional probability (corresponding to the optimal prediction of a Bayes classifier) if we know the probability of drawing the sample $\boldsymbol{x}$ in either of the two phases, that is, if we know $p(\boldsymbol{x} \mid \text{phase A})$ and $p(\boldsymbol{x} \mid \text{phase B})$.[29] Let us denote the physical parameter we vary to get from phase A to phase B as γ and assume that we sample phase A/B at distinct points $\{\gamma \in \text{phase A/B}\}$. Then, we have

$$p(\boldsymbol{x} \mid \text{phase A}) = \frac{1}{|\{\gamma \in \text{phase A}\}|} \sum_{\gamma \in \text{phase A}} p_\gamma(\boldsymbol{x}), \tag{3.20}$$

where $|\{\gamma \in \text{phase A}\}|$ is the number of distinct sampled points in phase A (and similarly for phase B). This shows that we can express $p(\boldsymbol{x} \mid \text{phase A/B})$ (and thus the class-conditional probability) based on the probability distribution $p_\gamma(\boldsymbol{x})$ underlying the physical system of interest sampled at distinct values of the tuning parameter γ. These probability distributions have a clear physical meaning, and we often know them completely or at least partially.[30] For example, when we analyze systems at thermal equilibrium at various temperatures, such as in the case of the Ising model (Section 3.1.1) or IGT (Section 3.1.2), we know that the underlying distribution is Boltzmann, Eq. (3.2). Similarly, when studying quantum phases, we may have access to the wave function of the quantum state that governs the measurement statistics.

[29] We might apply Bayes' rule in the same fashion to the task of classifying of images of cats and dots. In this case, we would need to know $p(\boldsymbol{x} \mid \text{cat})$ and $p(\boldsymbol{x} \mid \text{dog})$ to construct the class-conditional probabilities; quantities that are abstract and inaccessible and thus evade this discussion.

[30] In experimental scenarios, we may not be able to access the distribution directly. However, physicists strive to isolate and characterize the classical statistical ensemble or quantum state they realize in their experiment as best as possible.

From this perspective, the task of (optimal) phase classification boils down to characterizing the probability distributions underlying the physical system at hand. Phase transitions then manifest themselves as rapid changes in these probability distributions. This offers an alternative avenue for classifying data into distinct phases of matter (namely by estimating $p_\gamma(\boldsymbol{x})$ first rather than trying to estimate $p(\boldsymbol{x} \mid \text{phase A})$ and $p(\boldsymbol{x} \mid \text{phase B})$ directly) and sheds a different light on "traditional" NN-based phase classification. For a more complete overview of this probabilistic view on phase classification, see [169, 170].

3.6 Outlook and open problems

Over the last five years, there have been many works applying supervised and unsupervised phase classification algorithms, including supervised learning (Section 3.3), learning by confusion (Section 3.4.2), and the prediction-based method (Section 3.4.3), to models with well-known phases. However, only a few works applied unsupervised phase classification methods to experimental data. Moreover, the discovery of a novel phase of matter using unsupervised phase classification methods still remains to be demonstrated. This would constitute a major step toward the automation of scientific discovery.

While there has been significant progress regarding the interpretability of phase classification methods in recent years, we still lack a deeper understanding of these methods. In particular, it remains difficult to tell when and why a given method fails or succeeds [169]. With the goal of automated scientific discovery in mind and having demonstrated that phase classification methods are capable of dealing with a vast range of physical systems, addressing these gaps in knowledge and developing corresponding interpretability tools is of crucial importance.

Further reading

- Carleo, G. *et al.* (2019). *Machine learning and the physical sciences.* Rev. Mod. Phys. 91, 045002. An overview of the current state of the phase classification landscape is presented in section 4C [5].
- Neupert, T. *et al.* (2021). *Lecture notes: Introduction to machine learning for the sciences.* An introduction to fundamentals of ML and clustering algorithms for scientists [94].
- Molnar, C. (2019). *Interpretable Machine Learning: A Guide for Making Black Box Models Explainable.* An introductory book on interpretable ML [171].
- Jupyter notebook on phase classification [2].

4 Gaussian processes and other kernel methods

This section deals with the so-called kernel methods, of which support vector machines (SVMs) and Gaussian processes (GPs) are prominent examples. We point the interested reader to the exemplary and definitely non-exhaustive collection of [172, 173] for a deep dive into the foundations of kernel methods in ML and to [174] for a more high-level overview. These methods are particularly well suited in the case of low availability of labeled data. This usually happens when the creation of a large dataset is expensive in terms of money, time, effort, and so on. As a second advantage, the predictions of GPs are, *by construction*, accompanied by their uncertainties which other methods do not readily provide. We see how this property arises from the design choice of the models in Section 4.2.

But first, we have to introduce the notion of the *kernel* that is an integral part of all the methods discussed in this chapter. The introduction of the kernel allows us to extend the range of problems we are able to tackle substantially. Before properly defining the mathematical foundation of kernels, we start by providing some intuition on how to use them in practice based on the *kernel trick* and its implications. Afterward, we show how to extend the aforementioned methods via this kernel trick and discuss how to train each of them given data. For example, it turns out that GPs can be approached from an information-theoretic perspective. Furthermore, we explain how we can make use of concepts from information theory for a guided data acquisition procedure, as well as to select a good model among various possible ones. We end the chapter by showcasing the power of these methods in tackling quantum problems in Section 4.5.

4.1 The kernel trick

As we have seen in Section 2.4, simple approaches such as the linear regression model or the linear SVM have severe limitations with respect to the properties of data. They are applied to the input data *as is*, that is, they are bound to the given representation of the input data. To avoid confusion later on, we refer to this data space as the *input space*. In this input space, it can, for example, happen that the given input data is not linearly separable. One possible remedy consists of extending the classification power of the model by transforming the input data into an alternative *feature space*. In contrast to what we have said earlier in Chapter 1, we explicitly distinguish between the input and the feature spaces in the following.[1] Ideally, in this new representation, the data possesses a more convenient structure compared to the original representation. For example, data that was initially not linearly separable may be linearly separable in this new space. In particular, it is often useful to transform into a higher-dimensional feature space in which our data is now nested on a manifold that (ideally) possesses beneficial additional structure.

[1] Each element of a data point is still called a *feature* – in this chapter, we are however only interested in finding the most convenient representation of the data whose space we hence call the *feature* space.

One may think that this makes machine learning extremely costly (or even infeasible if the dimension of the feature space approaches infinity). However, we can make use of the fact that the predictions of our ML algorithms of interest are often formulated in terms of distances between data in the input space.

This is where the *kernel trick* comes in: Instead of explicitly transforming the data into the feature space and then calculating the distance therein, we start from the other end and provide a closed-form expression for the distance in terms of the data representation in the input space. This typically is much more efficient from a numerical perspective.

As we will see later, we can always associate a unique feature space with any valid distance function. Thus, we shift the focus from finding a suitable representation to choosing a suitable distance function. This trick is called the *kernel* trick because the kernel is the mathematical object we associate with such a feature space. As such, the kernel trick allows one to retain all the benefits of high-dimensional feature spaces at a manageable computational cost. Moreover, the mathematical foundation of kernels allows us, as we see in Section 4.1.1, to enrich our motivation with rigorous, analytical validity. Especially important from a practical point of view is the *representer theorem*: So far, we have set out to find a suitable transformation, that is, function to simplify our task at hand. However, it is unclear how to optimize over functions instead of parameters. The representer theorem endows us with both: In essence, it assures that the optimization over the function space is equivalent to optimizing the coefficients of a closed-form solution, which, in turn, allows us to devise feasible numerical optimization routines.

In the following, we start with an intuitive example to illustrate why and how the transformation into the feature space can be beneficial. Afterward, we properly introduce the mathematical notion of kernels, which gives us the analytical tools required to understand the representer theorem.

4.1.1 Intuition behind the kernel trick

To gain some intuition, let us consider a labeled two-dimensional dataset as depicted in Fig. 4.1. In this toy example, the black line indicates the underlying decision boundary, that is, the line that separates input data with different labels. In higher dimensions, the decision boundary generalizes to a hyperplane. In our example, one label refers to the center of the data cloud and the other to its outskirts. A label distribution is said to be separable if one can find at least one such hyperplane separating the two class sets. If, furthermore, this decision boundary is linear, the data is called linearly separable.[2] Clearly, our toy dataset is not linearly separable in the input space.

[2]Whether a given dataset actually is (linearly) separable or not is not easily detectable. In practice, we can at least run algorithms such as an SVM explained in Section 2.4.3, which are guaranteed to find the corresponding separating hyperplane if it exists.

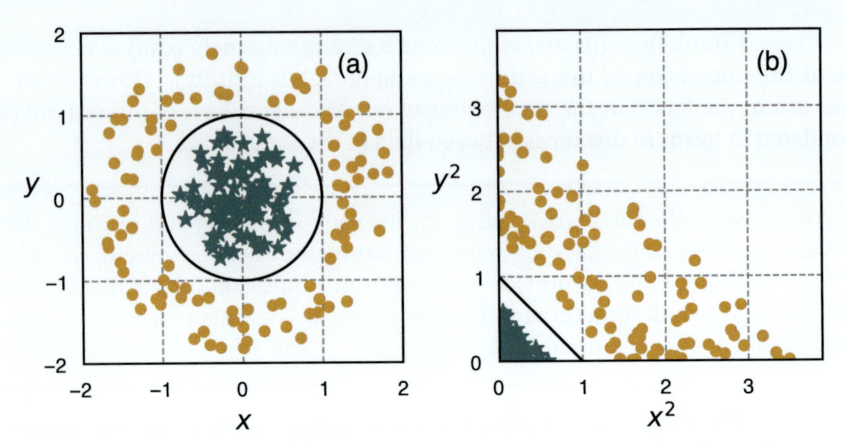

Figure 4.1 Toy example of a labeled two-dimensional dataset. The data points are labeled according to their position with respect to the decision boundary indicated by the black circle in (a). In this *input space*, such a dataset is not linearly separable. After a transformation of the input variables into a nonlinear *feature space*, however, the data becomes linearly separable as indicated by the black line in (b).

As a consequence, we cannot find a straight line that fully separates the two data classes by means of a simplistic linear classifier. Additionally, other linear methods such as PCA (see Chapter 3) fail to cluster this data.[3]

However, as shown in the right panel in Fig. 4.1, the dataset becomes linearly separable if we transform the data in the appropriate way. Here, we have applied the transformation $\phi : (x_1, x_2) \mapsto (x_1^2, x_2^2)$ to map the input data nonlinearly to the so-called feature space. Hence, the map ϕ is called the feature map. There are two important caveats: First, finding a useful feature map is a highly nontrivial task. In our toy example, the labeling procedure considers the data points in polar coordinates and uses the radial distance r to the origin as the label criterion. Using the connection $r^2 = x_1^2 + x_2^2$, we motivate our feature map ϕ, whose choice is by no means unique. Second, even if we had found a good feature map, it could be infinite-dimensional.[4] In that case, it would not be feasible to transform the data with the feature map. In our example, we only squared the input variables to achieve linear separability. However, there can be instances where the polynomial expansion has to be taken to infinite order. We encounter such an example where the most suitable feature space is infinite-dimensional later on when we come back to our toy example at the end of Section 4.2.2.

[3]As described in Chapter 3, PCA itself is not a clustering algorithm. However, we have also seen that it can be used to provide a low-dimensional representation of the data in which the data is split into different clusters.

[4]In a sense, we do the opposite of a dimensionality reduction method such as PCA in Section 3.2.1 – usually, we drastically increase the dimension of the feature space to rearrange the data most conveniently. Another difference is that we do not lose any information in the data by embedding them into a high-dimensional space.

Fortunately, it turns out that many ML algorithms can be expressed only in terms of inner products between data points $\boldsymbol{x}$ and $\boldsymbol{y}$.[a] As such, we do not need to consider the individual inputs $\boldsymbol{x}$ and their representation in feature space $\phi(\boldsymbol{x})$ explicitly. The *kernel trick* simply consists of exchanging the inner product $\langle \boldsymbol{x}, \boldsymbol{y} \rangle = \boldsymbol{x}^\top \boldsymbol{y}$ in the corresponding algorithms by the function $K(\boldsymbol{x}, \boldsymbol{y}) = \phi(\boldsymbol{x})^\top \phi(\boldsymbol{y})$.

[a] We remind the reader that these data points are, in general, vectors, that is, they may exist in a high-dimensional space; see also Section 1.5.

The function K can be efficiently evaluated, in particular, for an infinite-dimensional feature space to which ϕ maps and yields a single real nonnegative number regardless of the data under consideration or the dimension of the feature space.[5] As we see later, the function K is referred to as the kernel function. This way, the kernel trick allows for a feasible nonlinear extension of a variety of ML algorithms, such as ridge regression, SVMs, or PCA. We detail the mathematical foundation of kernels in Sections 4.1.2–4.1.5.

4.1.2 The function space as a Hilbert space

As we have sketched above, we want to find a suitable choice for our feature map, that is, search for a function that lives in some high-dimensional space of functions. A mathematical space is a set of elements that obey common rules specifying the relationship between them. As we want to search in a function space, it is reasonable to assume it to be a vector space. In such a space, we are allowed to add functions to each other or rescale them by a constant without leaving the function space. This assumption is necessary to express the unknown target function by a weighted sum of other known functions. Fortunately, tuning the weights is something that we are familiar with in ML problems. As a second ingredient, we require some distance measure or, equivalently, a measure of similarity between functions. This is achieved by equipping our vector space with an inner product $\langle \cdot, \cdot \rangle$, turning the mathematical space into a Hilbert space. Different choices of inner products lead to different notions of distance, that is, different spaces altogether. We will see why we need this similarity measure shortly.

Consider, for example, a space of real-valued *square-integrable* functions called the L^2-space. When equipped with an inner product, it is akin to the more intuitive Euclidean space. [6] Square-integrability refers to the fact that the integral over the full domain D remains finite, that is, $\int_{\mathbb{D}} f^2(\boldsymbol{x})\, d\boldsymbol{x} < \infty \; \forall f \in L^2$. The domain $\mathbb{D}$ is often

[5] In many cases, it is easier to work with kernels yielding infinite-dimensional feature spaces than with finite ones. Indeed, the finite dimensionality may often be a problem. We refer to Section 8.2.5 for quantum kernel methods and the references therein regarding this issue.

[6] Of course, not all functions we are interested in finding are necessarily members of this function space. Because we are always presented with a finite dataset, however, we usually do not care what happens very far outside this regime. Even though our target function might not be a member of this space, we should be able to find a member that resembles the target function in the region of interest, nevertheless. This is why we can restrict ourselves with the L^2-space in the first place.

given by $\mathbb{R}^m$ with some dimension m in typical ML scenarios as it corresponds to our input space. The inner product of an L^2-space can be defined as

$$\langle f, g \rangle = \int_{\mathbb{D}} f(\boldsymbol{x}) g(\boldsymbol{x})\, d\boldsymbol{x}, \tag{4.1}$$

which provides the notion of orthogonality as $\langle f, g \rangle = 0$, and the norm $\|f\|_{L^2}^2 = \langle f, f \rangle$. Since a Hilbert space is a vector space, one can find an orthogonal basis set $\{\phi_n(\boldsymbol{x})\}$ that spans the space. In the case of the L^2-space, it is of infinite (but countably infinite) dimension. This basis allows for any function $f \in L^2$ to be decomposed as

$$f(\boldsymbol{x}) = \sum_n a_n \phi_n(\boldsymbol{x}), \tag{4.2}$$

with expansion coefficients $\{a_n\}$.

Any two Hilbert spaces have the same geometric structure, regardless of their respective elements – this is the reason why we can draw the analogies between the L^2 and the intuitive Euclidean space in the first place. This can be made more formal by the representation theorem of Riesz [175], which essentially boils down to this: One can express certain linear functionals by means of the Hilbert space's inner product. The functional of interest is the *evaluation* of an arbitrary function f living in a Hilbert space $\mathcal{H}$ at any point $\boldsymbol{x} \in \mathbb{D}$, that is, $f \in \mathcal{H} \mapsto f(\boldsymbol{x}) \in \mathbb{R}$. This evaluation property will become essential once we are trying to learn a target function. Riesz' theorem further tells us that there exists a unique function $K_{\boldsymbol{x}}$ such that any function $f \in \mathcal{H}$ can be evaluated at $\boldsymbol{x}$ by its inner product with $K_{\boldsymbol{x}}$.

4.1.3 Reproducing kernel Hilbert spaces

As anticipated, we want to use the Riesz theorem to connect the function space to the evaluation of its members. That is, we require $K_{\boldsymbol{x}}$ to assert that point evaluation is of the form

$$f(\boldsymbol{x}) = \langle f, K_{\boldsymbol{x}} \rangle. \tag{4.3}$$

Because of Eq. (4.1) and with the definition $K_{\boldsymbol{x}} := K(\boldsymbol{x}, \cdot)$, this leads to the integral transform

$$f(\boldsymbol{x}) = \int_{\mathbb{D}} K(\boldsymbol{x}, \boldsymbol{x}') f(\boldsymbol{x}')\, d\boldsymbol{x}' \quad \forall f \in \mathcal{H}. \tag{4.4}$$

Here, K acts as the *kernel* of the integral transform. It can be seen that the kernel K of this transform must be symmetric, that is, $K(\cdot, \boldsymbol{x}) = K(\boldsymbol{x}, \cdot)$. Moreover, due to its role in Eq. (4.4), we call the kernel K a *reproducing* kernel as it faithfully reproduces any function f in our Hilbert space. If point evaluations of every function in a Hilbert space can be represented with a reproducing kernel, the Hilbert space is called a reproducing kernel Hilbert space (RKHS).

We will refer to $K(\boldsymbol{x}, \boldsymbol{x}')$ as the *kernel function*. Furthermore, we will restrict our discussion to functions $K(\boldsymbol{x}, \boldsymbol{x}')$ that are positive-semidefinite. Due to Mercer's theorem [176], any symmetric, positive-semidefinite function $K(\boldsymbol{x}, \boldsymbol{x}')$ can be represented as

$$K(\boldsymbol{x}, \boldsymbol{x}') = \sum_n \lambda_n \phi_n(\boldsymbol{x}) \phi_n(\boldsymbol{x}') \tag{4.5}$$

with nonnegative coefficients $\{\lambda_n\}$. Moreover, for an orthonormal basis set, the coefficients λ_n must be all equal to 1 to yield $K(\boldsymbol{x}, \boldsymbol{x}')$ satisfying Eq. (4.4). We can prove this by choosing $f = \phi_k$ for some integer k. Then from Eq. (4.4) and the decomposition of the kernel, we require that

$$\phi_k(\boldsymbol{x}) = \sum_n \lambda_n \phi_n(\boldsymbol{x}) \int_{\mathbb{D}} \phi_n(\boldsymbol{x}')\phi_k(\boldsymbol{x}')\, d\boldsymbol{x}' = \lambda_k \phi_k(\boldsymbol{x}). \tag{4.6}$$

Since k was chosen arbitrarily, we find that $\lambda_n = 1\ \forall n$ if and only if we select an orthonormal basis set. We can also convince ourselves that this kernel representation actually performs the integral transformation in Eq. (4.4) via

$$\begin{aligned}\int_{\mathbb{D}} K(\boldsymbol{x}, \boldsymbol{x}') f(\boldsymbol{x}')\, d\boldsymbol{x}' &= \sum_{n,k} a_n \phi_k(\boldsymbol{x}) \int_{\mathbb{D}} \phi_k(\boldsymbol{x}')\phi_n(\boldsymbol{x}')\, d\boldsymbol{x}' \\ &= \sum_n a_n \phi_n(\boldsymbol{x}) = f(\boldsymbol{x})\end{aligned} \tag{4.7}$$

as intended.

One may recognize Eq. (4.5) with $\lambda_n = 1\ \forall n$ as a basis-set representation of a delta-function $\delta(\boldsymbol{x} - \boldsymbol{x}')$. Obviously, this is not the most useful choice of the kernel function because, in ML, the task is generally to estimate the target function $f(\boldsymbol{x})$ at values of $\boldsymbol{x}$ different from the positions of the training points $\boldsymbol{x}_i$ by means of equations written in terms of the kernel functions $K(\boldsymbol{x}, \boldsymbol{x}_i)$. The question thus becomes: Can the above arguments be extended to any arbitrary function $K(\boldsymbol{x}, \boldsymbol{x}')$ that is symmetric and positive-semidefinite? In particular, can the reproducing kernel yielding Eq. (4.3) be defined for any symmetric, positive-semidefinite kernel function?

To answer this question, consider the eigenvalue decomposition of a positive-semidefinite function $K(\boldsymbol{x}, \boldsymbol{x}')$

$$\int_{\mathbb{D}} K(\boldsymbol{x}, \boldsymbol{x}')\phi_n(\boldsymbol{x}')\, d\boldsymbol{x}' = \lambda_n \phi_n(\boldsymbol{x}) \quad \forall n, \tag{4.8}$$

where $\lambda_n \geq 0$. Using Eq. (4.2), we can rewrite the inner product Eq. (4.1) as

$$\begin{aligned}\langle f, g\rangle &= \int_{\mathbb{D}} f(\boldsymbol{x}) g(\boldsymbol{x})\, d\boldsymbol{x} = \sum_{m,n} a_m b_n \underbrace{\int_{\mathbb{D}} \phi_m(\boldsymbol{x})\phi_n(\boldsymbol{x})\, d\boldsymbol{x}}_{=\langle \phi_m, \phi_n\rangle = \delta_{m,n}} = \sum_n a_n b_n \\ &= \sum_n \int a_m \phi_m(\boldsymbol{x})\phi_n(\boldsymbol{x})\, d\boldsymbol{x} \int b_k \phi_k(\boldsymbol{x}')\phi_n(\boldsymbol{x}')\, d\boldsymbol{x}' \\ &= \sum_n \langle f, \phi_n\rangle \langle g, \phi_n\rangle\,.\end{aligned} \tag{4.9}$$

However, for K to be a valid kernel function of an RKHS, it has to give rise to Eq. (4.3) as well. Plugging this into the first step of Eq. (4.9) together with Mercer's decomposition, we see that

$$\begin{aligned}\langle f, g\rangle &= \int f(\boldsymbol{x}) g(\boldsymbol{x})\, d\boldsymbol{x} = \int \langle f, K_{\boldsymbol{x}}\rangle\, g(\boldsymbol{x})\, d\boldsymbol{x} \\ &= \sum_n \lambda_n \int \phi_n(\boldsymbol{x}) \langle f, \phi_n\rangle\, g(\boldsymbol{x}) = \sum_n \lambda_n \langle f, \phi_n\rangle \langle g, \phi_n\rangle\,.\end{aligned} \tag{4.10}$$

Comparing the two previous results, Eqs. (4.9) and (4.10), we see a discrepancy in terms of the prefactors λ_n. To compensate for this, we can redefine the inner product as

$$\langle f, g \rangle_{\mathcal{H}} = \sum_{n=1}^{\infty} \frac{\langle f, \phi_n \rangle \langle g, \phi_n \rangle}{\lambda_n}, \tag{4.11}$$

which is equivalent to the previous inner product in Eq. (4.1) if and only if $\lambda_n = 1\, \forall n$. With this definition of the inner product, we have

$$\langle f, K_{\boldsymbol{x}} \rangle_{\mathcal{H}} = \sum_{n=1}^{\infty} \frac{\langle f, \phi_n \rangle \langle K_{\boldsymbol{x}}, \phi_n \rangle}{\lambda_n} = \sum_{n=1}^{\infty} \langle f, \phi_n \rangle \phi_n(\boldsymbol{x}) = f(\boldsymbol{x}). \tag{4.12}$$

Here, we have used Eq. (4.8) in the second step, and the last equality follows from the fact that the set $\{\phi_n\}$ forms a complete basis, that is, Eq. (4.2). This now fulfills Eq. (4.3) as intended and, furthermore, renders the Hilbert space $\mathcal{H}$ an RKHS, and shows that K, any arbitrary symmetric positive-semidefinite function of two arguments, is indeed a kernel function. We note that this definition of the inner product – which is unique for every kernel function – is crucial. It exemplifies what is known as the Moore–Aronszajn theorem [177], which states that every RKHS is associated with a unique positive-semidefinite kernel, and vice versa.

4.1.4 The representer theorem

If we were to represent our target function as a basis set expansion (Eq. (4.2)), determining the function would require finding an – in principle – infinite number of expansion coefficients. The representer theorem [178], which plays a central role in kernel methods for ML, allows one, however, to express the target function by a finite sum. Given a loss function $\mathcal{L}$ (including a regularization term) and n training samples $\{(\boldsymbol{x}_i, y_i)\}_{i=1}^{n}$, the theorem implies that

$$f^*(\boldsymbol{x}) := \arg\min_f \mathcal{L}\Big(\{(f(\boldsymbol{x}_i), y_i)\}_{i=1}^{n} \Big) = \sum_{i=1}^{n} a_i K(\boldsymbol{x}, \boldsymbol{x}_i), \tag{4.13}$$

that is, the loss is minimized by a function that can be written as a *finite* sum over the kernel function evaluated at one of the arguments set to the position of training data points. Moreover, one only requires the knowledge of the kernel function K – no explicit feature mapping is required.

The representer theorem guarantees that we can formulate the search for a function f^* that minimizes a specific loss function over an infinite-dimensional function space as a search over n kernel coefficients $\{a_i\}_{i=1}^{n}$. Thus, it significantly reduces the complexity of the minimization problem at hand, making it computationally tractable.

Finally, our mathematical efforts have come to fruition: The reproducing property of the kernel in Eq. (4.4) allows us to *implicitly* embed the input data in a (possibly) high-dimensional feature space in which we calculate the similarities to a test point $\boldsymbol{x}$. Because we are only required to calculate the similarity measure between data points, we do not lose efficiency here. We see how to practically do this in the following when we extend some of the models of Section 2.4 via the kernel trick in Section 4.2.

4.1.5 Consequences of the kernel trick

We have explored the mathematics behind kernels in some detail in particular, concerning the RKHS. Despite the perceived detour through Hilbert spaces and a redefinition of the inner product in the RKHS, this groundwork provides us with the necessary foundation for the theory of kernels: The reproducing feature of Eq. (4.4) is not a mere mathematical curiosity but has straightforward implications in terms of the representer theorem. Furthermore, the kernel formulation allows us to solve our initial problem (efficiently finding the unknown target function) by identifying the proper kernel. The kernel can be understood as a similarity measure between feature vectors representing data. This similarity between inputs can be directly calculated even if the underlying feature space is high- or even infinite-dimensional.

One important consequence of *implicitly* switching from the input to the feature space by means of the kernel trick is that we have to rethink our intuition of regularization: We have to perform the regularization of the learned function in the function space given by the RKHS as discussed in Section 4.1.4. For this, we have to start from the inner product in the RKHS, that is, Eq. (4.11). The coefficients λ_n correspond to the weights of the basis functions of the kernel and are nonnegative by construction. In particular, they depend on the actual choice of the kernel function K. For instance, our kernel decomposition could include zero entries for some of the basis functions. This is not an issue per se: The function f we are interested in might still lie entirely in the RKHS. Using Eq. (4.2), we can decompose it as $f(\boldsymbol{x}) = \sum_{n=1}^{\infty} a_n \phi_n(\boldsymbol{x})$. Its corresponding L^2-norm is $||f||_2^2 = \sum_n |a_n|^2$. Due to Eq. (4.11), this translates to a norm in the RKHS as

$$||f||_{\mathcal{H}}^2 = \sum_{n=1}^{\dim(\text{RKHS})} \frac{|a_n|^2}{\lambda_n}. \tag{4.14}$$

Thus, we see that we potentially run into trouble in the case of vanishing λ_n as our norm may diverge. It remains finite if and only if the corresponding function coefficient a_n is equal to 0 at the same time. This is the case if the function is entirely in the RKHS associated to the kernel function. Otherwise, when choosing the wrong kernel function, we cannot regularize our model and cannot expect to learn the unknown target function f entirely.

The theory of RKHS gives us an intuition on why certain choices of kernel function seem to work while others fail: The target function f has to fully lie in the RKHS uniquely defined by K. If not, our approach to learning the function is doomed to fail from the start. It is possible to develop strategies to build optimal kernels (which provide an optimal RKHS for a particular problem), as, for example, discussed in Section 4.4.2.

Table 4.1 Examples of kernel functions, where θ, θ_1, and θ_2 are free parameters [135].

Kernel function	Mathematical form
Linear	$K_{\text{LIN}}(\boldsymbol{x}, \boldsymbol{x}') = \boldsymbol{x}^{\intercal}\boldsymbol{x}' + \theta$
Radial basis	$K_{\text{RBF}}(\boldsymbol{x}, \boldsymbol{x}') = \exp\left(-\frac{1}{2\theta^2}\|\boldsymbol{x} - \boldsymbol{x}'\|^2\right)$
Matérn 5/2	$K_{\text{MAT}}(\boldsymbol{x}, \boldsymbol{x}') = \left(1 + \frac{\sqrt{5}}{\theta}\|\boldsymbol{x} - \boldsymbol{x}'\| + \frac{5}{3}\|\boldsymbol{x} - \boldsymbol{x}'\|^2\right) \times \exp\left(-\frac{\sqrt{5}}{\theta^2}\|\boldsymbol{x} - \boldsymbol{x}'\|\right)$
Rational quadratic	$K_{\text{RQ}}(\boldsymbol{x}, \boldsymbol{x}') = \left(1 + \frac{\|\boldsymbol{x}-\boldsymbol{x}'\|}{2\theta_1\theta_2^2}\right)^{-\theta_1}$

4.2 Kernel methods

In Section 4.1, we have presented the mathematical foundation of kernels. In short, we want to map our data to a feature space that possesses a more suitable structure for the task at hand. Instead of explicitly defining a feature map ϕ, we introduce a kernel function K, which provides a similarity measure between data points in the underlying feature space. As such, we exchange the problem of searching for a (potentially) high-dimensional feature map for finding an optimal kernel function, which is rigorously easier. This constitutes the *kernel trick*.

Kernel methods correspond to all classification and regression methods that take advantage of the kernel trick. The validity of these approaches is ensured by the representer theorem, see Section 4.1.4. The first step of every kernel method is to choose a kernel function. As discussed previously, any symmetric, positive-semidefinite function of two arguments can be used as a kernel function. Typically, one starts by assuming some functional form (see examples in Table 4.1). These functions are parametrized by a few parameters, such as θ, or θ_1 and θ_2 in these examples. Having chosen a particular functional form of the kernel function, one varies these parameters to find the best kernel in the corresponding functional ansatz class. This makes the optimization already easier as we now have to optimize over a set of parameters and not over a set of functions. Also note that given a set of kernels, there exist many transformations that yield new valid kernels. For example, any linear combination of kernel functions $\sum_i c_i K_i(\boldsymbol{x}, \boldsymbol{x}')$ with coefficients c_i constitutes a valid kernel. For a more exhaustive list of techniques for constructing new kernels, see Ref. [93]. We explicitly make use of these rules in Section 4.4.2, where we discuss how to construct good kernels systematically through compositional kernel search.

We turn to three prominent kernel methods in the remainder of this subsection: kernel ridge regression (KRR), SVMs, and GPs. As we focus on supervised kernel methods, we do not elaborate on a kernel extension for unsupervised methods such as PCA from Section 3.2 [117, 179].

4.2.1 Kernel ridge regression

Kernel ridge regression is an extension of ridge regression (presented in Section 2.4.1) to nonlinear regression problems [180]. The functional we want to minimize is very similar to the one of ridge regression, but this time, the model f lives in the RKHS $\mathcal{H}$ corresponding to the particular choice of the kernel function (however, note the use of MSE!):

$$\mathcal{L}_{\text{KRR}} = \sum_{i}^{n} (y_i - f(\boldsymbol{x}_i))^2 + \lambda \|f\|_{\mathcal{H}}^2 = \mathcal{L}_{\text{MSE}} + \mathcal{L}_{\text{reg}} \tag{4.15}$$

with a regularizing term introduced in Eq. (4.14). Here, f is not restricted to a linear function (as in linear ridge regression). Instead, f can, in principle, be arbitrary. As such, KRR is capable of building highly expressive models given an appropriate choice of kernel. Increasing the data efficiency of an ML problem, and consequently the accuracy of the resulting model given a fixed, finite dataset, translates to finding the optimal kernel. This is not immediately apparent from Eq. (4.15). To illustrate the role of kernels, recall that the model $f(\boldsymbol{x})$ can be written as the following sum over the training data:

$$f(\boldsymbol{x}) = \sum_{j=1}^{n} \alpha_j K(\boldsymbol{x}, \boldsymbol{x}_j) . \tag{4.16}$$

This formulation is an instance of the already discussed representer theorem, see Eq. (4.13). Apart from K, whose mathematical form we know (or assume), we also have here coefficients α_j of our kernel model $f(\boldsymbol{x})$ which we need to find. We can express Eq. (4.15) with Eq. (4.16) in matrix form:

$$\mathcal{L}_{\text{MSE}} = \sum_{i}^{n} (y_i - f(\boldsymbol{x}_i))^2 = (\boldsymbol{y} - \boldsymbol{K}\boldsymbol{\alpha})^\intercal (\boldsymbol{y} - \boldsymbol{K}\boldsymbol{\alpha}), \tag{4.17}$$

$$\mathcal{L}_{\text{reg}} = \lambda \|f\|_{\mathcal{H}}^2 = \lambda \boldsymbol{\alpha}^\intercal \boldsymbol{K} \boldsymbol{\alpha}, \tag{4.18}$$

where the matrix $\boldsymbol{K}$ is called the *kernel matrix*. It is a positive-semidefinite, square $n \times n$ matrix with elements $K(\boldsymbol{x}, \boldsymbol{x}')$ with training points $\boldsymbol{x}$ and $\boldsymbol{x}'$ belonging to the training set: $\{\boldsymbol{x}_1, \boldsymbol{x}_2, \boldsymbol{x}_3, \dots, \boldsymbol{x}_n\}$. The vector $\boldsymbol{y}$ represents the targets for the corresponding training input $\boldsymbol{x}$. Finally, if we set the derivative of the sum of these two components equal to 0, we can find a solution for $\boldsymbol{\alpha}$ which is:

$$\hat{\boldsymbol{\alpha}} = [\boldsymbol{K} + \lambda \mathbb{1}]^{-1} \boldsymbol{y}. \tag{4.19}$$

Given $\hat{\boldsymbol{\alpha}}$, we can write the estimator of the model $\hat{f}$ at a test point $\boldsymbol{x}^*$ as

$$\hat{f}(\boldsymbol{x}^*) = \boldsymbol{k}^\intercal(\boldsymbol{x}^*) \hat{\boldsymbol{\alpha}} = \boldsymbol{k}^\intercal(\boldsymbol{x}^*) [\boldsymbol{K} + \lambda \mathbb{1}]^{-1} \boldsymbol{y}, \tag{4.20}$$

where $\boldsymbol{k}(\boldsymbol{x}^*) = [k(\boldsymbol{x}^*)_i] = [K(\boldsymbol{x}^*, \boldsymbol{x}_i)]$. The analogous and thorough derivation of the kernel trick on the example of KRR is provided in Appendix B. Finally, we can see that the prediction of the output for an unseen input $\boldsymbol{x}^*$ can be written in terms of:

- The target vector $\boldsymbol{y}$,
- The kernel matrix $\boldsymbol{K}$, whose elements are the kernel function values $K(\boldsymbol{x}_i, \boldsymbol{x}_j)$, which takes advantage of the kernel trick,

- The column vector $\boldsymbol{k}(\boldsymbol{x}^*)$, whose elements are the kernel function values $K(\boldsymbol{x}^*, \boldsymbol{x}_i)$, which also takes advantage of the kernel trick,
- The regularization term of magnitude λ.

As anticipated earlier in the chapter, successfully applying KRR boils down to finding the appropriate kernel function K and tuning its corresponding hyperparameters, which – for KRR – is most often done by cross-validation. This should be contrasted with how the kernel function parameters are estimated for Gaussian process regression (GPR), discussed below.

4.2.2 Support vector machines

In Section 4.2.1, we have discussed how to use kernel methods for regression problems (in particular ridge regression). In this section, we show how one can use them for classification. In this context, the intuition behind the kernel approach is to embed the input space into the feature space in such a way that the data become linearly separable with a hyperplane (as described already in Section 4.1.1). The most common ML-classification method utilizing the kernel trick is SVMs [181].[7]

Support vector machines have been introduced already in Section 2.4.3 as geometric linear classifiers. Before we see how kernels enter SVMs, let us recall how linear SVMs work and rephrase the optimization problem that we have described in Section 2.4.3. The problem there is to find an optimal hyperplane separating data from different classes. The optimal hyperplane is defined as the one with the maximal distance between the hyperplane and the data points. In other words, we can say that all data points need to be at least at distance M away from the hyperplane. The data points that are separated from the hyperplane exactly by M, so are the closest to the hyperplane, become support points, $\boldsymbol{x}_{s,i}$. The classification problem boils down to finding such a hyperplane described by $\boldsymbol{\theta}$ that maximizes the margin between the hyperplane and support points $\boldsymbol{x}_{s,i}$ (see Fig. 2.5). As we can rescale the hyperplane in an arbitrary way, we can have $|\boldsymbol{\theta}| = 1/M$. Then maximizing a margin, becomes minimizing $\boldsymbol{\theta}$ that in turn comes down to minimizing the Lagrange function L in Eq. (2.41), which we restate here for readability:

$$L = \frac{1}{2}|\theta|^2 - \sum_{i=1}^{n} \alpha_i \left[y_i \left(\theta^{\mathsf{T}} \boldsymbol{x}_i + \theta_0 \right) - 1 \right] , \tag{4.21}$$

where the Lagrange multipliers α_i are chosen such that

$$\alpha_i \left[y_i \left(\theta^{\mathsf{T}} \boldsymbol{x}_i + \theta_0 \right) - 1 \right] = 0 \quad \forall i = 1, \dots, n \,. \tag{4.22}$$

As we have already discussed, α_i is nonzero (and positive) only for $\boldsymbol{x}_{s,i}$. In practice, rather than minimizing L, we go for the dual formulation of the problem, and we maximize a Lagrange dual, L_D, which provides the lower bound for L. L_D remains a quadratic program, similarly to L, as we have discussed in section 2.4.3. To express

[7] There is a variant of this approach designed for regression called support vector regression that is almost identical with KRR but minimizes a different form of the loss function.

the problem via L_D, we first take the derivative of L with respect to $\boldsymbol{\theta}$ and θ_0 and set it to zero. We arrive at:

$$\begin{aligned} \boldsymbol{\theta} &= \sum_{i=1}^{n} \alpha_i y_i \boldsymbol{x}_i \\ 0 &= \sum_{i=1}^{n} \alpha_i y_i \,. \end{aligned} \tag{4.23}$$

We can see that the coefficients $\boldsymbol{\theta}$ are given by the Lagrange multipliers α_i, which can be found numerically. When we plug these equations back into the Lagrange function in Eq. (4.21), we arrive at the Lagrange dual:

$$L_D = \sum_{i=1}^{n} \alpha_i - \frac{1}{2} \sum_{i=1}^{n} \sum_{j=1}^{n} \alpha_i \alpha_j y_i y_j \boldsymbol{x}_i^{\mathsf{T}} \boldsymbol{x}_j \quad \text{subject to } \alpha_i \geq 0\,. \tag{4.24}$$

Finally, to put kernels in the picture, we change the notation from $\boldsymbol{x}_i^{\mathsf{T}} \boldsymbol{x}_j$ to $\langle \boldsymbol{x}_i, \boldsymbol{x}_j \rangle$. For now, the SVM remains a linear model. To deal with nonlinearities in the input space, we can introduce a feature map, $\boldsymbol{x}_i \to \Phi(\boldsymbol{x}_i)$, which gives us

$$L_D = \sum_{i=1}^{n} \alpha_i - \frac{1}{2} \sum_{i=1}^{n} \sum_{j=1}^{n} \alpha_i \alpha_j y_i y_j \left\langle \Phi\left(\boldsymbol{x}_i\right), \Phi\left(\boldsymbol{x}_j\right) \right\rangle \,. \tag{4.25}$$

We finally see our kernel function, $K(\boldsymbol{x}_i, \boldsymbol{x}_j) = \left\langle \Phi\left(\boldsymbol{x}_i\right), \Phi\left(\boldsymbol{x}_j\right) \right\rangle$, appearing. In this kernel formulation, the margin we maximize is between the hyperplane and the support points *in the feature space* [71]. Therefore, the SVM problem boils down to maximizing L_D numerically to find the coefficients α_i, for example, using sequential minimal optimization [72]. The parameters of the kernel function K then need to be validated using, for example, the held-out validation set or cross-validation. As with NNs, this requires a retraining of the SVM for each trial choice of kernel parameters. Once both parameter sets are known, the hyperplane separating the classes in the typically high-dimensional space is also known. With the optimal hyperplane $\hat{f}$ we can then make predictions at an arbitrary test point $\boldsymbol{x}^*$:

$$\begin{aligned} \hat{f}(\boldsymbol{x}^*) = \boldsymbol{\theta}^{\mathsf{T}} \phi(\boldsymbol{x}) + \theta_0 &= \sum_{i} \alpha_i y_i \left\langle \phi\left(\boldsymbol{x}_i\right), \phi\left(\boldsymbol{x}_i\right) \right\rangle + \theta_0 \\ &= \sum_{i} \alpha_i y_i K\left(\boldsymbol{x}_i, \boldsymbol{x}\right) + \theta_0 \,. \end{aligned} \tag{4.26}$$

Finally, to turn this value into a class prediction, we take the sign of $\hat{f}$ as the corresponding class label. Note that the choice of the kernel function here matters, as discussed in the opening of Section 4.2. We visualize this problem in Fig. 4.2.

While we can put lots of effort into finding a kernel function that renders our problem linearly separable, we can also relax the problem by allowing some misclassification. Let us thus move to the *problems that are not linearly separable.* The derivations above still hold with one modification: we now allow a number of data points to be on the wrong side of the margin, as shown in Fig. 4.3. This modifies our constraint from Eq. (2.40) to $y_i(\boldsymbol{\theta}^{\mathsf{T}} \boldsymbol{x}_i + \theta_0) \geq 1 - \xi_i$, where $\xi_i = 0$ if the data point is on

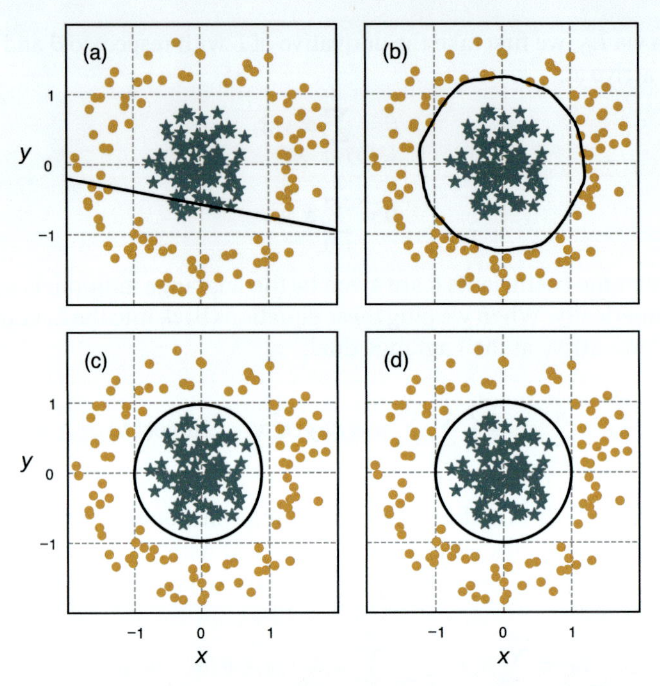

Figure 4.2 The kernel form makes a difference! The same data as in Fig. 4.1 is classified using an SVM with different kernel choices. The black line corresponds to each underlying decision boundary. (a) As the data is not linearly separable, a linear kernel does not work. (b) A polynomial kernel up to degree 50 does the trick at the expense of signs of overfitting. (c) Due to the rotation symmetry, an RBF kernel is best suited for classifying this dataset whose decision boundary closely resembles the actual underlying decision boundary of the data shown in (d).

the correct side of the margin. The variables ξ_i are often referred to as *slack variables*. We can incorporate the control over how "wrong" the hyperplane can be by adding another constraint, that is, $\sum_i \xi_i < C = \text{const}$, which also adds terms to the Lagrange function:

$$L = \frac{1}{2}|\theta|^2 - \sum_i \alpha_i \left[y_i \left(\theta^\top \boldsymbol{x}_i + \theta_0 \right) - (1 - \xi_i) \right] + C \sum_i \xi_i. \tag{4.27}$$

Now, we maximize the margin while minimizing the violation of the margin constraints. This loss function is still a quadratic program but now has also a largely increased number of optimization variables (one slack variable per data point). As previously mentioned, instead of minimizing L you can maximize L_D. Note that in this case, you get an additional regularization term with magnitude C. Large C allows for more misclassified data points but promotes simpler decision boundaries. This is because, in this case, the SVM focuses on a minimal number of relevant data points to draw a decision boundary. Since this number of relevant data points is usually much smaller than the total number of given training data points, this is referred to as a *sparse* solution. By contrast, a small C forces the model to better fit training data,

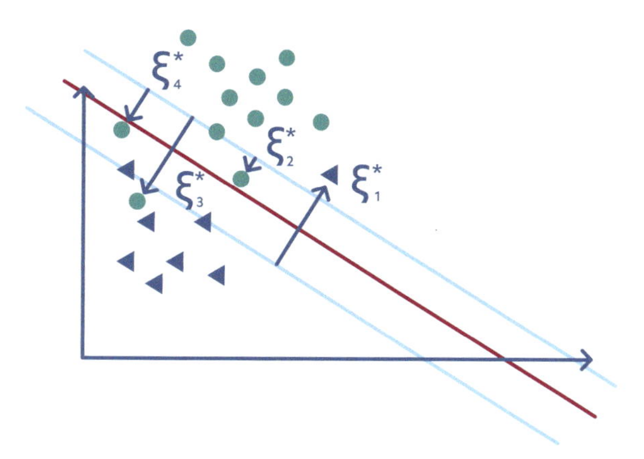

Figure 4.3 A linear SVM applied to a dataset that is not linearly separable. The SVM tries to minimize the total distance of all misclassified data points ξ_i called slack variables. $\xi_i \geq 1$ for misclassified $\boldsymbol{x}_i$ (here, $i = 1, 3, 4$) and $0 < \xi_i < 1$ for points on the correct side of the decision boundary but within the margin (here, $i = 2$).

sometimes at the expense of the validation data.[8] In this case, the solution is less sparse but can tend to be overfitted to the training data.

Finally, we can make a connection between SVMs and KRR. We started by saying that we need to minimize $|\boldsymbol{\theta}|^2$ subject to the following conditions:

$$\begin{gathered} y_i\left(\boldsymbol{\theta}^\mathsf{T}\boldsymbol{x}_i + \theta_0\right) \geq 1 - \xi_i \text{ with } \xi_i \geq 0 \\ \sum_i \xi_i < C = \text{const}\,. \end{gathered} \tag{4.28}$$

If we simply write

$$\xi_i \geq 1 - y_i\left(\boldsymbol{\theta}^\mathsf{T}\boldsymbol{x}_i + \theta_0\right) \tag{4.29}$$

and vary $\boldsymbol{\theta}$ and θ_0 to find the minimum of the following function:

$$L(\boldsymbol{\theta}, \theta_0) = \frac{1}{2}|\boldsymbol{\theta}|^2 + C\sum_i \left[1 - y_i\left(\boldsymbol{\theta}^\mathsf{T}\boldsymbol{x}_i + \theta_0\right)\right]_+ , \tag{4.30}$$

the problem is equivalent to minimizing the following function:

$$L(\boldsymbol{\theta}, \theta_0) = \frac{1}{2C}|\boldsymbol{\theta}|^2 + \sum_{i=1}^{n} \left[1 - y_i\left(\boldsymbol{\theta}^\mathsf{T}\boldsymbol{x}_i + \theta_0\right)\right]_+ , \tag{4.31}$$

which is the same as

$$L(\boldsymbol{\theta}, \theta_0) = \frac{1}{2}\lambda|\boldsymbol{\theta}|^2 + \sum_{i=1}^{n} \max\left[0, 1 - y_i f(\boldsymbol{x}_i)\right]. \tag{4.32}$$

[8] Beware of various definitions and conventions regarding regularization strength, particularly in SVMs. For example, in Scikit-learn, decreasing a hyperparameter C corresponds to more regularization.

Finally, let us compare it to the functional of KRR from Eq. (4.15): The regularization term is the same, the main difference is that KRR uses the squared error loss, while SVMs uses a function called the Hinge loss. The difference, of course, stems from the fact that we try to solve two different tasks, regression for KRR vs. classification for SVMs. It is possible, however, to modify SVMs such that they can be applied to regression tasks. In this case, KRR has the advantage of being computationally more efficient, especially for small datasets. The advantage of the modified SVMs, again, is their sparsity to yield potentially less overfitted solutions.

4.2.3 Gaussian processes

At this point, we have already covered powerful and general regression tools. We learned about the kernel trick and what it means to learn in feature spaces rather than in the input space. Moreover, we have seen how this is useful, particularly for high-dimensional feature spaces, and examined in more depth some tools that manifestly use the kernel trick to perform good learning tasks. In this section, we cover an additional tool, namely, Gaussian processes (GPs). Since we have already introduced powerful regression models such as KRR, a natural question to ask is: *Why do we need another regression model?* The short answer is that GPR does all that KRR offers but also provides the Bayesian uncertainty of the predictions. As we shall see, this can be used for Bayesian optimization (BO) [182]. GPR is also well suited for algorithms aiming to build kernels well aligned with data in a setting where data is very limited. In the remainder of this section, we describe what GPs are and how GPR works.

Consider the regression problem of finding a function $f : \mathbb{R}^m \to \mathbb{R}$. In the spirit of this chapter, we want to map the input to a suitable d-dimensional space using nonlinear functions ϕ_i. Then, as in Section 2.4.1, we apply a linear model, that is,

$$f(\boldsymbol{x}) = (\theta_0, \theta_1 \dots \theta_d)^\top (1, \phi_1(\boldsymbol{x}), \dots, \phi_d(\boldsymbol{x})) . \tag{4.33}$$

Here, each of the functions $\phi_i(\boldsymbol{x})$ is some parametrized nonlinear function of $\boldsymbol{x} \in \mathbb{R}^m$ such as

$$\phi_i(\boldsymbol{x}) = \tanh\left[w_i \sum_{j=1}^{m} x_j + b_i \right] . \tag{4.34}$$

Equation (4.33) can be visualized as the NN from Fig. 4.4. If all the weights of the network are fixed, the network maps any given data point $\boldsymbol{x}$ to a single value $f(\boldsymbol{x})$. However, if we now assume the parameters $\theta_0, \dots \theta_d$ as well as the hyperparameters of the activation functions $(w_i, b_i)_i$ to be samples of random variables, distributed according to some distribution, the output of the NN for a fixed $\boldsymbol{x}$ becomes a random variable itself. Such an NN becomes a Bayesian NN, where all tunable parameters of the NN have been promoted to random variables. As a consequence, we have to average over the parameter values to obtain, for example, the expectation value of $f(\boldsymbol{x})$ and similar quantities of interest. In Eq. (4.33), there are $d + 1 + 2d = 3d + 1$ many continuous random variables, θ_i, w_i, and b_i. Hence, taking averages requires us to solve a $(3d + 1)$-dimensional integral. This is just not feasible and potentially not possible analytically, depending on the distributions involved for each of the random

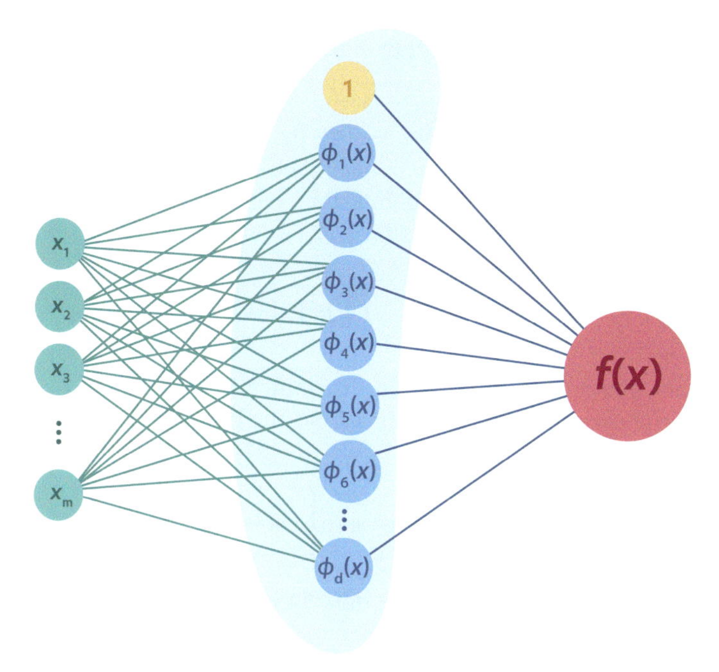

Figure 4.4 Sketch of a Bayesian NN. The features in the hidden layer (blue) are multiplied by random variables as shown in Eq. (4.33). This way, the NN is no longer deterministic, and its output $f(\boldsymbol{x})$ given the input $\boldsymbol{x}$ is itself a random variable.

variables. There is a way out of this conundrum, luckily: Assuming that all θ_i are i.i.d. random variables, the limit of $d \to \infty$, by the central limit theorem,[9] yields a normal distribution for the output [183]. Because the distribution always remains Gaussian for any given input $\boldsymbol{x}$, $f(\boldsymbol{x})$ is our first example of a GP. In general, a GP can be viewed as a normal distribution over functions. Its output is an instantiation of a random variable, distributed according to a multivariate density. The upshot of this preliminary example is that by incorporating randomness into the model, under the assumption of infinitely many independently sampled NN parameters, the output of the model remains Gaussian. The advantage of all this, compared to what we have previously analyzed, is that it is possible to obtain closed-form expressions for the log marginal likelihood and the predictive distribution later on.

Let us come back to our problem setting: We want to infer the function f that describes a given dataset. The data are generally noisy. We model this data noise by a Gaussian distribution with zero mean and variance σ^2:

$$y = f(\boldsymbol{x}) + \epsilon, \quad \epsilon \sim \mathcal{N}(0, \sigma^2). \tag{4.35}$$

For the sake of simplicity and clarity, we first consider only the linear model without any nonlinear feature map, that is, ϕ is just the identity map in this case, which reduces Eq. (4.33) to

[9]The central limit theorem states that the sum of many independent random variables is approximately normally distributed. For example, if you roll two six-sided dice multiple times, the sum of the obtained results converges to the Gaussian distribution centered at seven in the limit of an infinite number of rolls.

$$f(\boldsymbol{x}) = \theta^\mathsf{T} \boldsymbol{x}\,, \tag{4.36}$$

similar to Section 2.4.1. Again, this includes the bias term such that we have $d = m+1$ many random variables in this case. We assume that each data point $\boldsymbol{x}_i$ is independent of the others. For any given point, the model likelihood is expressed as $p(\boldsymbol{y}_i \mid \theta, \boldsymbol{x}_i)$. Due to the independence of the data points, for a labeled training dataset $\mathcal{D} = (\boldsymbol{X}, \boldsymbol{y})$ the joint likelihood reads (see also Section 2.3):

$$p(\boldsymbol{y} \mid \theta, \boldsymbol{X}) = \prod_{i=1}^{n} p(\boldsymbol{y}_i \mid \theta, \boldsymbol{x}_i)\,. \tag{4.37}$$

Because of our assumption of Gaussian distributed noise (see Eq. (4.35)), we can explicitly rewrite each term of the product from the right-hand side of the equation above as:

$$p(\boldsymbol{y}_i \mid \theta, \boldsymbol{x}_i) = \frac{1}{\sqrt{2\pi}\sigma} \exp\left[-\frac{(y_i - \theta^T \boldsymbol{x}_i)^2}{2\sigma^2}\right]. \tag{4.38}$$

Hence, the model likelihood given the data $\mathcal{D}$ is a product of Gaussians. Thus, Eq. (4.37) can be explicitly rewritten as:

$$p(\boldsymbol{y} \mid \theta, \boldsymbol{X}) = \prod_{i=1}^{n} p(\boldsymbol{y}_i \mid \theta, \boldsymbol{x}_i) = \frac{1}{(2\pi\sigma^2)^{n/2}} \exp\left[-\frac{|\boldsymbol{y} - \boldsymbol{X}\theta|^2}{2\sigma^2}\right]. \tag{4.39}$$

Our goal now is to use the Bayes' theorem to calculate the posterior over the weights θ (see Section 2.3). Using Eq. (2.19), we obtain

$$p(\theta \mid \boldsymbol{y}, \boldsymbol{X}) = \frac{p(\boldsymbol{y} \mid \theta, \boldsymbol{X}) p(\theta)}{p(\boldsymbol{y} \mid \boldsymbol{X})}, \tag{4.40}$$

where $p(\boldsymbol{y} \mid \boldsymbol{X})$ is a normalization constant not depending on θ and $p(\theta)$ represents our prior (again see Section 2.3). In general, one has freedom over choosing the prior. Nevertheless, it would always be better in practice to choose this prior wisely. Hence, one should choose it according to the prior knowledge of the problem at hand. For instance, when thinking about physical problems, one can use some context or prior knowledge of the system to set the prior appropriately. Of course, the better the prior is chosen, the more effective the model is. While these are useful guidelines in general, we completely discard them at this point and set the prior to be

$$\theta \sim \mathcal{N}(0, \boldsymbol{\Sigma}_d), \tag{4.41}$$

which is a joint normal distribution with zero mean and covariance matrix $\boldsymbol{\Sigma}_d$. In particular, under the assumption of independent parameters, $\boldsymbol{\Sigma}_d$ can be chosen to be a diagonal matrix. As we shall argue below, we are even able, without loss of generality, to choose $\boldsymbol{\Sigma}_d$ to be simply the identity matrix, as the specific choice of $\boldsymbol{\Sigma}_d$ (and hence also any prior knowledge of the problem at hand) can be rolled into the definition of the kernels. The choice of zero mean for the prior ensures that the covariance function of a GP is indeed the kernel function. We refer to Appendix C for further details on this.

Getting back to the posterior, we can collect all the θ-independent terms under a normalization constant A and rewrite the likelihood using Eqs. (4.37) and (4.40) as

$$\begin{aligned} p(\theta \mid \boldsymbol{y}, \boldsymbol{X}) &= \frac{p(\boldsymbol{y} \mid \theta, \boldsymbol{X}) p(\theta)}{p(\boldsymbol{y} \mid \boldsymbol{X})} \\ &= \frac{1}{A} \exp\left[-\frac{|\boldsymbol{y} - \boldsymbol{X}\theta|^2}{2\sigma^2}\right] \exp\left[-\theta^\top \boldsymbol{\Sigma}_d^{-1} \theta\right]. \end{aligned} \tag{4.42}$$

Rearranging the expressions in the exponents, the posterior becomes:

$$p(\theta \mid \boldsymbol{y}, \boldsymbol{X}) = \frac{1}{A} \exp\left[-\frac{1}{2}(\theta - \boldsymbol{\mu})^\top \boldsymbol{C}(\theta - \boldsymbol{\mu})\right], \tag{4.43}$$

where we used the following definitions

$$\boldsymbol{\mu} = (\boldsymbol{X}^\top \boldsymbol{X} + \sigma^2 \boldsymbol{\Sigma}_d^{-1})^{-1} \boldsymbol{X}^\top \boldsymbol{y} \tag{4.44}$$

$$\boldsymbol{C}^{-1} = (\boldsymbol{X}^\top \boldsymbol{X} + \sigma^2 \boldsymbol{\Sigma}_d^{-1})^{-1} \tag{4.45}$$

with $\boldsymbol{\mu}$ being the posterior mean and $\boldsymbol{C}^{-1}$ the posterior covariance matrix.[10] Finally, Eq. (4.43) gives us access to the analytical form for the distribution of the model parameters $p(\theta \mid \boldsymbol{y}, \boldsymbol{X})$ given the data $\mathcal{D}$. This, again, is a normal distribution with updated mean and covariance. These new values intrinsically possess the information about the training data as inferred from their analytical forms. Hence, $p(\theta \mid \boldsymbol{y}, \boldsymbol{X}) \sim \mathcal{N}(\boldsymbol{\mu}, \boldsymbol{C}^{-1})$. Once we have our updated distribution over our model's parameters, the next step is to predict the output y^* for a previously unseen data point $\boldsymbol{x}^*$. To do this, we need to multiply the posterior by the probability for y^* and integrate over all possible parameters. This yields:

$$\begin{aligned} p(y^* \mid \boldsymbol{x}^*) &= \int_{\mathbb{R}^d} p(y^* \mid \boldsymbol{x}^*, \theta) p(\theta \mid \boldsymbol{y}, \boldsymbol{X}) \, \mathrm{d}\theta \\ &\propto \int_{\mathbb{R}^d} \exp\left[-\frac{(y^* - \theta^\top \boldsymbol{x}^*)^2}{2\sigma^2}\right] \exp\left[-\frac{1}{2}(\theta - \boldsymbol{\mu})^\top \boldsymbol{C}(\theta - \boldsymbol{\mu})\right] \mathrm{d}\theta. \end{aligned} \tag{4.46}$$

It is easy to notice that this distribution is again going to be Gaussian. We can also analytically compute the conditional mean and variance[11]:

$$\hat{\mu} = \boldsymbol{x}^{*\top} \boldsymbol{\mu} = \boldsymbol{x}^{*\top} \sigma^{-2} \boldsymbol{C}^{-1} \boldsymbol{X}^\top \boldsymbol{y} \tag{4.47}$$

$$\hat{\sigma}^2 = \boldsymbol{x}^{*\top} \boldsymbol{C}^{-1} \boldsymbol{x}^*. \tag{4.48}$$

The mean can be used to make predictions, while the variance gives the uncertainty over such estimation. It is now time to compare these results with previously introduced methods:

$$\text{Linear regression: } \boldsymbol{x}^{*\top} \hat{\theta} = \boldsymbol{x}^{*\top} (\boldsymbol{X}^\top \boldsymbol{X})^{-1} \boldsymbol{X}^\top \boldsymbol{y} \tag{4.49}$$

$$\text{Linear ridge regression: } \boldsymbol{x}^{*\top} \hat{\theta} = \boldsymbol{x}^{*\top} (\boldsymbol{X}^\top \boldsymbol{X} + \lambda \mathbb{1})^{-1} \boldsymbol{X}^\top \boldsymbol{y} \tag{4.50}$$

$$\text{Linear GPR: } \quad \hat{\mu} = \boldsymbol{x}^{*\top} (\boldsymbol{X}^\top \boldsymbol{X} + \sigma^2 \boldsymbol{\Sigma}_d^{-1})^{-1} \boldsymbol{X}^\top \boldsymbol{y}. \tag{4.51}$$

[10] Even though we arrived at these results from a Bayesian perspective, similar results have already been developed in the geostatistics community in the 1960s. There, GPR is more often known as *kriging* [184].

[11] With $\boldsymbol{C} = [\sigma^{-2} \boldsymbol{X}^\top \boldsymbol{X} + \boldsymbol{\Sigma}_d^{-1}]$ already defined above.

This entire derivation, which we have been doing for the linear case, can easily be generalized to nonlinear regression. To this end, we revisit our Bayesian NN from Eq. (4.33). Here, we map $\boldsymbol{x} \mapsto \phi = \phi(\boldsymbol{x})$, embedding our input data with a feature map into a potentially high-dimensional space. It follows that $\boldsymbol{X} \mapsto \boldsymbol{\Phi} = \phi(\boldsymbol{X})$ and the conditional mean and variance can be derived accordingly. Substituting the terms appropriately and doing a little bit of math adjustments, under the simplification of unit variance $\boldsymbol{\Sigma}_d = \mathbb{1}$ from above, we obtain:

$$\hat{\mu} = \phi^{*\mathsf{T}}\boldsymbol{\Phi}^\mathsf{T}\left[\boldsymbol{\Phi}^\mathsf{T}\boldsymbol{\Phi} + \sigma^2\mathbb{1}\right]^{-1}\boldsymbol{y} \tag{4.52}$$

$$\hat{\sigma}^2 = \phi^{*\mathsf{T}}\phi^* - \phi^{*\mathsf{T}}\boldsymbol{\Phi}^\mathsf{T}\left[\boldsymbol{\Phi}^\mathsf{T}\boldsymbol{\Phi} + \sigma^2\mathbb{1}\right]^{-1}\boldsymbol{\Phi}\phi^* . \tag{4.53}$$

These expressions can be rewritten in terms of the kernel function K:

$$\hat{\mu} = \boldsymbol{k}^\mathsf{T}(\boldsymbol{x}^*)\,[\boldsymbol{K} + \sigma^2\mathbb{1}]^{-1}\boldsymbol{y} \tag{4.54}$$

$$\hat{\sigma}^2 = K(\boldsymbol{x}^*, \boldsymbol{x}^*) - \boldsymbol{k}^\mathsf{T}(\boldsymbol{x}^*)\left[\boldsymbol{K} + \sigma^2\mathbb{1}\right]^{-1}\boldsymbol{k}(\boldsymbol{x}^*) \tag{4.55}$$

where we used the following definitions:

$$\phi^{*\mathsf{T}}\phi^* = K(\boldsymbol{x}^*, \boldsymbol{x}^*) , \tag{4.56}$$

$$(\boldsymbol{\Phi}\phi^*)_i = \boldsymbol{k}(\boldsymbol{x}^*)_i = K(\boldsymbol{x}_i, \boldsymbol{x}^*) , \tag{4.57}$$

$$\left(\boldsymbol{\Phi}^\mathsf{T}\boldsymbol{\Phi}\right)_{ij} = \boldsymbol{K}_{ij} = K(\boldsymbol{x}_i, \boldsymbol{x}_j). \tag{4.58}$$

While choosing $\boldsymbol{\Sigma}_d = \mathbb{1}$ may appear as unnecessarily restrictive, we note that one can always rewrite the identities above such that they take the covariance matrix into account (e.g., Eq. (4.56) becomes $\phi^{*\mathsf{T}}\boldsymbol{\Sigma}_d\phi^{*\mathsf{T}} = K(\boldsymbol{x}^*, \boldsymbol{x}^*)$). Since the kernel function is to be defined anyway, the choice of $\boldsymbol{\Sigma}_d = \mathbb{1}$ does not lose generality. We detail in Appendix C that the kernel trick does not interfere with the Gaussian prior assumption of Eq. (4.41) for the parameters $\boldsymbol{\theta}$.

Thus, looking at the conditional mean from Eq. (4.54), one can directly see the analogy with the result of KRR from Eq. (4.20) where our regularization strength λ can be seen to correspond to the data noise assumption, parametrized by σ^2.[12] This comes to no surprise as the conditional mean from Eq. (4.54) corresponds to the MAP, Eq. (2.35) introduced in Section 2.4.1. However, GPR also yields the uncertainty of the prediction. Furthermore, we see, once again, the consequence of the representer theorem in Eq. (4.13) on the form of the conditional mean $\hat{\mu}$, that is, $\boldsymbol{\alpha} = (\boldsymbol{K} + \sigma^2\mathbb{1})^{-1}\boldsymbol{y}$.

4.2.4 Training a Gaussian process

In Section 4.2.4, we have seen what a GP is, how one constructs it, and how it allows one to obtain a closed-form expression for the conditional mean and the Bayesian uncertainty for the output over one (or more) unseen data point(s) $\boldsymbol{x}^*$. We understand that the performance of the GP strongly relies on the choice of the prior and on the amount of data the model is exposed to. So, at this point, one natural question

[12] The difference in the two constants actually only arises from our choice of the covariance matrix. If we had chosen $\boldsymbol{\Sigma}_d = \tau^2\mathbb{1}$ instead, the difference would vanish completely, and the two methods would yield the same estimator.

that might arise is: How do we train a GP? First and foremost, we need to choose an appropriate kernel function K, which defines the kernel matrix $\boldsymbol{K}_{ij} = K(\boldsymbol{x}_i, \boldsymbol{x}_j)$ accordingly. The parameters of this function (e.g., as in Table 4.1) are tuned to maximize the so-called marginal likelihood $p(\boldsymbol{y} \mid \boldsymbol{X})$. This name comes from the fact that this quantity is obtained from the Bayes' theorem Eq. (4.40) when marginalizing over the model parameters (i.e., taking the integral over $\boldsymbol{\theta}$). Here, our goal is to express the marginal likelihood of a GP in terms of the kernels. To this end, we choose the covariance matrix of the GP to be the kernel matrix such that $\mathrm{Cov}(\boldsymbol{x}, \boldsymbol{x}') = \boldsymbol{K}$[13], which turns out to be equivalent to the choice of the prior made previously $\boldsymbol{\theta}_{\mathrm{prior}} \sim \mathcal{N}(0, \boldsymbol{\Sigma}_d)$. A mathematical justification for this is provided in Appendix C. Our ultimate goal is to evaluate the marginal likelihood

$$p(\boldsymbol{y} \mid \boldsymbol{X}) = \int_{\mathbb{R}^d} p(\boldsymbol{y} \mid \boldsymbol{\theta}, \boldsymbol{X}) p(\boldsymbol{\theta} \mid \boldsymbol{X}) \, \mathrm{d}\boldsymbol{\theta} \,. \tag{4.59}$$

Given that the prior's covariance is the kernel function and the integrand is a product of two Gaussians, it is possible to express $p(\boldsymbol{y} \mid \boldsymbol{X})$ in terms of the kernel matrix $\boldsymbol{K}$. Working with the logarithm of the marginal likelihood, it follows from Appendix C that our objective of the training process is

$$\log p(\boldsymbol{y} \mid \boldsymbol{X}) = -\frac{1}{2} \boldsymbol{y}^{\intercal} \left(\boldsymbol{K} + \sigma^2 \mathbb{1} \right)^{-1} \boldsymbol{y} - \frac{1}{2} \log \left| \boldsymbol{K} + \sigma^2 \mathbb{1} \right| - \frac{n}{2} \log 2\pi \,. \tag{4.60}$$

When training a GP one aims to find the parameters of the kernel function that maximize the logarithm of marginal likelihood from Eq. (4.60). As can be seen, training a GP requires the inversion of the kernel matrix (and the calculation of its determinant), whose dimension is determined by the size of the training set. This already gives an intuition why GPs are the tool of choice in a regime of few data points where they can be very effective and relatively cheap. GP models can be difficult to train for problems with a lot of training data.

4.3 Bayesian optimization

In Section 4.2.3, we have discussed GPR and how to train a GP. Furthermore, we have shown that GPs yield a closed-form expression for the estimate of the output for a test data point $\boldsymbol{x}^*$, conditioned by a set of given data $\mathcal{D}$, in a similar fashion to KRR. Unlike KRR, GPR also comes with a prediction uncertainty. This is of a great relevance as it can be used for BO [182].

Bayesian optimization is a technique used for the optimization of expensive black-box functions where gradients cannot be easily computed or estimated (e.g., time-consuming experiments). Here, the term *expensive* is very important. Indeed, optimizing black-box functions is the general goal of a big set of ML models and techniques. Such optimization usually relies on efficient computation, arbitrarily large amounts of data, and so forth. However, this might not always be the case. Sometimes, we might face problems where the amount of data is rather limited, or the

[13] On the notation: when providing the inputs of the covariance matrix, we use Cov while when they are implicit we just use $\boldsymbol{C}$.

Algorithm 5 Bayesian optimization (BO)

Require: initial data set $\mathcal{D}_0 = \{(\boldsymbol{X}, \boldsymbol{y})\}$
Require: initial surrogate model GP trained on $\mathcal{D}_0$ with mean and variance $\hat{\mu}, \hat{\sigma}^2$
Require: acquisition function to be maximized
  for iteration $t < T_{\max}$ **do**
    Sample a set of candidate points $\boldsymbol{X}_{\text{cand}}$ ▷ batch optimization
    $\boldsymbol{x}^* \leftarrow \max_{\boldsymbol{x} \in \boldsymbol{X}_{\text{cand}}} a(\boldsymbol{x}, \hat{\mu}, \hat{\sigma})$
    $y^* \leftarrow f_{\text{BB}}(\boldsymbol{x}^*)$ at $\boldsymbol{x}^*$
    $\mathcal{D} \leftarrow \mathcal{D}_t \bigcup \{(\boldsymbol{x}^*, y^*)\}$ ▷ update data set
    Update the surrogate model's parameters $\hat{\mu}$ and $\hat{\sigma}^2$
  end for
  return $\hat{\mu}$ and $\hat{\sigma}^2$ ▷ prediction and uncertainty of the surrogate model

routine to extract additional experimental data is very expensive. In this context, the interplay between GP and BO becomes extremely important. It is also worth mentioning that contrary to the optimization methods used in most ML approaches we have encountered so far (in particular NNs), BO is a gradient-free method. Therefore, it is particularly well suited for functions that are very difficult or expensive to evaluate.

The way BO works in the context of GPR is that BO takes GPs *as surrogate models of the black-box function to be optimized.* Recalling the results of Section 4.2.4, GPR gives us access to a conditional prediction along with an estimate for its uncertainty. This said, the next important thing to notice is that BO is an iterative process. This process works as follows (see Algorithm 5 for the pseudo-code): The BO procedure starts with a few evaluations of the black box function at some random locations of the input space. We refer to this initial dataset as $\mathcal{D}_0 = \{(\boldsymbol{X}, \boldsymbol{y})\}$. These evaluations are used to train the first version of a GP. Hence, we can think of those as our training data.[14] Once we have our first surrogate model, we introduce the so-called acquisition function. The acquisition function is typically a function of both $\hat{\mu}$ and $\hat{\sigma}$ and essentially tells us where to perform the next evaluation x to maximize the knowledge we gain about the underlying black-box function. In the next section, we see what the acquisition function looks like. For now, we can just think of it to be an arbitrary function $a(\boldsymbol{x}, \hat{\mu}, \hat{\sigma})$. The prediction and its uncertainty are fixed given a surrogate GP prior. Thus, the acquisition function is only a function of a new candidate point $\boldsymbol{x}$. Our goal is now to find such a candidate point that is as informative as possible. As such, the next point to evaluate is determined by maximizing the acquisition function, that is, $\boldsymbol{x}^* := \max_{x \in D} a(x, \hat{\mu}, \hat{\sigma})$, where D is the domain of $\boldsymbol{x}$. Once the new target location $\boldsymbol{x}^*$ is found, the next step is to evaluate the black-box function such that $y^* = f_{\text{BB}}(\boldsymbol{x}^*)$. The result of the evaluation is appended to the training set for GP such that $\mathcal{D} = \mathcal{D}_0 \bigcup \{(\boldsymbol{x}^*, y^*)\}$ and a new, less uncertain, surrogate model is trained on the updated $\mathcal{D}$. From this point on, the iteration starts over: Every time we update the surrogate model, we have new predictions and uncertainties, hence a new acquisition function. At each step of the BO, a new point is thus added to $\mathcal{D}$, and the entire process goes on until a maximum number of iterations $T_{\max}$ is reached or some convergence

[14] Since we work in a Bayesian setting, the data noise is taken into account in Eq. (4.35), and all predictions are based on top of that assumption.

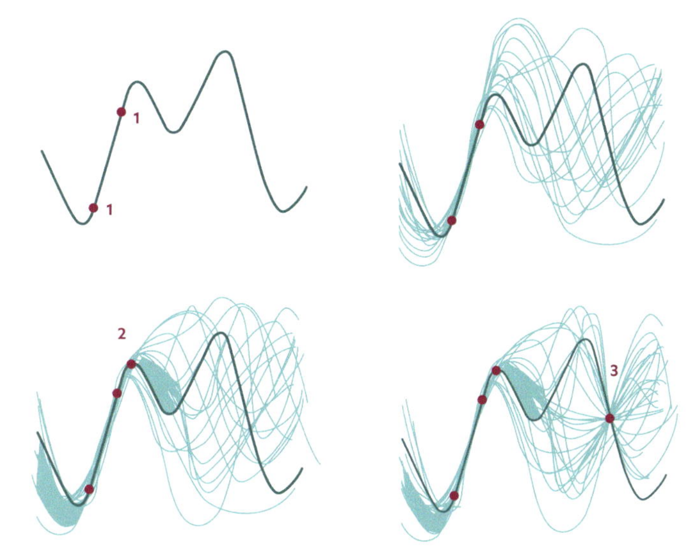

Figure 4.5 Example of how BO can be applied to GPR. A GP represents a surrogate model (light green lines) trained on an initial set of observations (red dots in the upper left panel). At each BO step, a new point is added to the previous set of observations such that the surrogate model becomes increasingly certain as more points are added. The dark green line represents the underlying black-box function.

criterion is met. The plot in Fig. 4.5 shows how three subsequent steps of BO result in an increasingly more certain surrogate model of the underlying black-box function (dark green line). This illustrates how BO can be used in the context of *active learning*, where the training dataset is built step-by-step with the aim of minimizing the number of training points while maximizing the information it contains. However, BO should not be confused with active learning as they serve different purposes. The former aims to optimize the target function with as few evaluations as possible. The latter, instead, tries to sample the input space as efficiently as possible to target more accurate prediction models.

Bayesian optimization with GPR has the following advantages over other optimization methods:

- smaller number of function evaluations,[a]
- gradient-free.

[a] A suitable choice for the kernel can be used to lower the number of function calls. Moreover, it is preferable to have smaller training sets for this method. Equation (4.54) shows that the kernel matrix needs to be inverted for each trial kernel, which adds a computational constraint.

The acquisition function. In the previous section, we have briefly described the idea of BO and how it operates combined with GPs. In this context, we have introduced the acquisition function. This quantity is very important as it represents a mathematical technique that guides the exploration of the entire parameter space during the BO routine. We have previously defined the acquisition function as a general function of $\boldsymbol{x}$, the surrogate's prediction and its uncertainty. There are different kinds of acquisition functions, and most of the time, the choice is problem-dependent. However, most importantly, its mathematical form should always incorporate the trade-off between exploration and exploitation. In other words, the goal of the acquisition function is to evaluate the usefulness of the next data location to look at to achieve the maximization of the surrogate model of our black box function and, thus, to approximate the target function with lower uncertainty. As such, the ultimate goal in BO is to find the next point to evaluate by maximizing such acquisition function. One example, commonly used and easy to interpret, is the Upper Confidence Bound (UCB):

$$a_{\mathrm{UCB}}(\boldsymbol{x}, \hat{\mu}, \hat{\sigma}) = \hat{\mu}(\boldsymbol{x}) + \beta\hat{\sigma}(\boldsymbol{x}), \tag{4.61}$$

where $\beta \geq 0$ is an arbitrary parameter that ideally should be tuned during the optimization routine. Here, the first term drives the exploitation, while the second drives the exploration. In the remainder, we refer to those as the *exploitation* and the *exploration* terms, respectively.

Depending on the value of β, the exploration term might dominate in the maximization. By looking at Eq. (4.61), it is immediately clear that a new candidate point with the higher variance is preferred as the model rewards the evaluation of currently unexplored regions of the domain. That is not surprising as the model seeks to explore what it does not know yet. With respect to the mean, according to the UCB, higher values for the mean are preferred. That is because, by definition, we are seeking for an upper bound, hence enhancing sampling in the upper quartile of the surrogate model. In other words, in the extreme case where $\beta \gg 0$, the exploration dominates, hence regions of higher variance are preferred (see Fig. 4.6(a)).

When instead $\beta \to 0$, the acquisition function becomes far more conservative, hence samples aggressively around the best solution, that is, exploiting the region where the surrogate model feels confident as visible in Fig. 4.6(c). Figure 4.6(b) shows a good balance between the exploration and the exploitation. Hence, for two candidate points with comparable predicted mean, the one with higher uncertainty is preferred. As a consequence, the acquisition function, at least at the beginning of the optimization, prefers to explore rather than exploit.

Moreover, looking at the analytical form from Eq. (4.54) (which appears as the first term in Eq. (4.61)), the acquisition function might not always be easy to maximize (minimize) in practice. Therefore, one needs to leverage efficient numerical optimization routines. As acquisition functions are highly nonconvex, what is done in practice is to do *batch optimization*. At each BO step, starting points $\boldsymbol{x}_{\mathrm{cand}}$ are randomly sampled over a specified domain D. Then, one takes the best one of the sampled points (that maximizes the acquisition function) as the actual candidate. Other prominent examples of widely used acquisition functions are: expected improvement, noisy expected improvement, and probability of improvement. For a more detailed overview of other types of acquisition functions, we refer to [185].

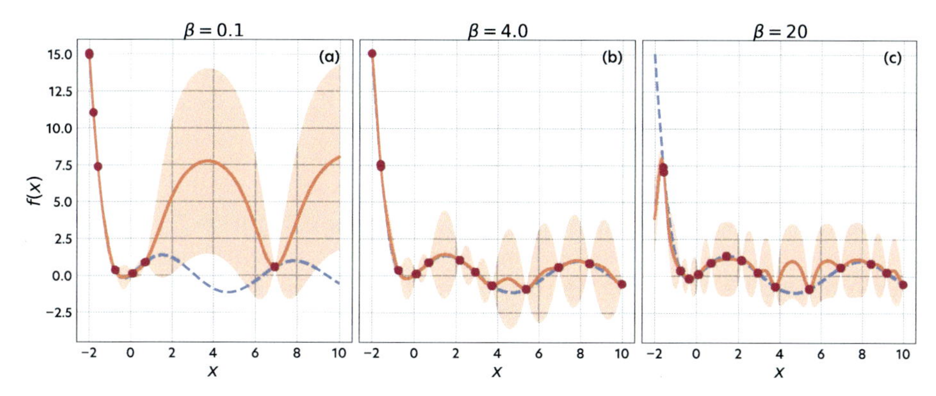

Figure 4.6 Selection of new candidate points via Bayesian optimization using the Upper Confidence Bound acquisition function. The target function is represented by the blue dashed line. The solid orange line is the surrogate model (GP), while the orange shading represents its uncertainty. (a) The plot exhibits exploitative behavior, that is, the most selected points are around the peak(s). (c) Contrarily, in this plot, the parameter choice for β heavily enforces exploration. As such, new sampled points (red dots) are evenly distributed through some part of the domain (e.g., $x \sim 2$ would require more exploitation). (b) The plot shows a tradeoff between exploration and exploitation: The sampled candidates are well distributed across the entire domain, thus approximating the target function efficiently even around the peak(s) and boundaries.

4.4 Choosing the right model

Having introduced the powerful toolbox of kernels, a natural question arises: Suppose we are given a set of "noisy" data points, forming a data set $\mathcal{D}$, and two distinct models $\mathcal{M}_1$ and $\mathcal{M}_2$ possibly based on two different kernels, which one should we choose? This is the central question behind *model selection*. To answer this question, we again take a Bayesian approach (see Section 2.3). Applying Bayes' theorem from Eq. (2.19) to each model $\mathcal{M}_i$ yields

$$p(\mathcal{M}_i \mid \mathcal{D}) = \frac{p(\mathcal{D} \mid \mathcal{M}_i)p(\mathcal{M}_i)}{p(\mathcal{D})}. \tag{4.62}$$

We combine the expressions of the two models to obtain

$$\frac{p(\mathcal{M}_1 \mid \mathcal{D})}{p(\mathcal{M}_2 \mid \mathcal{D})} = \frac{p(\mathcal{D} \mid \mathcal{M}_1)}{p(\mathcal{D} \mid \mathcal{M}_2)} \frac{p(\mathcal{M}_1)}{p(\mathcal{M}_2)}. \tag{4.63}$$

If we have no prior knowledge of the model performance, we must set the priors for the two models to be the same. In this case, the ratio of the posterior probabilities $P(\mathcal{M}_i \mid \mathcal{D})$ is equal to the ratio of the prior probabilities $P(\mathcal{M}_i)$ times the so-called Bayes factor

$$\frac{p(\mathcal{D} \mid \mathcal{M}_1)}{p(\mathcal{D} \mid \mathcal{M}_2)}. \tag{4.64}$$

Equation (4.63) gives us a first answer to our question: In a Bayesian framework, the ratio of posterior probabilities can be used to decide which model is superior given the data at hand, that is, the model with the larger posterior probability is superior. In scenarios where we do not know anything about the data, we can set the prior probabilities equal to each other, which leaves us with the Bayes factor

$$\frac{p(\mathcal{M}_1 \mid \mathcal{D})}{p(\mathcal{M}_2 \mid \mathcal{D})} = \frac{p(\mathcal{D} \mid \mathcal{M}_1)}{p(\mathcal{D} \mid \mathcal{M}_2)}. \tag{4.65}$$

To calculate the Bayes factor in Eq. (4.64), we need to compute $p(\mathcal{D} \mid \mathcal{M}_i)$ for each model which can be viewed as *marginal likelihood*, that is, a likelihood function in which all variables except the type of the model have been marginalized (integrated out). Let us define the likelihood as $p(\mathcal{D} \mid \boldsymbol{\theta}, \mathcal{M}_i)$, such that the marginal likelihood can be obtained as

$$p(\mathcal{D} \mid \mathcal{M}_i) = \int_{\mathbb{R}^d} p(\mathcal{D} | \boldsymbol{\theta}, \mathcal{M}_i) p(\boldsymbol{\theta} \mid \mathcal{M}_i) \, d\boldsymbol{\theta}, \tag{4.66}$$

where we integrate over the distribution of model parameters $\boldsymbol{\theta} \in \mathbb{R}^d$ given by $p(\boldsymbol{\theta} \mid \mathcal{M}_i)$. Unfortunately, marginal likelihoods are typically hard to compute as they involve high-dimensional integrals. Choosing a kernel with d parameters results in a d-dimensional integral for its marginal likelihood.

Having encountered this problem, let us take a step back: When we train a model, we minimize a loss function (or equivalently, we maximize the log-likelihood). Therefore, why not simply choose the model that gives the lowest loss or largest likelihood? Intuitively, this leads to *overfitting*. This intuition is formalized by the *bias-variance trade-off* (see Section 2.2). In particular, the bias-variance trade-off makes it clear that the ideal model realizes an optimal balance between the training error and the model complexity. Rather than choosing the model that results in the lowest loss during training, we thus need to take its complexity into account.

4.4.1 Bayesian information criterion

A computationally tractable criterion for model selection which takes model complexity into account is the Bayesian information criterion (BIC) [186] defined as

$$\text{BIC} = -2\max(\ell) + d\log(n), \tag{4.67}$$

where $\max(\ell)$ is the maximum of the log likelihood, n is the number of training points, and d is the number of model parameters. The lower the BIC, the better the model. Clearly, the BIC reflects the trade-off between bias, here given by $\max(\ell)$, and the model complexity as measured by $d\log(n)$. Moreover, it turns out that the BIC approximates the logarithm of the marginal likelihood in the large n-limit [187]:

$$\log p(\mathcal{D} \mid \mathcal{M}_i) \approx \log p(\mathcal{D} \mid \boldsymbol{\theta}^*, \mathcal{M}_i) - \frac{d}{2}\log(n), \tag{4.68}$$

where $\boldsymbol{\theta}^*$ are the model parameters that maximize the likelihood. This expression reveals that the model selection criterion given in Eq. (4.63) based on the Bayesian approach does indeed take the model complexity into account. Moreover, we see that the criterion can be used to estimate the posterior probability of a model $\mathcal{M}_i$ as

$$p_i = \frac{\exp\left(-\frac{1}{2}\text{BIC}_i\right)}{\mathcal{N}}. \tag{4.69}$$

Here, the normalization constant $\mathcal{N} = \sum_i e^{-\mathrm{BIC}_i/2}$ ensures that each model $\mathcal{M}_i$ is assigned a valid probability p_i to enable comparability. As such, the BIC gives us a tractable way to select models according to the criterion given in Eq. (4.63). In fact, BIC is asymptotically consistent as a model selection metric: Given a family of models, including the model underlying the data, the probability that BIC correctly selects the model underlying the data approaches one as $n \to \infty$.[15]

Inspired by these findings, we can adapt the criterion to GPR based on the log marginal likelihood, which is optimized during training (at fixed kernel parameters) and the number of kernel parameters in the GPR. This criterion is computationally tractable and thus allows one to select between different kernels in the regression task using GPs.

4.4.2 Kernel search

Choosing the right kernel is crucial when using a kernel-based method, as we have seen, for example, for the performance of SVMs for different kernels in Fig. 4.2. When performing GPR in a *naive* manner, we simply select a fixed kernel from a set of conventional kernels such as listed in Table 4.1. We then optimize their hyperparameters by maximizing the marginal likelihood during training. There are now several possible routes toward achieving a more accurate model. Clearly, we may improve the model accuracy by providing more training points. However, keeping the number of training points low is one of the main advantages of GPR compared to other methods and constituted our main initial motivation. At a fixed number of training data, the result from GPR can only be improved through a better kernel. Moreover, while BO is guaranteed to converge, the exact number of iterations may vary drastically. The choice of a good kernel can significantly speed up the convergence of BO.

The construction of good kernels ultimately boils down to a (possibly high-dimensional) optimization problem [189]. This happens, for example, when constructing a good kernel through optimization of the kernel hyperparameters itself. The key challenge is posed by the fact that the parametric form of the kernel must be proposed by the user itself. This is a nontrivial task that relies on trial and error – even for experts. In [135, 189, 190], the kernel learning problem was reframed as a search tree problem (see Fig. 4.7): The space of parametric forms of kernels is constructed as a tree that can be searched systematically in an automated fashion, where the powerful BIC is used for the kernel selection and new kernels are proposed via composition.

We start by selecting each kernel from a set of conventional kernels and training a GP for each of them on the same dataset. Then, we select the one that achieves the lowest value of BIC as given by Eq. (4.67) (highlighted in blue). This kernel serves as the base kernel for the subsequent round, where it is combined with the various

[15]While the BIC criterion approximates the log marginal likelihood in the large n-limit, it can still be applied as a heuristic model selection criterion at low values of n and can be confirmed empirically to often still yield good results. There exist many other model selection criteria (see [187] for a review), a popular one being the Akaike information criterion [188] which closely resembles the BIC. The crucial difference between the BIC and many other methods is its asymptotic consistency. One may question the importance of asymptotic consistency due to the fact that the ground-truth model typically is not present in the candidate set of models in practice.

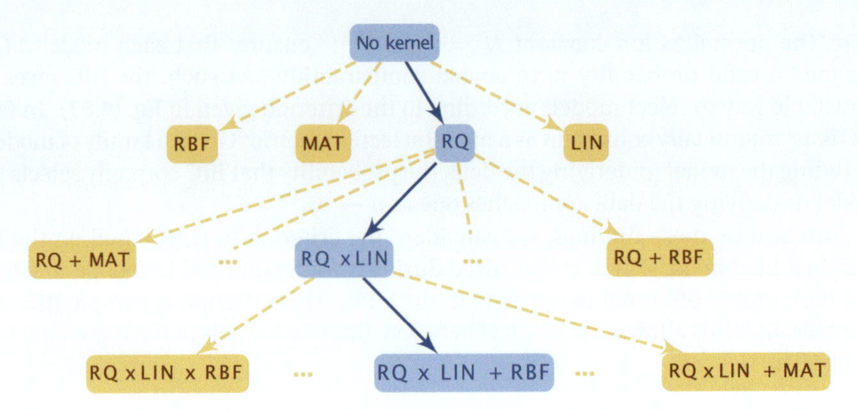

Figure 4.7 Illustration of the search tree behind the algorithm for the optimal kernel construction in GPR. It utilizes the BIC for the model selection introduced in Eq. (4.67). For an overview of possible kernel functions and corresponding abbreviations, see Table 4.1. Adapted from [135].

kernels from the starting set to create new candidate kernels by forming products or combining them linearly. Again, the best one is selected according to the BIC, and the process is repeated. The complexity of the model, that is, of the composite kernel, increases as one progresses in the search tree. Eventually, increasing the kernel complexity further leads to overfitting and, hence, does not improve the BIC value compared to the kernel of the previous round, and the algorithm is stopped. The algorithm can also be stopped prematurely if the number of kernel parameters becomes large, and the associated training simply takes too long to be practical. Other than greedily searching the tree, the reformulation of the kernel construction as a search tree problem opens up the possibility for more advanced strategies that could yield better kernels more efficiently [191, 192].

4.5 Applications in quantum sciences

In the previous sections, we have motivated GPs and BO as powerful methods that together allow us to build expressive ML models. Importantly, they are equipped with an intrinsic measure for uncertainty and can be trained using a small amount of training data. In this section, we discuss how these two methods can be useful in the context of quantum sciences, as sketched in Fig. 4.8. In particular, GP and BO can be used to tackle inverse problems, extrapolate in Hamiltonian parameter spaces, and increase the accuracy of quantum dynamics calculations.

4.5.1 Inverse problems

As explained in Section 4.3, BO is very useful when you need to optimize black-box functions that are expensive to evaluate. This property proves extremely useful in inverse quantum problems aiming at finding a theoretical description of the system by experimentally measuring its observables. The idea is related to a popular exper-

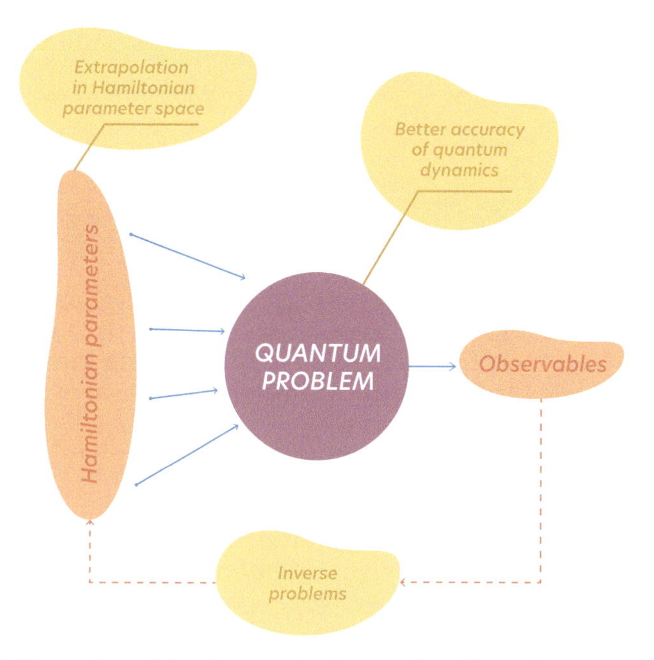

Figure 4.8 Illustration of the three main classes of problems in quantum sciences (marked in yellow) that have been successfully tackled with BO and GPs.

imental approach known as *optimal control*. The optimal control approach aims to design external field parameters that yield the desired quantum dynamics. It is usually achieved by a feedback loop, which iteratively modifies experimental parameters such that they yield system dynamics advancing to the target one.

We can imagine applying a similar feedback loop for the inverse quantum problems. It would consist of iterative modifications of parameters of the theoretical description (such as Hamiltonian parameters) till the observables predicted theoretically agree with those measured experimentally. However, solving the iterative inverse quantum problem is challenging. Each iteration requires an additional run of theoretical calculations, for example, the numerical solution of the Schrödinger equation, which is time-consuming. The optimization itself is also difficult as we do not explicitly know the range of parameters that needs to be explored. Finally, the curse of scaling of the Hilbert space dimension with the complexity of the quantum systems definitely does not help. How to make it more feasible? Both inverse quantum problems and optimal control become easier when the expensive black box (either the experimental setup or the theoretical calculations) is replaced by a trained surrogate ML model such as a GP. Finally, instead of a blind search for the optimal parameters, we can employ BO.

As a practical example of the inverse problems solved with BO and GPs, let us consider the application to *scattering experiments*. The outcomes of such experiments are determined by the microscopic interactions between scattered particles. We have a quantum theory that describes these interactions and can predict the outcome of such scattering events. Therefore, our aim, in the case of an inverse problem, may

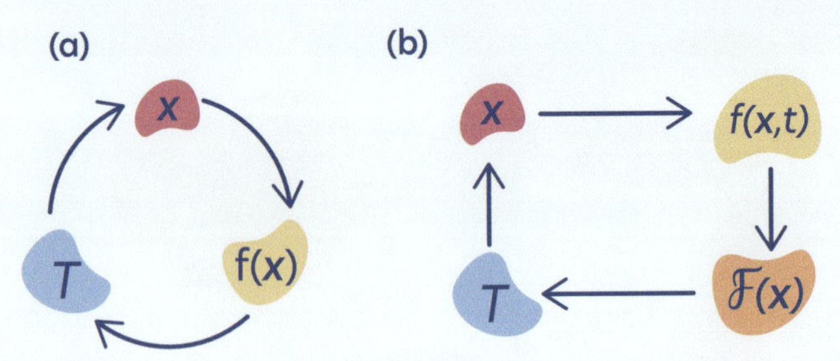

Figure 4.9 Examples of feedback loops whose optimization becomes feasible when implemented with BO and GPs. (a) $\boldsymbol{x}$ corresponds to PES, $f(\boldsymbol{x})$ is quantum scattering calculations taking the PES as an input, and T is the difference between reaction probabilities calculated by $f(\boldsymbol{x})$ and measured in the experiment across various collisional energies. A search for the optimal $\boldsymbol{x}$ would require minimization of T via optimization of $f(\boldsymbol{x})$. It becomes feasible when we surrogate $f(\boldsymbol{x})$ with one GP and $\boldsymbol{x}$ with another GP and apply BO. (b) $\boldsymbol{x}$ are Hamiltonian parameters, $f(\boldsymbol{x}, t)$ is the time-dependent observable f (e.g., molecular orientation or alignment), $\mathcal{F}(\boldsymbol{x})$ is the time-dependent Schrödinger equation, and T is the difference between calculated and measured time-dependent observable f. When $\mathcal{F}(\boldsymbol{x})$ is surrogated by a GP, BO is used to minimize T and find $\boldsymbol{x}$ of the underlying Hamiltonian.

be to infer these microscopic interactions from the experiment. More concretely, the authors of [194] aimed to recover a global potential energy surface (PES)[16] governing the chemical reaction $H + H_2 \longrightarrow H_2 + H$, using as few experimental measurements of the reaction rates (depending on the constituents' translational energy) as possible. The feedback loop that needed to be solved is presented in Fig. 4.9(a). First, they trained a GP to surrogate a quantum scattering theory on a series of PESs and predicted reaction rates. Second, they modeled the PES with another GP. Finally, they used BO to find the three-dimensional PESs, recovering the measured reaction rates. Only eight iterations of BO (where every iteration rebuilds the PES completely) were required to reach the accuracy of conventional approaches! Moreover, in this case, a traditional approach of building a PES requires around 8 700 points – their GP was modeled based only on 30 points![17] This impressive scaling is presented in Fig. 4.10. As a result, they successfully surrogated two complex models (PES and quantum scattering calculations using PES as an input) with two GPs trained on a much smaller number of data points than needed to build the original complex models. They also used this approach for a six-dimensional PES of $OH + H_2$, where BO beat the traditional approach with 290 points compared to 17 000 points.[18]

[16] A PES describes interactions between some particles. As a result, it models landscapes of chemical reactions, which can be used to predict reactive pathways and final products. Traditionally, it is constructed as an analytic fit to many, usually costly ab initio quantum-chemical calculations of the potential energy for reactants for various relative positions.

[17] Remember, these are not any 30 points, but points indicated by BO as needed for the optimal description.

[18] The number of points needed for efficient BO scales roughly like 10 times the number of PES dimensions. We can raise an interesting point, which is, how are we even sure that we faithfully reproduce the PES if we build it only from reaction probabilities? It may happen that we capture the reactive chemical channels accurately, but the remaining parts of the surface are unconstrained and as a result may be non-

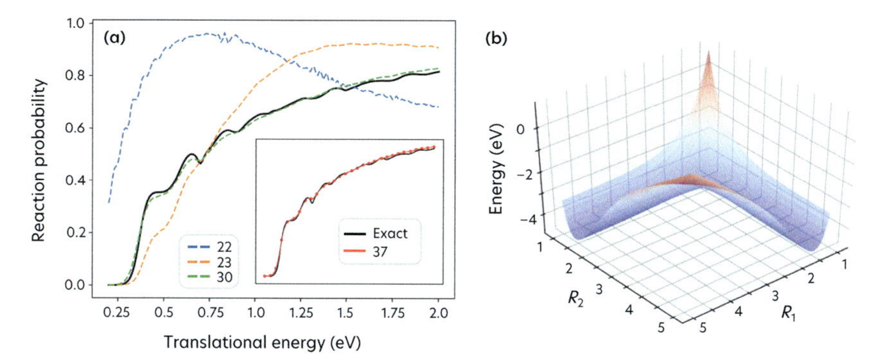

Figure 4.10 (a) The reaction probability for the $H_2 + H \longrightarrow H + H_2$ reaction as a function of the collision energy. The black solid curve represents calculations from [193] based on the surface with 8 701 ab initio points. The dashed blue/orange/green/red curves are calculations based on the GP PES obtained with 22/23/30/37 ab initio points. (b) GP model of the PES for the H_3 reaction system constructed with 30 ab initio points. R_1 and R_2 are the distances between atoms 1 and 2 and atoms 2 and 3, respectively. Adapted from [194] under the CC BY 3.0 DEED license.

Another example of an inverse quantum problem is the task of inferring molecular properties from time-dependent observables. Authors of [195] tackled the reconstruction of molecular polarizability tensors from the observed time evolutions of the orientation or alignment signals of SO_3 and propylene oxide induced by strong laser pulses. The feedback loop that was solved is drawn in Fig. 4.9(b). They used a GP with a vector output whose elements corresponded to a prediction of a chosen observable (orientation or alignment) in a different time step. The GP was trained to surrogate the numerical integration of the time-dependent Schrödinger equation given the Hamiltonian parameters. Interestingly, the authors showed what we discussed already in Section 4.4: that a proper choice of the kernel can result in a two times faster convergence of BO. Analogous approaches were used for the reconstruction of scattering matrices of molecules from molecular hyperfine experiments [196] and for optimizing the reaction conditions of an organic chemistry experiment [197].

4.5.2 Improving quantum dynamics, physical models, and experiments

Gaussian processes and BO can also be used for transfer learning in the context of quantum dynamics calculations. These are typically very difficult, and one quickly has to rely on approximations. The authors of [198] proposed to apply GPs to correct such approximate quantum calculations for computing cross-sections for molecular collisions. The idea is to train a model on a small number of exact results and a large number of approximate calculations, resulting in ML models that can generalize exact quantum results to different dynamical processes.

---

physical. One can argue that we ultimately do not need a complete faithful reproduction of the underlying PES. We only need a PES that allows us to accurately predict what we are interested in, here reaction probabilities. Note, however, that if we take a PES built from a particular set of observables and we use it to calculate another observable, the result may be wrong.

Moreover, as the minimization of any function using BO bypasses the need for computing gradients [182], successful applications of BO include optimization of parameters of physical models. Most models do not have a closed-form solution and conventionally have to be approximated numerically using finite differences. For example, [199, 200] showed that BO could efficiently optimize density functional models to improve their accuracy and minimize the energy of the Ising model [201]. Furthermore, BO was used to generate low-energy molecular conformers [202, 203], tuning the parameters of various models used to simulate *cis–trans* photoisomerization of retinal in rhodopsin [204], and the optimization of lasers [205–207]. BO has also been impactful in material science in chemical-compounds screening [208–213] and optimization of experimental setups [214–219].

4.5.3 Extrapolation problems

The second class of problems that seems suitable for GPs are extrapolation tasks: Given some function values for data points in one regime, the goal is to accurately predict the function values of data points in different regimes. This section touches upon two possible applications that are (1) learning PES from a possibly smallest number of ab initio calculations in one regime and (2) predicting the existence of quantum phases without knowledge of the full phase diagram.

An example of a successful extrapolation in the case of PES learning was shown in [191] where authors studied the six-dimensional PES of H_3O^+. They trained GP models on 1 000 ab initio geometries from a low-energy regime (up to $\approx 7\,000\ \mathrm{cm}^{-1}$) and checked that the model predictions in higher energy regimes match the full calculations with a high level of accuracy. If you doubt it, this result can be reproduced using the published code and data [2]. It gets better! You can get similarly accurate extrapolations from a GP model trained on 5 000 molecular geometries of a 51-dimensional problem of a protonated imidazole dimer, which contains 19 atoms [220]. The scaling of the extrapolation accuracy with respect to the number of training points seems to be even more favorable for large molecules: high accuracy was reached already for 1 000 randomly sampled geometries of the 57-dimensional aspirin.[19]

Another example of extrapolation, this time in the space of Hamiltonians, is the task of inferring properties of other phases of a system given knowledge of one particular phase. In [135], the authors proposed to train a GP on one phase of the system and expected the model to predict phase transitions and properties of other phases. Let us start by discussing how they achieved this for the mean-field Heisenberg spin model in the nearest-neighbor approximation. They trained the GP on the free energy of the system in the high-temperature regime, where the average spin magnetization is zero, far from the phase transition point. Then the trained GP was asked to extrapolate within the low-temperature regime, and it predicted correctly both the location of

[19]More precisely, the test energy mean absolute error was 0.177 kcal/mol. Is this error small? Are these PESs accurate enough for modern spectroscopic applications? Spectroscopy discerns two kinds of accuracies. The spectroscopic accuracy is 1 cm^{-1}, while the chemical accuracy is 1 kcal/mol $\approx$ 350 cm^{-1}. Modern spectroscopic applications need a spectroscopic accuracy, and this result has an error that is 60 times larger than spectroscopic accuracy – so no, it is not yet good enough. But it is more than enough, for example, for simulations of molecular dynamics or reactions, especially at room temperature [221].

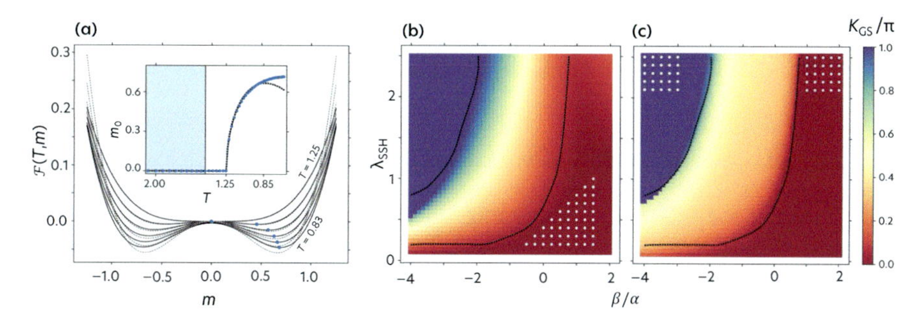

Figure 4.11 GPs for extrapolation to non-seen quantum phases. (a) The mean-field Heisenberg spin model in the nearest-neighbor approximation, where black dots are mean-field results and blue dots are GP predictions. GP was trained on the high-temperature shadowed regime of the data. (b) and (c) Generalized lattice polaron model with different GP training regimes (marked with white dots). In both cases, GP correctly predicted the phase transitions (color map) as compared to the quantum calculations (black lines). Adapted from [135].

the phase transition and the free energy (and consequently nonzero magnetization) of the system as presented in Fig. 4.11(a).

The authors also applied this approach to a much more complex system[20] whose Hamiltonian can be written in the following generic form:

$$H = H_0 + \alpha H_1 + \beta H_2 , \tag{4.70}$$

where α and β are tunable parameters along which phase transitions occur. They trained a GP in some parameter regime of the Hamiltonian and were able to successfully extrapolate to others.[21] This approach proves to be useful for such a class of Hamiltonians for another reason. Usually, we are able to easily compute or measure the eigenspectrum in certain limits of α and β but not at arbitrary points within the parameter space. We can then train GPs in these limits and can expect them to extrapolate successfully to other parameter regimes where the direct calculation is more difficult. Finally, in the same system, the authors of [194] studied the importance of choosing the kernel. They compared the results from the original work [135] obtained for kernels found with the BIC as described in Sections 4.4.1 and 4.4.2 with the results obtained for kernels with the same complexity (that is, at the same search tree level, see Fig. 4.7) but chosen at random. Predictions of such GPs were much worse and were prone to overfitting, which stresses the power of the BIC as a selec-

[20] The studied system was a generalized lattice polaron model [222] describing an electron in a one-dimensional lattice with $N \to \infty$ sites coupled to a phonon field. The interaction between an electron and a phonon field was a combination of two qualitatively different terms: the Su–Schrieffer–Heeger electron–phonon coupling and the breathing-mode model with the Holstein coupling.

[21] How is this even possible? The intuition behind it is that the evolution of physical properties that are given to the ML model as input should somehow reflect the fact that there is a phase transition. The model probably picks up on the prevalent correlations within one phase, and it observes that these correlations change when crossing to other phases.

tion criterion for kernels. The appropriate choice of the kernel is, therefore, crucial as it determines how far the model can accurately extrapolate.

4.5.4 Bayesian optimization of variational quantum algorithms

Another suitable application for BO is within the context of near-term quantum computing, where a computational advantage is sought by the use of noisy intermediate-scale quantum (NISQ) devices; see also Section 8.2.4. One popular strategy is variational quantum eigensolvers (VQEs): These algorithms are hybrid quantum-classical algorithms that are suited for finding the ground state of a given Hamiltonian. Applications of VQEs can be found in several domains, see, for example, [223–227]. In these algorithms, parameter optimization happens classically. Treating the quantum circuit as a parametrized black box, BO, hence, presents itself as a gradient-free optimization tool.

Recent works have demonstrated the resource efficiency of BO in optimizing VQEs [228–230], that is, the number of calls of the quantum algorithm. In particular, [230] showed that kernel methods are a natural choice for this task by incorporating physical prior knowledge directly into the kernel itself. To achieve this, the authors derive a new general type of kernel with the same functional form as the target black-box function one seeks to minimize, for example, using BO. Thus, optimizing the parameters of a variational quantum circuit is aided by the prior knowledge one has about the quantum circuit at hand. In addition to that, the authors introduce a new type of acquisition function. The pivotal feature of this acquisition function is to look for the next point to measure by using the level of confidence the model has with respect to the current choice of parameters. Should the confidence of the GP for a given candidate point be high, that is, small variance, it is likely to be skipped. Therefore, leveraging a sequential minimal optimization scheme [231, 232], that is, sequentially optimizing one parameter after the other, this new acquisition function achieves better convergence to the global optimum. The results shown in [230] show that the novel optimization scheme introduced therein is capable of outperforming the state of the art [231] (at the time of writing). While the kernel in [230] has been proposed within the context of finding the ground state of a given Hamiltonian, it can be combined with any other acquisition function and used for any other optimization task. Similarly, the novel acquisition function serves as a general framework suited for any other optimization tasks when combined with the kernel mentioned above.

4.6 Outlook and open problems

- While GPs successfully surrogate PESs and need a much smaller number of ab initio calculations, it is challenging to reach the spectroscopic accuracy with this approach. What is stopping us from achieving such accuracy levels with GPs? The major limitation is the number of training data. In practice, it is often observed that the error during learning eventually decreases by a factor of $1/n$, where n is the number of training points. As such, the number of training points required to reach a level of accuracy on the order of 10 cm^{-1} for a

57-dimensional surface is still manageable. However, reaching spectroscopic accuracy requires an excessive amount of training data. In particular, the size of the training dataset grows beyond the regime where GPs are useful [233]. The high-accuracy limit may be obtained if one incorporates some knowledge of the system into the kernel, however, how to do that remains an open question.

- In Section 4.5, we have presented how BO and GPs can be used to tackle optimization of expensive setups, where gradients are not accessible. Such is also the case of quantum NNs or VAEs. Therefore, this approach may prove useful in the optimization of a quantum model!
- An interesting research direction is combining the power of automatic differentiation (AD), described in Section 7.1, and kernel methods. Already, AD has played a major role in developing more robust kernel functions for GP models. For example, [234] showed that by maximizing the log marginal likelihood, Eq. (4.60), one could jointly optimize the weights and biases of a deep NN combined with any parameter of a standard kernel function. A more recent work [235] also showed that learning the composition of kernels is differentiable under the AD framework, and more complex kernels could be parametrized. Currently, there are two main ecosystems for GPs based on AD, `GPytorch` [236, 237], and `GPflow` [238].
- The training procedure of KRR could also be differentiated using AD bypassing the need of using a cross-validation scheme [239].
- With the advent of quantum extensions of classical ML methods for near-term quantum devices, there are several paths on how to encode a data point $\boldsymbol{x}$ in a Hilbert space as $|\boldsymbol{x}\rangle$. As a consequence, the kernel function has to be promoted to its quantum version. Interestingly, there is a provable advantage of such kernels based on measurement results of the quantum state [240]. We give a bit more detail in Section 8.2.

Further reading

1. Rasmussen, C. E. & Williams, C. K. I. (2006). *Gaussian Processes for Machine Learning.* The MIT Press. The standard go-to reference on kernel methods and GPs in particular [241].
2. Bishop, C. M. (2006). *Pattern Recognition and Machine Learning.* Another go-to reference for GPs [93].
3. Krems, R. V. (2019). *Bayesian machine learning for quantum molecular dynamics.* Phys. Chem. Chem. Phys. 21(25), 13392–13410. Discusses various applications of GPs for quantum molecular dynamics [242].
4. Jupyter notebook allowing to faithfully reproduce a six-dimensional PES with a GP and BO including optimal kernel search using the BIC criterion for the H_3O^+ [2].

5. Vargas-Hernández, R. A., & Krems, R. V. (2020). *Physical extrapolation of quantum observables by generalization with Gaussian processes.* Lect. Notes Phys., 968, 171–194. In-depth review of possible applications of GPs and BO for extrapolation problems in quantum sciences [243].
6. Huang, H. et al. (2021). *Provably efficient machine learning for quantum many-body problems.* Science 377(6613). It introduces quantum-measurement-inspired kernels for a provable advantage of kernel methods over classical methods that do not use measurement data [244].

5 Neural-network quantum states

In the early days of quantum mechanics, it soon became clear that approximation methods would be needed to solve most relevant real-world problems [245]. Indeed, in most cases, the Schrödinger equation cannot be exactly solved for systems of more than a few interacting particles. This came to be referred to as the *quantum many-body problem*. In this chapter, we show how neural networks (NNs) have been introduced to tackle this problem [246], in a variety of applications, including ground-state and quantum dynamics of interacting quantum systems. For simplicity, we mainly focus our discussion on spin systems and discuss applications to fermions [247] and bosons [248] only toward the end.

According to the axioms of quantum mechanics, the state of an isolated quantum system is encoded into a complex-valued vector of probability amplitudes commonly known as the wave function. In the case of a single spin-$\frac{1}{2}$, the wave function in the computational z-basis $\hat{Z}$ is $|\Psi\rangle = C_\uparrow|\uparrow\rangle + C_\downarrow|\downarrow\rangle$. The coefficients $C_\uparrow$ and $C_\downarrow$ are the complex probability amplitudes of the spin being aligned along ($C_\uparrow$) or opposite to ($C_\downarrow$) the direction of the computational basis, and they are subject to the normalization condition $|C_\uparrow|^2 + |C_\downarrow|^2 = 1$. For many-body quantum systems of N spins, where N can be any large number from tens to the order of the Avogadro number $\sim 10^{23}$, the number of coefficients in the wave function scales as 2^N. Following up on the spin example, the wave function can be expressed as follows:

$$\begin{aligned} |\Psi\rangle &= C_{\uparrow\uparrow\cdots\uparrow}|\uparrow\uparrow\cdots\uparrow\rangle + C_{\uparrow\uparrow\cdots\downarrow}|\uparrow\uparrow\cdots\downarrow\rangle + \cdots + C_{\downarrow\downarrow\cdots\downarrow}|\downarrow\downarrow\cdots\downarrow\rangle \\ &= \sum_{\sigma_1,\sigma_2,\cdots,\sigma_N} C_{\sigma_1,\sigma_2,\ldots,\sigma_N}|\sigma_1\rangle \otimes |\sigma_2\rangle \otimes \cdots \otimes |\sigma_N\rangle, \end{aligned} \tag{5.1}$$

where the $|s\rangle = |\sigma_1\rangle \otimes |\sigma_2\rangle \otimes \cdots \otimes |\sigma_N\rangle$ are the basis vectors of the Hilbert space that describes the N spin system, and $C_{\sigma_1,\sigma_2,\ldots,\sigma_N}$ are their associated amplitudes.

The *quantum many-body problem* originates from the exponential scaling of the number of the basis elements, which leads to an exponential computational complexity in the system size. In particular, the memory required to naively store the wave function of just 60 spins is $16 \cdot 2^{60} \approx 18$ exabytes, about 500 times more than what is available on the world's largest supercomputer as of 2022.

Nevertheless, while the Hilbert space of many-body quantum systems is exponentially large, physically relevant states are typically confined to a corner of the Hilbert space that is of limited dimension. For instance, many physical Hamiltonians only contain *local* interactions, which significantly constrains the form of the associated many-body wave functions.

> The main idea behind variational methods is to find a computationally efficient representation of the physically relevant quantum states within the Hilbert space of interest.

Variational methods circumvent the issue of an exponential complexity by encoding the complex amplitudes of the wave function onto a parametrized function (often called the *ansatz*), which depends on a set of parameters θ. If the number of parameters is polynomial in the system size, the state can be efficiently stored with limited computational resources. In general, the variational state $|\Psi_\theta\rangle$ can be expanded onto the computational basis as

$$|\Psi_\theta\rangle = \sum_{s=1}^{2^N} \Psi_\theta(\boldsymbol{s})|s\rangle, \tag{5.2}$$

where $\Psi_\theta(\boldsymbol{s}) = \langle s|\Psi_\theta\rangle$ denotes the probability amplitude corresponding to the state $|s\rangle$. The task is then to find the parametrization θ that best describes our desired quantum state of interest, such as the ground state of a given Hamiltonian.

5.1 Variational methods

Even when using variational states, computing expectation values can still be of exponential complexity since one must perform sums over all the basis elements of the Hilbert space for these calculations. Among the variational states that are practically usable, there are two possible approaches that distinguish two families of variational ansätze: those that can be used to compute expectation values exactly with a polynomial cost, and those that do so only approximately, with an accuracy improvable at a polynomial cost in system size. In the former, the only source of error in the expectation value of observables comes from the truncation of (exponentially large) regions of the Hilbert space, limiting its ability to represent wave functions. In the latter, an additional source of error typically comes from sampling, although it does not necessarily add a systematic error and can be improved upon for an additional computational cost.

The third category consists of parameterized quantum states whose cost for computing expectation values scales exponentially with system size. In practical applications, for example, in the case of tensor networks (TNs) in two dimensions, approximate algorithms for computing expectation values are introduced. Strictly speaking, however, these are not variational methods, as we cannot compute expectation values to arbitrary accuracy in polynomial time, and they introduce a systematic bias that goes beyond the pure variational error.

5.1.1 Variational states with exact expectation values

In the first kind of variational states, we mainly encounter locally constrained ansätze, for which mean-field and matrix product states (MPSs) are notable examples.

Mean-field Ansatz. Mean-field states are one of the simplest variational quantum states. With these, we model our variational wave function by the mean-field approximation, that is, as the tensor product of single-spin wave functions

$$\begin{aligned}|\Psi_\theta\rangle &= |\phi_1(\theta_\uparrow^{(1)},\theta_\downarrow^{(1)})\rangle \otimes |\phi_2(\theta_\uparrow^{(2)},\theta_\downarrow^{(2)})\rangle \otimes \cdots \otimes |\phi_N(\theta_\uparrow^{(N)},\theta_\downarrow^{(N)})\rangle \\ &= \bigotimes_{i=1}^{N} |\phi_i(\theta_\uparrow^{(i)},\theta_\downarrow^{(i)})\rangle,\end{aligned} \tag{5.3}$$

where $|\phi_i\rangle$ are the single-spin wave functions at site i. They are subject to the orthogonality condition $\langle\phi_i|\phi_j\rangle = \delta_{ij}$, with δ_{ij} denoting the Kronecker delta. This way, $|\phi_i\rangle$ has only two coefficients corresponding to the probability amplitudes of the spin being up or down, which we take as variational parameters

$$|\phi_i\rangle = \theta_{\uparrow}^{(i)}|\uparrow\rangle + \theta_{\downarrow}^{(i)}|\downarrow\rangle \tag{5.4}$$

$$\left|\theta_{\uparrow}^{(i)}\right|^2 + \left|\theta_{\downarrow}^{(i)}\right|^2 = 1\,, \tag{5.5}$$

resulting into $2N$ complex parameters in total, that is, $\theta = \left\{\theta_{\uparrow}^{(i)}, \theta_{\downarrow}^{(i)} \middle| i = 1, 2, \dots, N\right\}$.

With this family of wave functions, we can compute expectation values of quantum Hamiltonians exactly. This is a consequence of the fact that we can exploit the tensor product structure of our wave function to simplify the expectation values over many-body states to the expectation over the corresponding single-body ones. For example, the expectation value of the σ_i^x Pauli operator acting on the ith site can be obtained as $\langle\Psi_\theta|\sigma_i^x|\Psi_\theta\rangle = \langle\phi_i|\sigma_i^x|\phi_i\rangle$. The calculation is straightforward, as $|\phi_i\rangle$ is a two-dimensional vector and σ_i^x is a 2×2 matrix.

Tensor network states. However, mean-field states are not able to capture correlations between local degrees of freedom. Tensor network states (TNSs) are a family of quantum states that improve upon such a limitation, and a subset of TNSs also allow to compute expectations exactly. One of the most broadly used TNSs with this property are matrix product states (MPSs), which predominate in the study of one-dimensional systems.

Let us consider the coefficients $C_{\sigma_1,\sigma_2,\dots,\sigma_n}$, defined in Eq. (5.1). We can consider $C_{\sigma_1,\sigma_2,\dots,\sigma_N}$ as a tensor with N indexes, which we can always express as the contraction of tensors A^{σ_i}, such that:

$$C_{\sigma_1,\dots,\sigma_N} = \sum_{\alpha,\beta,\dots,\gamma} A_{\alpha,\beta}^{\sigma_1} A_{\beta,\delta}^{\sigma_2} \dots A_{\gamma,\alpha}^{\sigma_N}, \tag{5.6}$$

where the maximal dimension of the Greek indices $\alpha, \beta, \dots$ is the bond dimension χ. This way, an exact representation of $C_{\sigma_1,\dots,\sigma_N}$ requires an exponentially large number of parameters. This means that the bond dimension, χ, must increase exponentially with N. The idea of the MPS ansatz resides in the truncation of the dimension of the indices of the tensors A^{σ_i}. With the truncation, we reduce the number of parameters of our ansatz to be $\mathcal{O}(dN\chi^2)$, where d is the local Hilbert space dimension, for example, two for a spin-1/2 particle. We usually truncate the bond dimension in an elegant and controlled way using the singular value decomposition of the tensors A^{σ_i}, which has a strict connection with the maximal entanglement entropy the MPS can contain,[1] as we discuss in Section 5.2.3.

As a final remark, let us mention an important algorithm proposed by S. White for the energy minimization of variational quantum states, known as density matrix renormalization group [249]. This algorithm is particularly well suited for MPSs, and their combination is the current state-of-the-art technique to compute the ground-state wave function of one-dimensional systems. However, the description of this

[1]The bond dimension is, in fact, the rank of the Schmidt decomposition of the quantum state.

algorithm falls out of the scope of this book. We refer to [250] for a complete review of the use of MPS and to [251] for a review of methods based on tensor networks (TNs).

5.1.2 Variational states with approximate expectation values

The second family of variational states we encounter are known as *computationally tractable states* [252].

Variational ansätze must satisfy two conditions to be computationally tractable:

- Amplitudes for arbitrary single basis elements $\Psi_\theta(\boldsymbol{s}) = \langle s|\Psi_\theta\rangle$ can be computed efficiently.
- It is possible to efficiently generate samples s from the Born distribution $P(s) = \frac{|\langle s|\Psi_\theta\rangle|^2}{\langle\Psi_\theta|\Psi_\theta\rangle}$.

If these two conditions are met, we can efficiently estimate expectation values of arbitrary k-local operators, and the statistical error due to the stochastic sampling can be rigorously controlled by increasing the number of samples. Therefore, the computational time to compute expectation values is polynomial in both the symstem's size and accuracy.

A k-local operator is an operator that contains terms acting on at most k local quantum numbers at the same time. For instance, a nearest-neighbor Hamiltonian is a 2-local Hamiltonian because it contains terms acting on two qubits.

In general, given a variational state $|\Psi_\theta\rangle$, we can obtain the expression for the expectation value of an operator $\hat{O}$ as follows:

$$
\begin{aligned}
\langle\hat{O}\rangle &= \frac{\langle\Psi_\theta|\hat{O}|\Psi_\theta\rangle}{\langle\Psi_\theta|\Psi_\theta\rangle} && (5.7)\\
&= \frac{\sum_{s,s'}\langle\Psi_\theta|s\rangle\langle s|\hat{O}|s'\rangle\langle s'|\Psi_\theta\rangle}{\sum_s|\langle\Psi_\theta|s\rangle|^2} && (5.8)\\
&= \frac{\sum_s\langle\Psi_\theta|s\rangle\frac{\langle s|\Psi_\theta\rangle}{\langle s|\Psi_\theta\rangle}\sum_{s'}\langle s|\hat{O}|s'\rangle\langle s'|\Psi_\theta\rangle}{\sum_s|\langle\Psi_\theta|s\rangle|^2} && (5.9)\\
&= \frac{\sum_s|\langle\Psi_\theta|s\rangle|^2\sum_{s'}\langle s|\hat{O}|s'\rangle\frac{\langle s'|\Psi_\theta\rangle}{\langle s|\Psi_\theta\rangle}}{\sum_s|\langle\Psi_\theta|s\rangle|^2}, && (5.10)
\end{aligned}
$$

where we have added two identities of the form $\sum_s |s\rangle\langle s| = \mathbb{1}$ in the numerator and one in the denominator. Then, we have multiplied by $\frac{\langle s|\Psi_\theta\rangle}{\langle s|\Psi_\theta\rangle}$ in the numerator.[2] We identify two main terms:

$$P(\boldsymbol{s}) = \frac{|\langle \Psi_\theta | s\rangle|^2}{\sum_s |\langle \Psi_\theta | s\rangle|^2} \tag{5.11}$$

$$O_{\text{loc}}(\boldsymbol{s}) = \sum_{s'} \langle s|\hat{O}|s'\rangle \frac{\langle s'|\Psi_\theta\rangle}{\langle s|\Psi_\theta\rangle}, \tag{5.12}$$

where $O_{\text{loc}}(\boldsymbol{s})$ is the so-called *local estimator* of $\hat{O}$. Therefore, we can write the quantum expectation value of an observable $\hat{O}$ as the statistical expectation value of its local estimator O_{loc} over the probability distribution $P(\boldsymbol{s})$:

$$\langle \hat{O} \rangle = \sum_s P(\boldsymbol{s}) O_{\text{loc}}(\boldsymbol{s}) = \langle O_{\text{loc}}(\boldsymbol{s}) \rangle_P . \tag{5.13}$$

Let us stress that these calculations only hold for operators with the property that the number of states $\boldsymbol{s}'$ such that $|\langle s|\hat{O}|s'\rangle| \neq 0$, for arbitrary $\boldsymbol{s}$ is at most polynomial in the number of spins. For example, it is easy to convince oneself that k-local operators satisfy this property. Conversely, evaluating $O_{\text{loc}}(\boldsymbol{s})$ would not be tractable, given that the sum over s' in Eq. (5.12) would be over an exponential number of elements.

The procedure described above allows to obtain a controlled, stochastic estimate of the expectation values by directly sampling a series of states, $\boldsymbol{s}^{(1)}, \boldsymbol{s}^{(2)}, \dots, \boldsymbol{s}^{(M)}$, from $P(\boldsymbol{s})$, and approximating $\langle \hat{O} \rangle$ with the following arithmetic mean:

$$\langle \hat{O} \rangle \approx \frac{1}{M} \sum_{i=1}^{M} O_{\text{loc}}(\boldsymbol{s}^{(i)}). \tag{5.14}$$

The statistical error associated with such an estimate is $\varepsilon = \sqrt{\sigma^2/M}$, and it is bounded as long as the variance σ^2 of O_{loc} is finite. For example, when $\hat{O}$ is a k-local spin operator with bounded coefficients, its variance is strictly finite since it can be shown that $\sigma^2 = \langle \hat{O}^2 \rangle - \langle \hat{O} \rangle^2$.[3] Therefore, the error in the estimate of expectation values decreases as $\varepsilon \sim 1/\sqrt{M}$, which allows us to reach arbitrary accuracy in the estimation by increasing the number of samples M, given that $\lim_{M\to\infty} \varepsilon = 0$. However, generating a set of samples according to the Born distribution, $\{\boldsymbol{s}^{(i)}\} \sim P(\boldsymbol{s})$, is in general a nontrivial computational task in the case where the variational ansatz, $\Psi_\theta(\boldsymbol{s})$, is parameterized by an efficiently computable, yet arbitrary function. One of the most commonly adopted strategies to sample from $P(\boldsymbol{s})$ is through Markov chain Monte Carlo (MCMC) methods, including the Metropolis–Hastings method, which generate a sequence of correctly distributed samples $\boldsymbol{s}^{(i)}$.

[2] Notice that this manipulation is always valid, since amplitudes with $\langle s|\Psi_\theta\rangle = 0$ never appear in the summation over s.

[3] It is also simple to prove that, when $|\psi_\theta\rangle$ approaches an eigenstate of $\hat{O}$, the variance vanishes. Consequently, considering $\hat{O} = \hat{H}$, the statistical error vanishes as we approach the ground (or any excited) state.

Metropolis–Hastings methods construct a Markovian stochastic process which satisfies the *detailed balance* relation for the target probability distribution

$$P(\boldsymbol{s})\mathcal{T}(\boldsymbol{s} \to \boldsymbol{s}') = P(\boldsymbol{s}')\mathcal{T}(\boldsymbol{s}' \to \boldsymbol{s})\,, \tag{5.15}$$

where $\mathcal{T}(\boldsymbol{s}^{(i)} \to \boldsymbol{s}^{(i+1)})$ is the probability that the state $\boldsymbol{s}^{(i)}$ at step i transitions to the state $\boldsymbol{s}^{(i+1)}$ at the following step. As the process is Markovian, the transition probability at every step depends exclusively on the current configuration. The detailed balance condition ensures that regardless of the initial configuration $\boldsymbol{s}^{(0)}$, the sequence eventually converges to the correct distribution $P(\boldsymbol{s})$ in the long time limit.

One possible choice of the transition probability $\mathcal{T}$ is given by the *Metropolis–Hastings algorithm* [253]. The main idea is to express $\mathcal{T}$ in terms of a local transition kernel T and an acceptance probability A such that

$$\mathcal{T}(\boldsymbol{s} \to \boldsymbol{s}') = T(\boldsymbol{s} \to \boldsymbol{s}')A(\boldsymbol{s} \to \boldsymbol{s}')\,. \tag{5.16}$$

This way, we split the global stochastic process into the product of two local subprocesses that we can compute efficiently. For instance, it is very easy to find a normalized local transition kernel that allows us to modify only a few degrees of freedom, like flipping a single spin in a given configuration. Conversely, it is hard to find a normalized global kernel that would act on all spins.

The acceptance probability to go from a configuration $\boldsymbol{s}$ to $\boldsymbol{s}'$ through a local transition is defined as

$$A(\boldsymbol{s} \to \boldsymbol{s}') = \min\left(1, \frac{P(\boldsymbol{s}')T(\boldsymbol{s}' \to \boldsymbol{s})}{P(\boldsymbol{s})T(\boldsymbol{s} \to \boldsymbol{s}')}\right). \tag{5.17}$$

Notice that the normalization of the Born probabilities cancels out, giving the expression

$$\frac{P(\boldsymbol{s}')}{P(\boldsymbol{s})} = \left|\frac{\langle s' \mid \Psi_\theta\rangle}{\langle s \mid \Psi_\theta\rangle}\right|^2, \tag{5.18}$$

which allows us to consider unnormalized variational ansätze. Additionally, if the variational state is computationally tractable, the transition probability also has a tractable complexity, provided it only acts on the basis elements.

Choosing a valid transition rule $T(\boldsymbol{s} \to \boldsymbol{s}')$ is not trivial, and we must take special care in the case of systems with symmetries. For example, if the total magnetization along the direction of the computational basis is known, we might want to fix it and use a transition rule that does not project the Markov chain outside of a certain region. In general, a computationally expensive yet effective choice for the transition kernel is to use the Hamiltonian itself:

$$T(\boldsymbol{s} \to \boldsymbol{s}') = \frac{\left|\langle s|\hat{H}|s'\rangle\right|(1 - \delta_{\boldsymbol{s},\boldsymbol{s}'})}{\sum_{'\neq s}\left|\langle s|\hat{H}|s'\rangle\right|}\,, \tag{5.19}$$

which is known as the Hamiltonian transition rule [246].

This way, with the Metropolis–Hastings algorithm, starting from a random configuration $\mathbf{s}^{(0)}$, we can sample from $P(\mathbf{s})$ by iteratively proposing local modifications $\mathbf{s}'$ according to $T(\mathbf{s} \to \mathbf{s}')$, and accepting them according to $A(\mathbf{s} \to \mathbf{s}')$.

Nonetheless, this sampling procedure is imperfect, and it can fail to converge for a reasonable number of iterations if the sampled distribution is too complex. In addition, the procedure suffers from the fact that the samples are correlated since we flip spins iteratively. See Algorithm 6 for further details.

Algorithm 6 Metropolis–Hastings algorithm

```
s ← uniform∈ [1, 2^N]                          ▷ sample initial state uniformly at random
for i = 1 to M do
    propose s' according to T(s → s')
    A ← P(s')T(s'→s) / P(s)T(s→s')              ▷ calculate acceptance probability
    ξ ← uniform∈ [0, 1]
    if ξ ≤ A then
        s ← s'                                  ▷ update state
    end if
end for
```

5.2 Representing the wave function

Now that we have seen how to compute the quantities of interest using parametrized quantum states, let us dive into how to devise expressive variational states in practice. The main idea is that we need to represent high-dimensional functions with a parametrization that is flexible and general enough to describe physical systems while involving only a polynomial amount of parameters.

Traditionally, researchers have relied on physically inspired variational ansätze. The Jastrow wave function [254, 255] stands out as one of the most successful and widely used ones. It is based on the assumption that two-body interactions are the most physically relevant, and it assigns a trainable potential to every interacting pair. Formally,

$$\Psi_\theta(\mathbf{s}) = e^{-\frac{1}{2}\sum_{i\neq j}\theta_{ij}\sigma_i\sigma_j}, \tag{5.20}$$

where the sum runs over all possible spin pairs, and θ_{ij} are the parameters encoding pairwise spin correlations. Therefore, for a system of N spins, the resulting wave function has $\mathcal{O}(N^2)$ parameters. Moreover, in translationally invariant systems, the parameters θ_{ij} can be made dependent exclusively on the distance between i and j, resulting in a reduced number, $\mathcal{O}(N)$, of parameters.

Artificial neural networks (ANNs) have taken over more traditional ansätze to approximate the wave function itself [246]. This family of variational states is known as neural quantum states (NQSs). For instance, we can write a parametrized wave function as a feed-forward NN. In this case, $\Psi_\theta(\boldsymbol{s})$ corresponds to the output of an NN that takes the configuration $\boldsymbol{s}$ as input in the form of a vector.

In a feed-forward NN of depth D, every layer l consists of a nonlinear activation function $\mathrm{g}^{(l)}$ that acts, component-wise, on a vector resulting from applying the weight matrix $\boldsymbol{W}^{(l)}$ to the output of the previous layer. This way, it is possible to write the variational state as the composition of operations $\mathrm{g}^{(l)} \cdot \boldsymbol{W}^{(l)}$, where "$\cdot$" indicates pointwise operation, such that

$$\Psi_\theta(\boldsymbol{s}) = \mathrm{g}^{(D)} \cdot \boldsymbol{W}^{(D)} \ldots \mathrm{g}^{(2)} \cdot \boldsymbol{W}^{(2)} \mathrm{g}^{(1)} \cdot \boldsymbol{W}^{(1)} \boldsymbol{s}\,. \tag{5.21}$$

Hence, the output is a scalar, complex or real, representing the probability amplitude of configuration $\boldsymbol{s}$.

From a mathematical perspective, these ansätze are of great interest given that NNs are subject to universal representation theorems [75], as we explain in Section 2.4.4. According to Eq. (2.46), we could represent the many-body wave function with a polynomial number $\mathcal{O}(N^2)$ of one-dimensional nonlinear functions, with N denoting the number of spins.

However, these results hold for arbitrary nonlinear functions, Φ_q, $\phi_{q,p}$ in Eq. (2.46), that must be appropriately found to represent the target function. In practice, NNs use a fixed nonlinear activation function, and we can only adjust the number of operations. In these cases, the number of neurons does not have a strict polynomial scaling, and it can be, in the worst case, exponential in N [76]. Nevertheless, the state-of-the-art results in computer vision and natural language processing [256–258] should be sufficient motivation to employ similar techniques to represent quantum states. Note that the NN representation of quantum states does not preserve the Hilbert space structure, which means that for two NN representations $|\psi_1\rangle$ and $|\psi_2\rangle$, it is not possible to construct a valid wave function $|\psi\rangle = |\psi_1\rangle + |\psi_2\rangle$ represented by an NN of the same size as the ones representing $|\psi_1\rangle$ and $|\psi_2\rangle$ by simply adding up the parameters together, as the ansatz is generally nonlinear.

5.2.1 Restricted Boltzmann machines

NQSs were first introduced using restricted Boltzmann machines (RBMs) [246]. RBMs are shallow models featuring two fully connected layers: a *visible* layer, consisting of N units, and a *hidden* layer, consisting of M units. A scheme of an RBM architecture is presented in Fig. 5.1. The wave function amplitudes of an RBM ansatz are given by:

$$\Psi_\theta(\boldsymbol{s}) = \sum_{\boldsymbol{h}} e^{\boldsymbol{b}_v^\dagger \boldsymbol{s} + \boldsymbol{b}_h^\dagger \boldsymbol{h} + \boldsymbol{h}^\dagger \boldsymbol{W} \boldsymbol{s}}, \tag{5.22}$$

where $\boldsymbol{s}$ and $\boldsymbol{h}$ represent the visible and hidden units, respectively, and the parameters $\theta = \{\boldsymbol{b}_v, \boldsymbol{b}_h, \boldsymbol{W}\}$ represent the visible and hidden biases and the weight matrix, respectively. In the NN picture, the RBM is a single-layer nonlinear feed-forward NN, with the visible units serving as inputs and the exponential serving as the activation

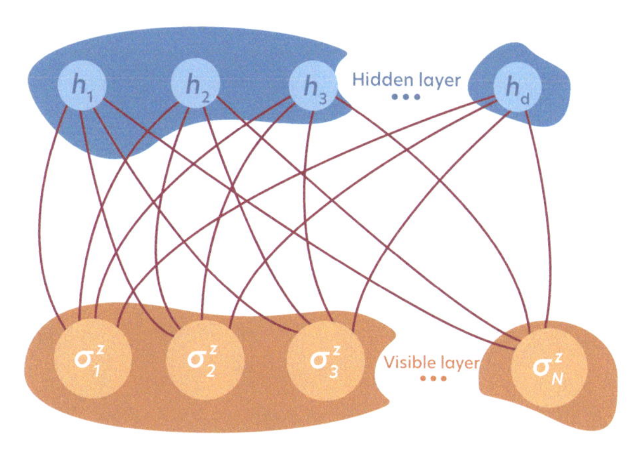

Figure 5.1 Pictorial representation of a restricted Boltzmann machine that represents the wave function of an N-spin system, with $\boldsymbol{s} = (\sigma_1, \sigma_2, \dots, \sigma_N)$ and $\boldsymbol{h} = (h_1, h_2, \dots, h_d)$ the hidden units.

function. While it is common to have biases for the hidden layer (see Section 2.4.4), RBMs also have somewhat unusual biases connected to the input values, which is explained in the next paragraph.

By construction, RBMs are designed in such a way that computing the summation over hidden units, as in Eq. (5.22), can be done analytically. To see this, we can rewrite Eq. (5.22) in a tractable form considering binary hidden units $h_i \in \{-1, 1\}$, leading to

$$\Psi_\theta(\boldsymbol{s}) = e^{\boldsymbol{b}_v^\dagger \boldsymbol{s}} \prod_{i=1}^{M} 2\cosh\left(\boldsymbol{b}_{h,i} + \boldsymbol{W}_{i\cdot}\boldsymbol{s}\right), \tag{5.23}$$

where $\boldsymbol{b}_{h,i}$ and $\boldsymbol{W}_{i\cdot}$ denote the ith hidden bias and weight matrix row, respectively. To treat spin systems, the visible units will represent the N physical spins. Thus the input of the RBM is simply the spin configuration $\boldsymbol{s}$. In this way, we obtain an analytical expression to evaluate the amplitude for a given spin configuration and thus represent the full wave function with this ansatz. One can also interpret the hidden units as M hidden spins, and in this picture, the RBM can be thought of as an interacting spin model with interaction strengths $\boldsymbol{W}_{ij}$. Moreover, we can treat an RBM as a model with an associated energy depending on its parameters, input, and hidden spin values. This is known as an energy-based model and explains why input biases are present in Eq. (5.22). In fact, the RBM is equivalent to a Hopfield network, a type of spin glass [259]. For more details on this view, see [260].

Being the first to be introduced in this context, most of the early works about NQSs employ RBMs, but other architectures have been systematically explored in more recent years. The capacity of RBMs and its relationship to quantum entanglement has been examined in various works [261, 262]. An extension of this architecture, the deep RBM, has also been introduced to solve more complex problems [263], which consists of stacking more than two fully connected layers.

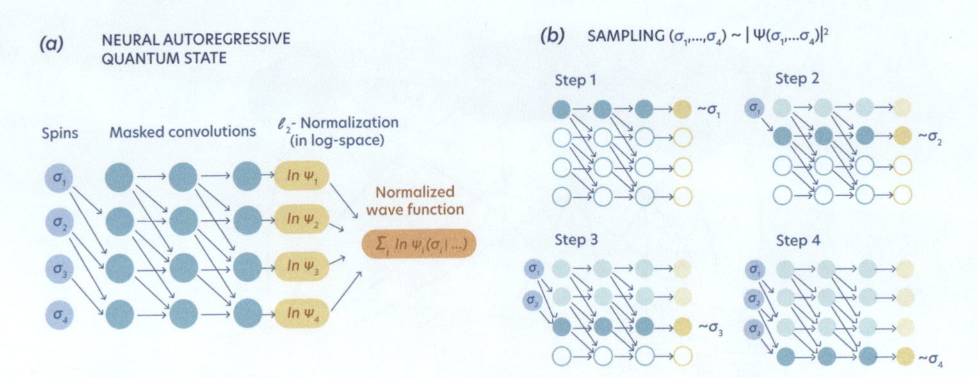

Figure 5.2 Example of an ARNN quantum state for four spins. (a) Pictorial representation of the network. The arrows representing the model's weights are skewed to ensure that the conditional structure of the output probability distribution is not broken. These layers are "masked" due to some connections being deleted. (b) Sampling algorithm. One samples consecutive spins using direct sampling on the conditional probabilities at each step. Adapted from [92].

5.2.2 Autoregressive and recurrent neural networks

Autoregressive neural networks (ARNNs), as presented in Section 2.4.6, can also be used for constructing NQSs, as introduced in [92] and later applied to both quantum [264] and classical problems [88]. Their main advantage is that their Born probability distribution is normalized, allowing for direct (autoregressive) sampling, which is easier to parallelize than MCMC. A pictorial representation of both the network and the sampling algorithm is presented in Fig. 5.2.

Analogously to Eq. (2.47), we express the many-body wave function in terms of a product of conditional complex amplitudes:

$$\Psi_\theta(\boldsymbol{s}) = \prod_{i=1}^{N} \phi_i(\sigma_i \mid \sigma_{i-1}, \dots, \sigma_1), \tag{5.24}$$

which is subject to the normalization condition $\sum_\sigma \left|\phi_i(\sigma \mid \sigma_{i-1}, \dots, \sigma_1)\right|^2 = 1$. With this architecture, we can compute expectation values by directly sampling state configurations instead of building a Markov chain through the Metropolis–Hastings algorithm, for example (see Algorithm 6). We sample state configurations by iteratively sampling one spin after the other: We start sampling the first spin σ_1 from the reduced probability distribution $|\phi(\sigma_1)|^2$. Then, we sample the second one σ_2 according to the conditional probability distribution $|\phi_2(\sigma_2 \mid \sigma_1)|^2$, then the next one $|\phi(\sigma_3 \mid \sigma_2, \sigma_1)|^2$, and so on until σ_N. This sampling procedure is embarrassingly parallelizable.[4]

[4]We can use the intermediate conditional probabilities to draw samples for a low computational cost, for example, use the probabilities for $N-1$ spins and sample from the last one, to obtain new samples; with MCMC we cannot do this.

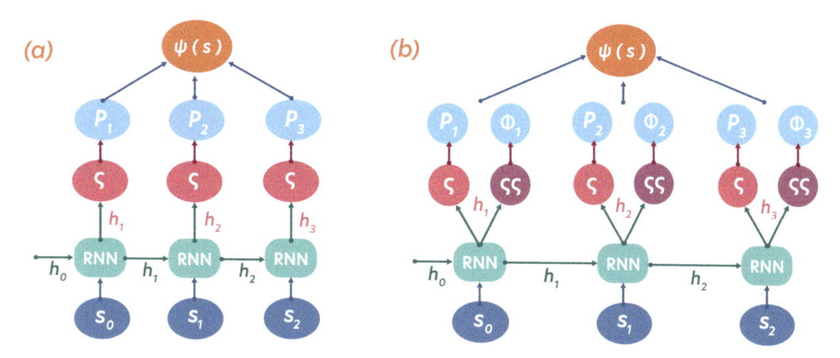

Figure 5.3 Pictorial representation of an RNN architecture for NQS. Panel (a) is for real-valued wave functions, which can be relevant for a certain class of problems, and panel (b) is for complex-valued wave functions. In both schemes, a local spin configuration s_i and a hidden vector h_i are fed into an RNN cell, which performs a nonlinear transformation. Then an activation function (ς, for softmax and/or $\varsigma\varsigma$, for softsign) is applied to obtain the final probability and/or phase corresponding to the configuration. In the end, the probabilities (and phases) are combined to obtain the final wave function amplitudes $\psi(\boldsymbol{s})$.

This sampling procedure yields independent, identically distributed samples. Conversely, MCMC methods may suffer from highly correlated consecutive samples,[5] which is problematic for complex probability distributions, for example, that are far from Gaussian. Consider a quantum state that spans several separated regions in the Hilbert space, where the probability is concentrated. In this case, Markov chains generally remain stuck in one of the regions, given that it must take several penalizing steps to travel from one to another, resulting in a highly inaccurate sampling. By contrast, the direct sampling procedure can seamlessly draw spin configurations belonging to all the regions according to the probability distribution, yielding much better samples.

While the first autoregressive models used in quantum physics were built from masked dense or convolutional layers, mimicking the so-called PixelNet architecture [265], recurrent neural networks (RNNs) were later introduced [266]. RNNs, inspired by natural language processing models, are generative models. We can draw a simple analogy between correlations in sentences, with their elements living in a large "word space," and spin configurations. Considering spin systems and supposing some hidden structure, quantum states are correlated, and their base elements are elements of the Hilbert space. Following this analogy, Hibat-Allah *et al.* introduced RNN wave functions [266], obtaining impressive results even for frustrated systems. An example of such an architecture is shown in Fig. 5.3. Clearly, many different NN architectures can work. A plethora of different architectures have been implemented as NQSs in

[5]MCMC methods such as the Metropolis–Hastings algorithm generally rely on performing modifications to the spin configurations to sample subsequent states. Therefore, this process could yield highly correlated consecutive samples that may have a negative impact on the results. To compute expectation values, we need to estimate the autocorrelation time to draw uncorrelated samples from the resulting chain. Moreover, when approaching a phase transition points, such methods suffer from critical slowing down, making the sampling of uncorrelated configurations unfeasible in many situations.

recent years, such as CNNs [267] and group CNNs [268], which can conveniently implement certain symmetries, as we describe in more detail in Section 5.2.4.

5.2.3 Capacity and entanglement

As we show in Sections 5.1 and 5.2, there is a whole plethora of methods to represent quantum many-body wave functions. For instance, only in NQSs, we already encounter substantial differences between ansäzte based on different NN architectures. Hence, a natural question arises regarding their expressive capacity and how they compare to each other.

Tensor networks have been a recurrent tool to perform this kind of studies, provided that they are well established and characterized, and they constitute a theoretical language to study quantum many-body phenomena. For this reason, there has been a significant community effort to study the relationship between TNSs and NQSs [262, 269, 270], which provides insight about the expressive capacity of NQSs [271]. Following the first introduction of NQSs implementing RBMs [246], early works focused on finding direct relationships between various kinds of RBM-based states and TNSs [262, 269]. It has been proven that NNs can efficiently approximate, in logarithmic space-complexity, all efficiently contractible TNs with arbitrary precision. Therefore, for every TNS there exists an equivalent NQS of polynomial size. Conversely, there are quantum states that can be efficiently described by NQSs, whose representation in terms of TNSs requires an exponential amount of parameters. Hence, TNSs are a subset of NQSs [270], as depicted in Fig. 5.4.

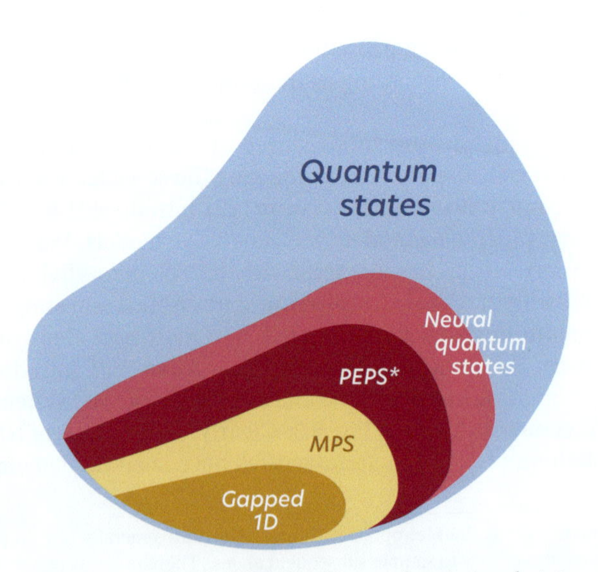

Figure 5.4 Expressive capacity of different classes of variational states, as explicitly proven in [270] by mapping TNSs to NQS. PEPS* refers to a subclass of projected entangled pair states, a generalization of MPSs. Adapted from [270].

As a measure of expressive capacity, we often rely on the entanglement that the different ansätze can capture. For instance, the mean field ansatz is, by construction, a product state [recall Eq. (5.3)]. Hence, it cannot capture entanglement, while TNSs and NQSs do not have such strong local limitations. This way, TNSs and NQSs have higher expressive capacity than the mean field ones.

More precisely, we study the entanglement scaling captured by the different ansätze. In a generic quantum many-body system with density matrix ρ, the entanglement entropy is defined as

$$S(\rho) = -\operatorname{Tr}\left[\rho \log_2 \rho\right] , \tag{5.25}$$

which is zero for any pure state. Let us consider a partition of the system in two subsets: I and its complementary O, as well as the reduced density matrix $\rho_I = \operatorname{Tr}_O[\rho]$. In general, ρ_I represents a mixed state, which can have nonzero von Neumann entanglement entropy. For a generic quantum state, the entanglement entropy of ρ_I grows with the volume of the cut. Thus, it corresponds to a *volume-law* scaling. NQSs can efficiently capture such scaling with architectures ranging from very basic shallow ones, such as RBMs [261], to more modern and deeper approaches, such as CNNs or RNNs [271]. Some traditional ansätze, such as the Jastrow wave function [see Eq. (5.20)], can also capture volume-law entanglement.[6]

However, there is a subclass of states in which the entanglement entropy grows, at most, as the boundary area between two regions. This is known as *area-law* scaling, and it is a property of ground states of local and gapped Hamiltonians [272]. Due to their local nature, TNSs can efficiently capture area-law entanglement [273]. For instance, in a one-dimensional chain, the area of the cut between two subsystems is constant, meaning that the entanglement entropy is a constant, and not an extensive quantity in the infinite volume limit. For an MPS with bond dimension χ, the von Neumann entanglement entropy of any possible bipartition of the system is bounded from above as $S \leq \mathcal{O}(\log_2 \chi)$, thus making the MPS ansatz an excellent candidate to study one-dimensional systems.

We can understand most differences between the ansätze at an intuitive level by, simply, looking at how they are built. In Fig. 5.5, we provide a pictorial representation of the different connections that some ansätze can draw in a bidimensional system. Clearly, the MPS ansatz, depicted in Fig. 5.5(b), is the most locally restricted one, as it can only account for nearest-neighbor connections in a snake-like pattern. This effectively limits the entanglement that MPS can capture. The RNN ansatz, illustrated in Fig. 5.5(d), while it is limited to parse the state in the same pattern as the MPS, it has the freedom to account for additional information, allowing it to capture richer correlations.

By contrast, other ansätze such as the Jastrow or RBM wave functions, respectively illustrated in Fig. 5.5(a) and (c), can draw connections between arbitrary sites. The Jastrow ansatz can account for all possible pairs in the system, regardless of the distance. Then, the RBM ansatz is a generalization of the Jastrow by means of an auxiliary hidden layer of variable size. Through the hidden neurons, the ansatz is no longer limited to pairs, and it can actually consider up to all-to-all connections. This nonlocal character allows them to capture volume-law entanglement.

[6]The Jastrow ansatz is, indeed, a specific case of RBM wave function with $N(N-1)/2$ hidden neurons [269].

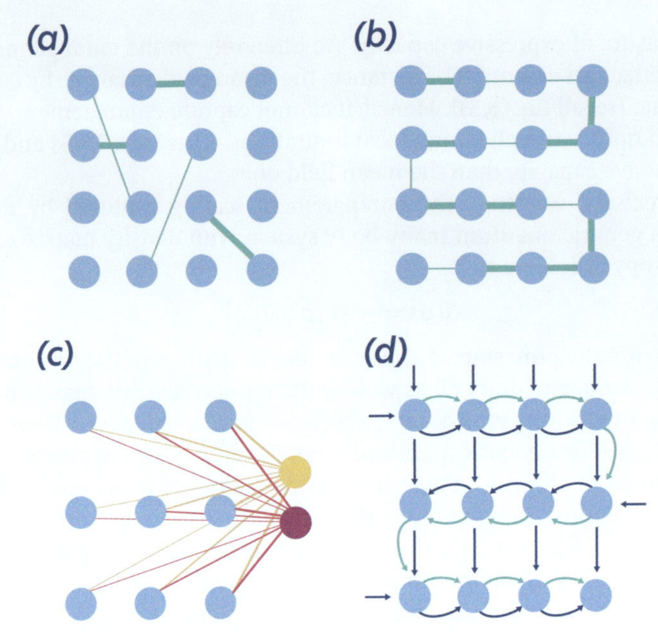

Figure 5.5 Schematic representation of various ansätze inspired by [269, 274]. (a) The Jastrow ansatz draws connections between all possible pairs of sites. (b) The MPS ansatz draws connections between nearest-neighbor sites along a line. (c) The RBM ansatz connects all the sites to every hidden neuron, illustrated in different colors. (d) The RNN ansatz processes the state sequentially, following the green arrows. The dark blue arrows indicate the flow of information within the model. Arrows without a starting site correspond to free parameters.

5.2.4 Implementing symmetries

Encoding symmetries in NQSs allows us to reduce the number of parameters in the NNs, restricting the region of the Hilbert space that our ansatz can cover to a subspace of interest, thus improving the accuracy of the results. Let us first explain what we mean by symmetry in this context. Consider a group of linear transformations: If the Hamiltonian is invariant under those transformations, meaning that they all commute with the Hamiltonian, then the Hamiltonian is symmetric under that group. Some of the most common symmetries in lattice models are the translation symmetry, the rotation symmetry in two or higher dimensions, the inversion or reflection symmetries, and all the compositions of those.

It is possible to show that if the Hamiltonian commutes with a set of operators $\mathcal{T} = \{\hat{T}_k\}_{k=1}^{K}$, its ground state must also be left invariant under those transformations. Therefore, the amplitude for two configurations $|s\rangle$ and $|s(k)\rangle = \hat{T}_k\,|s\rangle$ must be invariant for any $\hat{T}_k$: $\Psi_\theta(\mathbf{s}) = \Psi_\theta(\hat{T}_k\mathbf{s})\,\forall k$.[7] One way to introduce symmetries in our NQS is to take, as output, the sum of the ansatz evaluated on the set of symmetry-invariant

[7]Up to a phase on the right-hand side, but let us ignore it for convenience.

configurations $\{s(k)\}$. This way, the output is invariant by construction. However, we have not improved the performance of our model with this approach.

A more efficient approach is to build a dense layer at the beginning of the NQS model that fulfills the symmetry condition [267]. We can use this technique to encode any symmetry group isomorphic to a polynomially large permutation group. This usually comprises the set of all *lattice symmetries* (translations, rotations, reflections, and so on), global discrete symmetries, such as a global spin-flip, but it cannot deal with continuous symmetries, such as SU(2). For instance, we can implement translation symmetries through a convolution with a kernel as wide as the system itself. Since the convolution is translationally invariant by definition, it's easy to see that the output of the layer is symmetry-invariant.

In the case of RBMs, we can rearrange the terms of Eq. (5.22) to make it invariant under the elements of a symmetry group. Let us denote the transformation of local spins as $\sigma_j(k) = \hat{T}_k \sigma_j$. We can write our symmetry-invariant amplitude as:

$$\Psi_\theta(\mathbf{s}) = \sum_{\mathbf{h}} \exp\left(\sum_{f=1}^{\alpha} b_v^f \sum_{k=1}^{K} \sum_{j=1}^{N} s_j(k) + \sum_{f=1}^{\alpha} b_h^f \sum_{k=1}^{K} h_{f,k} + \sum_{f=1}^{\alpha} \sum_{k=1}^{K} h_{f,k} \sum_{j=1}^{N} W_j^f s_j(k) \right), \tag{5.26}$$

where we have explicitly written the matrix products as sums. The important point here is that $\boldsymbol{b}_v^f, \boldsymbol{b}_h^f$ are now vectors in a feature space with $f = 1, \ldots, \alpha$, and the matrix $\boldsymbol{W}^f$ is now of size $\alpha \times N$.[8] If we consider translational invariance, the corresponding symmetry group is made of N translation operators. In this case, $\boldsymbol{W}^f$ can be seen as a kernel acting over configurations to which we have applied the translation operators.

There are, in fact, many ways to directly encode symmetries in NQSs. For more details, we refer to [275] for general feedforward networks, or [267] for an example with CNNs.

5.2.5 Limitations

Similar to many ML methods, NQSs suffer from an interpretability problem, as we have discussed extensively in Section 3.5 for generic ML approaches. However, there has been substantial progress since the seminal paper from Carleo and Troyer [246]. For instance, a recent work introduced an interpretable RBM ansatz, in which the authors add some correlation terms to the expression of the probability distribution given by Eq. (5.22). With this, one can look at the magnitude of the trained parameters to understand which correlations are more important for the given physical problem [276].

Another route to gain further understanding of NQSs is through the mapping of NQS architectures to other known ansätze, such as TNSs. By exploiting this idea, works have shown NQSs to be capable of describing volume-law states, as opposed to TNSs, as we show in Section 5.2.3. In terms of expressive capacity,

[8] Note that this expression is equivalent to Eq. (5.22) with $M = K \times \alpha$ hidden variables.

NQSs can efficiently represent the ground states of one-dimensional gapped Hamiltonians, all the TNSs that are efficiently contractable in classical computers, and volume-law states [270]. Furthermore, there have been found exact NQS representations of several interesting phases of matter, such as topological states and stabilizer codes [261, 263, 269, 277–281]. However, not all quantum states can be efficiently represented in terms of NQSs. For instance, we cannot represent random states, since they do not have structure.

Another important aspect is choosing the right NN architecture and training strategy for the problem. For instance, we may be interested in implementing certain symmetries, as we have discussed in Section 5.2.4. However, on a given problem, a certain NQS ansatz may be well-suited for the task, but the training procedure can fail numerically. Some works have analyzed the training procedure involving stochastic reconfiguration [282]. Others have found that states involved in the dynamics of non-integrable systems are not representable by various architectures, but their entanglement structure can be recovered, hinting at a different limit from the built-in limitation on entanglement in TN-based ansätze [283].

These findings, along with state-of-the-art results, point toward a superior expressive power of NQS over existing simulation methods, but many research routes have to be taken to fully understand their capabilities, much like many ML methods discussed in this book.

5.3 Applications

In this section, we present various applications of NQS, ranging from the ground-state search to quantum-state tomography, featuring real-time dynamics, quantum circuits, and fermionic systems. In addition to presenting how the methods described previously apply to such problems, we provide results for each application and compare them to other state-of-the-art methods. By doing this, we hope to show both the potential and versatility of NQS approaches, which is still a young field of research.

5.3.1 Finding the ground state

As common in many ML tasks, we define a loss function $\mathcal{L}$ that depends on the trainable parameters of the NN. In this situation, this corresponds to the variational energy, that is, the expectation value of the Hamiltonian in the variational state:

$$\mathcal{L}(\boldsymbol{\theta}) = E(\boldsymbol{\theta}) = \langle \Psi_{\boldsymbol{\theta}} | \hat{H} | \Psi_{\boldsymbol{\theta}} \rangle. \tag{5.27}$$

This choice of the loss function is naturally introduced since it follows from the variational principle in quantum mechanics.

The *variational principle* states that given a Hamiltonian $\hat{H}$, the energy $E(\boldsymbol{\theta})$ of a variational wave function $|\Psi_{\boldsymbol{\theta}}\rangle$ is greater or equal than the exact ground-state energy, that is,

$$E(\boldsymbol{\theta}) = \frac{\langle \Psi_{\boldsymbol{\theta}} | \hat{H} | \Psi_{\boldsymbol{\theta}} \rangle}{\langle \Psi_{\boldsymbol{\theta}} | \Psi_{\boldsymbol{\theta}} \rangle} \geq E_0. \tag{5.28}$$

Therefore the energy is a valid loss function, as the *lower* the expectation value of the energy, *the better the approximation is.*[a]

[a] We stress that the principle is only valid when computing expectation values exactly. When the energy is computed as a stochastic average, its estimated average can be lower than the exact energy. Nevertheless, as discussed previously, the increase in the number of samples and the use of an efficient sampling approach systematically reduce fluctuations below the exact energy.

In fact, having a loss function strongly rooted in a principle of physics is crucial since it also allows us to compare different methods. By looking at the variational energy, we can, for example, understand how a method performs at solving a given problem: If the resulting approximate ground-state energy is significantly lower than what was found by alternative techniques, we can be reasonably sure that the solution found is of better quality. Following the general discussion on expectation values of operators, the variational energy can be stochastically approximated as

$$E(\theta) \approx \frac{1}{M} \sum_{i}^{M} E_{\mathrm{loc}}(\boldsymbol{s}^{(i)}), \tag{5.29}$$

where E_{loc} is the local estimator and is defined as $E_{\mathrm{loc}}(\boldsymbol{s}) = \sum_{\boldsymbol{s}'} \langle s|\hat{H}|s\rangle \frac{\langle s'|\Psi\rangle}{\langle s|\Psi\rangle}$. We aim to minimize this loss function by means of gradient-based optimization algorithms. The energy gradients can also be written in terms of expectation values[9]

$$\begin{aligned} \frac{\partial E(\theta)}{\partial \theta_p} &= 2\,\mathrm{Re}\left[\langle E_{\mathrm{loc}}(\boldsymbol{s}) O_p^*(\boldsymbol{s})\rangle - \langle E_{\mathrm{loc}}(\boldsymbol{s})\rangle\langle O_p^*(\boldsymbol{s})\rangle\right] && (5.30)\\ &= 2\,\mathrm{Re}\left[\langle (E_{\mathrm{loc}}(\boldsymbol{s}) - \langle E_{\mathrm{loc}}(\boldsymbol{s})\rangle) O_p^*(\boldsymbol{s})\rangle\right], && (5.31) \end{aligned}$$

where we have assumed that the parameters are real[10] and that θ_p is the pth parameter of the NQS. The diagonal operator $\hat{O}_p$ is defined as

$$O_p(\boldsymbol{s}) = \frac{\partial}{\partial \theta_p} \log\langle s|\Psi_\theta\rangle = \langle s|\hat{O}_p|s\rangle. \tag{5.32}$$

We also remark that the expression used in Eq. (5.31) has the form of a covariance and therefore is particularly stable with respect to sampling noise. Most notably, when the wave function is close to the exact ground state, statistical fluctuations in the local energy are suppressed, implying that also statistical fluctuations of the gradients are small because of the covariance structure.

[9] Computationally speaking, one does not need to store in memory the full Jacobian matrix $O_p(\boldsymbol{s})$ but can compute this gradient directly through the vector-Jacobian product (reverse-mode differentiation) of the vector $v = E_{\mathrm{loc}}(\boldsymbol{s}) - \langle E_{\mathrm{loc}}(\boldsymbol{s})\rangle$) and the Jacobian $O_p(\boldsymbol{s})$. This approach considerably lowers the memory and computational cost. For more details, see Section 7.1.

[10] The requirement of real parameters is not actually necessary. For complex parameters, the expression is very similar, though care has to be taken to consider non-holomorphic ansätze. Note that many common ansätze, particularly most autoregressive ones, are not holomorphic. Discussion on this can be found in the appendix of [284].

The learning algorithm is thus straightforward. First, we initialize the weights $\theta^{(0)}$. Next, at each step a sequence of M configurations is sampled according to the Born distribution: $P(\boldsymbol{s};\theta^{(s)}) \sim \boldsymbol{s}^{(1)} ... \boldsymbol{s}^{(M)}$. This can be done with a Markov chain or with direct sampling techniques as explained above.

The next step is to compute the mean of the local energy $E(\theta)$, which gives us the estimate of the expectation value of the Hamiltonian. Additionally, the gradients can also be calculated as shown in Eq. (5.31). For the last step, we can use a gradient-based optimizer of our choice, to update the parameters for the next step, that is, $\theta_p^{(s+1)} = \theta_p^{(s)} - \eta \frac{\partial E(\theta)}{\partial \theta_p}$ for vanilla gradient descent where η is the learning rate.

The procedure is repeated until it converges to a minimum of the energy landscape. Here, there is no training dataset as the approach is not based on any supervised learning method. The presented task is in fact to determine the optimal (unknown) wave function by drawing samples from the associated Born distribution and using an NN to model the state itself. These steps are summarized in Algorithm 7. Note that this algorithm is not the most commonly used, as it is less accurate than imaginary-time evolution, which is presented in Section 5.3.3.

Algorithm 7 Ground state search with NQS

```
Initialize θ randomly
for i = 1 to n_steps do
    Generate M samples according to some algorithm (usually a Markov chain)
    Calculate the gradient of the energy ∂E(θ)/∂θ_p
    Update parameters as θ_j ← θ_j − η∂E(θ)/∂θ_j (or with a more advanced update
rule)
end for
return Optimized parameters θ
```

5.3.2 Real-time evolution

Neural quantum states can also be used to variationally perform real-time evolution [285] through a procedure known as time-dependent variational Monte Carlo (t-VMC) [246, 267, 286, 287]. This is of particular interest for nonequilibrium quantum dynamics of closed, interacting quantum systems. Studying these problems enables one to understand critical properties, entanglement spectra, and many other physical quantities of interest in complex many-body quantum systems. The problem one wants to solve is to integrate the time-dependent Schrödinger equation ($\hbar = 1$ in the following) in time, using a parametrized wave function $|\Psi_\theta(t)\rangle$:

$$\mathrm{i}\frac{d\,|\Psi_\theta(t)\rangle}{dt} = \hat{H}\,|\Psi_\theta(t)\rangle\,, \tag{5.33}$$

that is, find the correct form of $|\Psi_\theta(t)\rangle\ \forall t$. Expanding Eq. (5.33) at first order in δ and taking the inner product with $\langle s|$, we obtain:

$$\Psi_\theta(t+\delta)(\boldsymbol{s}) = 1 - \mathrm{i}\delta\langle s|\hat{H}|\Psi_\theta(t)\rangle + O(\delta^2) \tag{5.34}$$

$$= 1 - \mathrm{i}\delta E_{\mathrm{loc}}(\boldsymbol{s}) + O(\delta^2), \tag{5.35}$$

where we used $E_{\rm loc}(\boldsymbol{s})$ as defined in Eq. (5.29) in Section 5.3.1. To obtain a good variational approximation of the state at the next time step, $t+\delta$, it is natural to define the cost function $\mathcal{L}(\tilde{\theta})$:

$$\mathcal{L}(\tilde{\theta}) = \text{dist}\left(|\Psi_{\tilde{\theta}}\rangle, |\Psi_{\theta}(t+\delta)\rangle\right), \tag{5.36}$$

with θ the variational parameters at the previous time step, and $\tilde{\theta}$ variational parameters to be determined. The loss function can be minimized analytically, if the time step is sufficiently small. One starts by noticing that $\tilde{\theta} = \theta + \delta\dot{\theta} + \mathcal{O}(\delta^2)$. One can therefore expand the variational state $|\Psi_{\tilde{\theta}}\rangle$ at first order and take its inner product with $\langle s|$, much like we did for Eq. (5.34):

$$\Psi_{\theta+\tau\dot{\theta}}(\boldsymbol{s}) = \left(1 - \delta\dot{\theta}\partial_{\theta}\Psi_{\theta}(\boldsymbol{s})\right)\Psi_{\theta}(\boldsymbol{s}) + \mathcal{O}(\delta^2). \tag{5.37}$$

We need to consider a distance measure between the two states $|\Psi\rangle$ and $|\phi\rangle$ which can be efficiently sampled. There is a certain freedom in this choice, which can lead to slightly different variational principles. For an extensive discussion of these issues, see [285]. By considering the infidelity, keeping in mind that for many NQS architectures the quantum states are unnormalized, we have[11]:

$$\text{dist}\left(|\Psi\rangle, |\phi\rangle\right) = 1 - \frac{\langle\phi|\Psi\rangle\langle\Psi|\phi\rangle}{\langle\phi|\phi\rangle\langle\Psi|\Psi\rangle}. \tag{5.38}$$

By plugging Eqs. (5.37) and (5.34) into the distance of Eq. (5.38), minimizing it, and keeping the leading terms in δ one obtains an equation giving the time derivative of the variational parameters $\dot{\theta}$, enabling high-order integration methods such as Runge–Kutta integration:

$$\boldsymbol{S}\dot{\theta} = -\mathrm{i}\boldsymbol{f} \tag{5.39}$$

with the *quantum geometric tensor* $\boldsymbol{S}$ and the vector $\boldsymbol{f}$, whose elements are given by:

$$S_{pp'} = \langle O_p^* O_{p'}\rangle - \langle O_p^*\rangle\langle O_{p'}\rangle \tag{5.40}$$

$$f_p = \langle E_{\rm loc} O_p^*\rangle - \langle E_{\rm loc}\rangle\langle O_p^*\rangle \tag{5.41}$$

with the O_ps given by Eq. (5.32) and $E_{\rm loc}$ is the local energy. The vector $\boldsymbol{f}$ is the gradient of the local energy with respect to the variational parameters and, in analogy with classical mechanics, it is often called the vector of *forces*. The spectrum of the geometric tensor instead encodes the (linearized) curvature of the variational space, akin to the Hessian discussed in Section 4.5. For a full derivation and an in-depth discussion of the time-dependent variational principles, see [285]. The spectrum of $\boldsymbol{S}$ has been extensively studied in the case of ground state optimization with RBMs [282], where it has been connected to the different regimes of considered physical system. In practice, solving the linear system Eq. (5.39) implies either using an iterative solver (e.g., conjugate gradient) or a direct solver (e.g., QR factorization). An important practical numerical issue is that the matrix $\boldsymbol{S}$ is often singular. Some techniques have been

[11] Rigorously, one should consider the Fubini–Study metric, but taking this distance leads to the same equations.

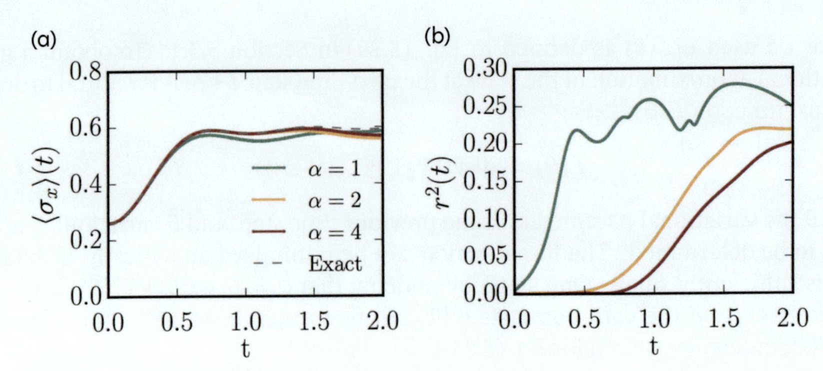

Figure 5.6 Critical quench dynamics with an RBM, preparing the system in the ground state of $\hat{H}_{\text{TFI}}$ for $h_i/J = 1/2$, then suddenly quenching to $h_f/J = 1$. This excites many eigenstates of the system at criticality (which exhibit long-range correlations, making the dynamics difficult to capture). (a) Average magnetization along x for different values of the density of hidden neurons α of the RBM. (b) Integrated error, systematically reduced by increasing α. Taken from [246].

found to regularize $\boldsymbol{S}$ and obtain more stable dynamics [267, 288]. In all cases, since only stochastic averages for both $\boldsymbol{S}$ and $\boldsymbol{f}$ are available, stable, and accurate long time dynamics are still a challenge for NQS [288]. As an example, in Fig. 5.6 we show the quench dynamics of a one-dimensional spin chain, subject to the Ising Hamiltonian with a transverse field:

$$\hat{H}_{\text{TFI}} = -J \sum_j \hat{\sigma}^z_j \hat{\sigma}^z_{j+1} + h \sum_j \hat{\sigma}^x_j. \tag{5.42}$$

Here, J is the nearest-neighbor coupling and h is the transverse field strength. This model exhibits a second-order phase transition in one dimension at $h = J$, which separates a ferromagnetic (for $J > 0$, or antiferromagnetic for $J < 0$) phase from a paramagnetic phase, with all spins aligned along the transverse field for $h \gg J$. The critical quench dynamics can be investigated by preparing the system in an eigenstate of the Hamiltonian for some value of $h = h_i$, then suddenly switching the Hamiltonian parameters to $h_f/J = 1$. As seen in Fig. 5.6, an RBM captures the dynamics up to about $Jt = 1.5$, and increasing the number of hidden layers α systematically improves the precision. As mentioned, more recent results have also been obtained using a CNN on a two-dimensional system, whose dynamics are a challenge for TN methods [267].

5.3.3 Imaginary-time evolution

The first-order optimization scheme presented in Section 5.3.1 to estimate the ground state of many-body systems can be improved to yield more accurate results. For this purpose, it is useful to consider an imaginary-time evolution through Wick's rotation $t \to i\tau$:

$$|\psi_\theta(\tau)\rangle = \exp(-\tau \hat{H}) |\psi_\theta(0)\rangle, \tag{5.43}$$

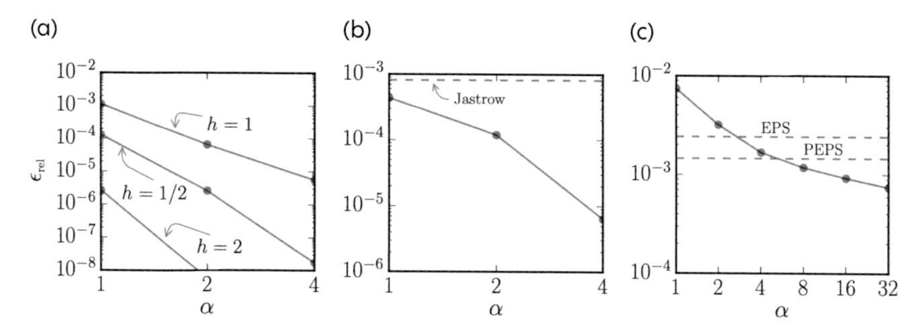

Figure 5.7 Relative error of the ground state search using imaginary time evolution with respect to the exact solution. (a) Relative error for the ground state of the transverse-field Ising model, for various values of h/J as a function of the number of $\alpha = M/N$, M being the number of hidden units for $N = 80$ spins. (b) Same plot for the ground state of the antiferromagnetic Heisenberg model, compared to another variational ansatz, the Jastrow wave function. (c) Comparison with state-of-the-art TN results, showing that NQS perform at least as good. Taken from [246].

where $\hat{H}$ is the Hamiltonian and τ is a real number. It can be shown that $\lim_{\tau\to\infty} |\phi(\tau)\rangle = |\phi_0\rangle$, with $|\phi_0\rangle$ being the exact ground state of the Hamiltonian and provided $|\langle\psi_\theta(\tau)|\phi_0\rangle| \neq 0$. Furthermore, it can be shown that the convergence of imaginary-time evolution toward the exact ground state is exponentially fast with τ, thus offering a systematic way to find the ground state. Analogously to real-time evolution, imaginary-time evolution can also be performed variationally. This leads to the same type of equation as Eq. (5.39):

$$\boldsymbol{S}\dot{\boldsymbol{\theta}} = -\boldsymbol{f}, \tag{5.44}$$

with the factor i missing due to the form of the exponent in Eq. (5.43). Hence, a very similar procedure is obtained as for the real-time evolution in which we can update the weights according to the update given by Eq. (5.44). To summarize both real and imaginary time, the algorithm for variational time evolution is given in Algorithm 8. In the case of imaginary-time evolution, the algorithm is typically modified in such a way that the S matrix is regularized by adding a constant, $\Lambda > 0$, proportional to the identity: $\boldsymbol{S} \to \boldsymbol{S} + \Lambda\boldsymbol{I}$. In this case, one recovers the stochastic reconfiguration method, as originally introduced by Sorella [289, 290]. In Fig. 5.7, results are shown for imaginary-time evolution performed on the transverse-field Ising model and the Heisenberg model [246]. These results show two important features: (i)

Algorithm 8 Real (ξ = i) or imaginary ($\xi = 1$) time evolution algorithm for NQS

```
θ ← random initialization
for i = 1 to n_steps do
    Calculate S_pp' and F_p
    Get θ̇ by inverting the equation Σ_p' S_pp' θ̇_p' = −ξF_p (and possibly regularizing)
    Update θ using an ODE integrator
end for
```

using an RBM ansatz, the relative error can be systematically reduced by increasing $\alpha = M/N$, with M the number of hidden units, and (ii) the results achieve a higher precision than state-of-the-art TN methods.

5.3.4 Fermionic systems

Fermions constitute one of the two fundamental types of elementary subatomic particle. As such, fermionic systems are ubiquitous in many-body quantum physics, high-energy physics, as well as chemistry. However, the classical simulation of fermionic systems is difficult because fermionic operators obey anticommutation relations, which constrain the wave functions of fermionic systems to be antisymmetric under particle exchange. Due to the infamous sign problem [291], all the known quantum Monte Carlo methods become extremely expensive computationally for fermionic systems.

Variational approaches with NQS for fermions may be divided into two classes: (i) using the first quantization, one may impose antisymmetry on the wave function by constructing it as a Slater determinant, or (ii) going to the second quantization and mapping the fermionic many-body Hamiltonian to a spin Hamiltonian. The first approach builds on a formulation of the problem in terms of a continuous state space. Many impressive results for realistic systems have been obtained by employing such an approach [292–294]. For further reading, we recommend [295] which is a recent review on the topic. In the following, we will focus our discussion on the second approach based on the second quantization.

As explained, a convenient approach to simulate fermionic systems is based on mapping the fermionic degrees of freedom to spins. A generic protocol is the Jordan–Wigner transformation, which enables us to map fermionic problems to interacting spin problems. Note that there are many other possible transformations that have mostly been developed in the context of quantum simulation, such as the Bravyi–Kitaev encoding [296]. Historically, this technique has been used to solve spin models [297]. Here we do the opposite: We map fermionic operators to spin operators to use NQSs and the techniques presented throughout the chapter to solve the corresponding many-body problem. This approach does not suffer from the sign problem because the antisymmetry is directly encoded in the terms of the Hamiltonian. However, ultimately the approach is limited by the difficulty of the resulting spin problem that may include complicated, nonlocal interactions.

Let us consider the creation and annihilation fermionic operators acting on site j, $\hat{c}_j^\dagger$ and $\hat{c}_j$, respectively. The Jordan–Wigner transformation prescribes:

$$\hat{c}_j = \left(\prod_{k=1}^{j} \hat{\sigma}_k^z \right) \hat{\sigma}_j^- , \tag{5.45}$$

$$\hat{c}_j^\dagger = \left(\prod_{k=1}^{j} \hat{\sigma}_k^z \right) \hat{\sigma}_j^+ , \tag{5.46}$$

where $\hat{\sigma}_j^\pm = \frac{1}{2}\left(\hat{\sigma}_j^x \pm i\hat{\sigma}_j^y\right)$ denote the spin raising and lowering operators. The first term in the transformation provides a phase that can be ± 1 depending on whether the number of occupied fermionic modes is even or odd in sites $k = 1, \dots, j$. We can conveniently rewrite this term using the relations $\hat{\sigma}_j^z = 2\hat{\sigma}_j^+\hat{\sigma}_j^- - 1$ and $\hat{\sigma}_j^+\hat{\sigma}_j^- = \hat{c}_j^\dagger\hat{c}_j = n_j$. This ensures that the resulting operators fulfill fermionic anticommutation relations.

For example, using this transformation, we can map a Hamiltonian describing free fermions in one dimension

$$\hat{H} = -\frac{1}{2}\sum_j \hat{c}_j\hat{c}_{j+1}^\dagger + \hat{c}_j^\dagger\hat{c}_{j+1} \tag{5.47}$$

to an interacting spin Hamiltonian of the form

$$\hat{H} = -\frac{1}{2}\sum_j \hat{\sigma}_j^+\hat{\sigma}_{j+1}^- + \hat{\sigma}_{j+1}^+\hat{\sigma}_j^-. \tag{5.48}$$

In this form, we can implement all the methods described throughout Chapter 5.

The main issue with this transformation is that it does not generalize well to arbitrary dimensions. In higher dimensions, the Jordan–Wigner transformation results in a nonlocal spin Hamiltonian which cannot be tackled with most standard techniques. Different mappings for fermionic degrees of freedom that work in higher dimensions have been proposed. These are not general mappings but instead are tailored to specific problems. For example, we can map local Hamiltonians in more than one dimension to local bosonic Hamiltonians for certain specific gauge theories [298, 299]. Another approach to avoid nonlocal spin Hamiltonians in high dimensions based on using an ancillary system has been considered in [300]. In this case, the degrees of freedom of the ansatz are separated into a main system and an auxiliary system. By doing so, one can build a local spin Hamiltonian from a local fermionic Hamiltonian at the expense of having a larger Hilbert space.

Calculating the electronic structure of molecules is a timely fermionic problem that is important for applications in chemistry [247]. It is one of the oldest instances of a quantum many-body problem first mentioned by Dirac in 1929 [245]. In this context, one is generally interested in finding the ground-state energy as a function of some physical parameter, such as the distance between two nuclei for a diatomic molecule. This way, by looking at the minimum of the energy, one can find out what the stable geometry of the molecule of interest is. Usually, the interacting fermionic Hamiltonian is defined on a lattice and takes the following form:

$$\hat{H} = \sum_{i,j} t_{ij}\hat{c}_i^\dagger\hat{c}_j + \sum_{i,j,k,l} U_{ijkl}\hat{c}_i^\dagger\hat{c}_j^\dagger\hat{c}_k\hat{c}_l, \tag{5.49}$$

where t_{ij} is a single-body hopping term, U_{ijkl} is a two-body interaction strength, and $\hat{c}_i$ ($\hat{c}_i^\dagger$) is a fermionic annihilation (creation) operator for mode i.[12] The Jordan–Wigner transformation changes this Hamiltonian to the form

$$\hat{H} = \sum_r a_r \mathbf{S}_r, \tag{5.50}$$

[12] We can formulate it in a real-space or a momentum-space basis, which we leave unspecified for the sake of generality.

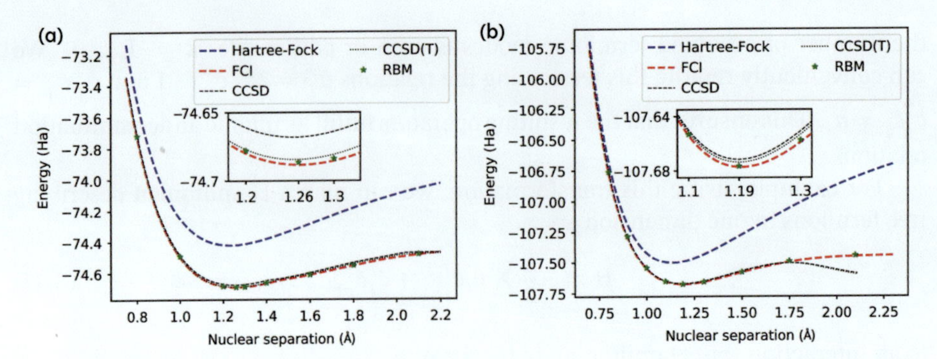

Figure 5.8 Ground-state energies of a (a) C_2 and (b) N_2 molecule measured in Hartree (Ha) as a function of nuclear separation, given by various techniques compared with results obtain by an RBM ansatz with $M = 40$ hidden units. CCSD(T): coupled-cluster approaches, FCI: full-configuration interaction. Taken from [247] under the CC BY 4.0 DEED license.

where a_r are scalar coefficients and $\mathbf{S}_r$ are Pauli strings composed of elements of the set of single-qubit operators $\{\mathbb{1}, \hat{\sigma}^x, \hat{\sigma}^y, \hat{\sigma}^z\}$. In other words, we now have an interacting spin problem and, while the resulting Hamiltonian is not necessarily k-local, it can be shown that the local energy can still be estimated efficiently. Therefore, these mappings are amenable to variational searches using NQS.

Another recent improvement in the field is the construction of explicit autoregressive ansätze, as presented in Section 2.4.6, for fermions [301]. Here, authors consider a basis set of spin-orbitals consisting of M spatial orbitals, each existing for upward and downward spins, that is, spanned by $|\mathbf{x}_k\rangle = \left|x_k^{1\uparrow}, x_k^{1\downarrow}, x_k^{M\uparrow}, x_k^{M\downarrow}\right\rangle$. Each spatial orbital can be treated as a single unit that can take on four possible values, denoted v_k^i. Therefore, the logarithm of the corresponding wave function coefficients takes on the following form:

$$\ln \psi_k = \sum_{j=1}^{M} \ln |\psi_j(v_k^j \mid v_k^1 \dots v_k^{j-1})| + \mathrm{i}\phi(v_k^1 \dots v_k^M). \tag{5.51}$$

This ensures a proper normalization of the wave function provided the conditional amplitudes ψ_i are normalized, therefore a direct sampling scheme which leads to improved results.

Finally, in Fig. 5.8, we show physical results obtained using NQS for fermionic systems. The dissociation curves (ground-state energies) as a function of the nuclear separation for the molecules C_2 and N_2 provided by various numerical methods are displayed. By using an RBM ansatz and a simple Jordan–Wigner transformation, one is able to recover results that are competitive with recent full configuration interaction calculations, which demonstrates the versatility and power of NQSs.

5.3.5 Classical simulation of quantum circuits

Another promising direction for NQSs is the classical simulation of quantum circuits, which we introduce in Section 8.2.1. Indeed, current classical simulation methods for large quantum circuits (of the order of at least 50 qubits) rely on TN methods that are explicitly restricted by entanglement. In particular, TNSs cannot capture volume-law entanglement scaling, which quickly arises in quantum circuits, whereas certain NQS architectures such as deep CNNs can [270]. In this context, NQSs can be investigated in two somewhat orthogonal directions: One could use quantum circuits to probe the limits of their capacity and trainability, and NQSs can be used to push the classical simulation limits of quantum hardware.

Let us consider a quantum circuit defined by a set of D gates $\mathcal{G} = \{\hat{G}_i\}_{i=1}^{D}$, each gate being defined by a unitary operator $\hat{G}_i$. After each gate, the variational state must be updated so as to capture the application of the previous gate. The following variational distance must therefore be minimized for each gate:

$$\mathcal{L}(\tilde{\theta}) = \operatorname{dist}\left(|\Psi_{\tilde{\theta}}\rangle, \hat{G}_i|\Psi_{\theta}\rangle\right), \tag{5.52}$$

with $\tilde{\theta}$ the parameters to be optimized, θ the parameters of the previous variational state, and $\hat{G}_i$ the unitary operator corresponding to gate G (for instance, for a NOT gate, $\hat{G}_i = \hat{\sigma}^x$). One can apply this procedure for each gate G and obtain the output state at the end of a circuit, after D optimizations. Note that $\hat{G}_i$ must be a k-local gate, or else the minimization procedure cannot be carried out (this is reminiscent of ground state search). This is rarely a problem, since universal gate sets can be constructed with only single- and two-qubit gates. With this condition, minimizing Eq. (5.52) closely resembles the ground state optimization. One can develop this expression using the infidelity, see Eq. (5.38), and obtain:

$$\mathcal{L}(\tilde{\theta}) = 1 - \langle G_{\text{loc}}(\theta, \tilde{\theta})\rangle_{|\Psi_{\tilde{\theta}}|^2} \langle G_{\text{loc}}(\tilde{\theta}, \theta)\rangle_{|\Psi_{\theta}|^2} \tag{5.53}$$

$$\text{with } G_{\text{loc}}(\theta, \tilde{\theta})(s) = \sum_{s'} \frac{\langle s|\hat{G}_i|s'\rangle}{\Psi_{\tilde{\theta}}(s)} \Psi_{\theta}(s'), \tag{5.54}$$

where we have defined G_{loc} as a local estimator, similarly to the procedure described for a ground state search. As often in ML, the minimization of $\mathcal{L}(\tilde{\theta})$ may be inaccurate, which reduces the overall fidelity of the simulation. Using an RBM architecture, however, not all gates need to be approximated through minimization of the loss above. Some gates can be applied "analytically," that is, it is possible to find the exact update on the parameters of the network so as to match the applied gate. In general, it is impossible to realize these exact updates for all gates of a universal gate set, or else one could simulate any quantum circuit with a RBM with infinite precision by obtaining the exact parameter update for each gate. For example, let us consider a Z gate acting on spin s_j defined by the operator $\hat{G} = \hat{\sigma}^z_j$. The action of such an operator on a basis state $|s\rangle$ is simply $\hat{G}\,|s\rangle = (-1)^{s_j}\,|s\rangle$. The RBM parameters before the gate are defined as $\theta = (\boldsymbol{b}_v, \boldsymbol{b}_h, \boldsymbol{W})$, and the parameters after the gate are defined as $\tilde{\theta} = (\tilde{\boldsymbol{b}}_v, \tilde{\boldsymbol{b}}_h, \tilde{W})$. The parameter update is given by the solution of the following equation (with C a constant):

$$\Psi_{\tilde{\theta}}(s) = C\,\langle s|\hat{G}|\Psi_{\theta}\rangle \tag{5.55}$$

$$e^{\tilde{b}_{v,j}s_j} = C(-1)^{s_j} e^{b_{v,j}s_j}, \tag{5.56}$$

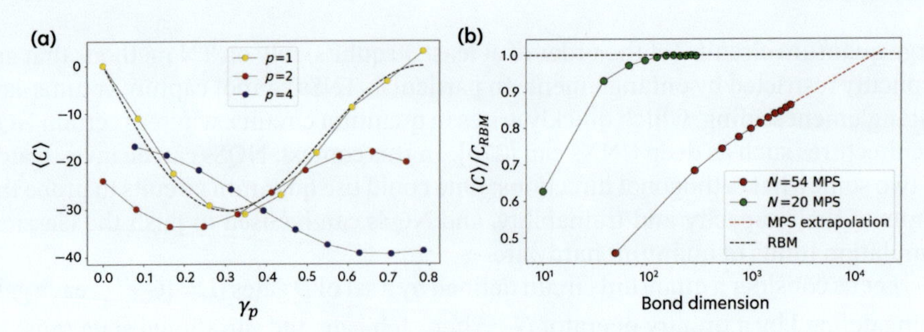

Figure 5.9 Results for the classical simulation of QAOA for a three-planar graph. Panel (a) shows $\langle C \rangle$, the approximated cost function for various values of p, the depth of the quantum circuit. The dashed line corresponds to the exact simulation of the $p = 1$ quantum circuit which the RBM simulation accurately reproduces for this task. (b) Estimation of the required bond dimension in an MPS simulation of the QAOA circuit to match the accuracy of the RBM. For $N = 54$ qubits, the required bond dimension is of about 10^4, which amounts to using billions of parameters, whereas the RBM uses a few hundred. Taken from [303] under the CC BY 4.0 DEED license.

which is simply $\tilde{b}_{v,j} = b_{v,j} + i\pi$ for $C = 1$. The simplification in the previous equation is due to the fact that this gate acts trivially on the other parts of the RBM amplitude, defined in Section 5.2.1. Details of how to apply other gates analytically can be found in [302] and [303]. In this last reference, authors classically simulate the circuit corresponding to the quantum approximate optimization algorithm (QAOA) [304][13] using an RBM ansatz. This quantum algorithm enables one to access the solution of a certain class of combinatorial optimization problems. The corresponding circuit, which is quite shallow, can be implemented on current hardware [305]. In Fig. 5.9, one can see that the results obtained by simulating the quantum circuit with the RBM ansatz closely match the result of the exact simulation, enabling one to find the solution of the optimization problem for large systems. Authors also estimate a significant advantage over TN methods.[14] Indeed, in the left panel of Fig. 5.9, one can see that the required bond dimension required to reach the same results as that of the RBM would quickly become dauntingly large when using an MPS.

Alternatively, the authors of [306] have proposed to simulate quantum circuits using a transformer architecture. A transformer is a deep learning model that adopts the mechanism of self-attention, differentially weighting the significance of each part of the input data [307]. Using this framework, a practical algorithm to simulate quantum circuits using a transformer ansatz responsible for the most recent breakthroughs in natural language processing was introduced in [306]. This framework allows for the simulation of circuits that build Greenberger–Horne–Zeilinger and linear graph states of up to 60 qubits.

[13]QAOA is a variational quantum algorithm designed to tackle combinatorial optimization problems.
[14]Based on an extrapolation of numerical simulation data.

5.3.6 Open quantum systems

The idea of using NNs to represent quantum states was also applied to open quantum systems. An open quantum system is a physical system that interacts with an environment, for example, an array of atoms interacting with an electromagnetic field. Rather than describing the full system + environment ensemble, one is generally only interested in the properties of the system (in the example above, the atoms) and one only keeps an effective description of its interaction with the environment (the field). This description enables one to understand effects such as decoherence. In the Born–Markov approximation, the time evolution of an open quantum system is given by the Lindblad master equation [308]:

$$\partial_t \hat{\rho} = -\mathrm{i}[\hat{H}, \hat{\rho}] + \sum_{i=1}^{D} \left(\hat{J}_i \hat{\rho} \hat{J}_i^\dagger - \frac{1}{2}\{\hat{J}_i^\dagger \hat{J}_i, \hat{\rho}\} \right) \equiv \mathbf{L}[\hat{\rho}], \tag{5.57}$$

where $\hat{\rho}$ is the system density operator, $\hat{H}$ is the system Hamiltonian ($\hbar = 1$), and $\hat{J}_i$ are so-called jump operators, that describe the system–environment interaction. We have also defined $\mathbf{L}$, the Liouvillian, which is to open quantum systems what the Hamiltonian is to closed quantum systems (up to an i) – their time evolution generator. Many works focus on finding the steady state that corresponds to the state $\hat{\rho}$ which satisfies $\mathbf{L}[\hat{\rho}] = 0$, or the dynamics of particular systems, which means one must in general integrate Eq. (5.57) in time, analogously to nonequilibrium dynamics of closed systems. A first difficulty one faces is finding a correct representation for $\hat{\rho}$ in terms of an NN. Indeed, a density matrix is harder to represent than a wave function because it has to be Hermitian, semi-positive, and of trace one. In fact, a general method to encode a density matrix into arbitrary NNs has still not been found. The key point of these works is that one can always purify a density matrix, and write its elements as

$$\langle s | \hat{\rho} | s' \rangle = \sum_{s'} \Psi(s, s') \Psi^*(s, s'), \tag{5.58}$$

with $\Psi(s, s')$ the purification that belongs to the joint Hilbert space composed of the system and an imaginary ancilla (whose Hilbert space is of at least the same dimension as the system's Hilbert space). With an RBM architecture, one can encode the purification $\Psi_\theta(s, s')$ with an NN, and the RBM architecture enables one to trace out the ancilla analytically spins s' without explicitly performing the summation which in general requires exponentially many operations. For more details, see [309–312].

A second more recent approach proposes to view the density matrix as a probability distribution over positive operator-valued measures (POVMs), and represent the resulting distribution using models employed in general density estimation, such as RNNs and ARNNs. In this formalism, the density matrix is simply written as:

$$\hat{\rho}_\theta = \sum_a p_\theta(a) \hat{M}_a, \tag{5.59}$$

with $\hat{M}_a$ POVMs that belong to a chosen complete set of POVMs. This could, for example, be all the operators composed as tensor products of the Pauli operators and the identity. In this picture, one simply needs to encode the probability distribution $p_\theta(a)$ with an NN. This POVM-based representation is motivated by the fact that the density matrix can be viewed as an ensemble of 4^N measurements, which is naively how experimentally one performs tomography to reconstruct the density matrix. This method alleviates the constraint on using an RBM for open systems but does not guarantee positivity of the density matrix, which can lead to unphysical states. However, a certain number of results using the POVMs encoding are promising [313, 314], and understanding in which regime they work best is a key research direction.

Finding the steady state(s) of open quantum systems is both challenging, due to the daunting size of the Liouvillian one would need to diagonalize (4^N for a spin system of N spins), and interesting, for example, for the study of dissipative phase transitions [315]. Once a parametrization $\hat{\rho}_\theta$ of the density matrix is constructed, one can simply minimize the following cost function:

$$\mathcal{L}(\theta) = \langle \hat{\rho}_\theta \mathbf{L}^\dagger \mathbf{L} \hat{\rho}_\theta \rangle \,. \tag{5.60}$$

The obtained state corresponds to the zero eigenvalue is zero of the Liouvillian, which is the steady state. For details about the procedure and how to retrieve the gradients, see [309–312]. As one can see in Fig. 5.10, this method has been applied to the dissipative version of the transverse-field Ising model, with good results for both a POVM approach and an RBM approach. In the former case, the expressive power of the network is higher, but the positivity of the density matrix is not enforced. A clear picture of when each approach fails or succeeds is still lacking and is an important research direction. The dynamics of open quantum systems is not described in detail here, but stochastic reconfiguration can also be used for open systems in both the RBM [312] and POVM [314] approaches.

5.3.7 Quantum-state tomography

The future quantum technologies are fueled by quantum resources such as coherence, entanglement, or Bell nonlocality. One of the main challenges is the experimental certification of such properties for a given unknown quantum state [316–327]. Information about quantum resources is encoded in the density matrix of the state, which can only be reconstructed based on finite-statistic measurements – this process is known as quantum-state tomography [328–338]. Density matrix reconstruction is a challenging task – with increasing system size, the number of required measurements scales exponentially. The field of quantum-state tomography has been entered by artificial NNs proposing supervised deep-learning approaches [339–344]. The following paragraphs introduce basic concepts of NNs-assisted quantum-state tomography.

Quantum-state tomography is the process of density matrix reconstruction based on finite-statistic measurement data. Due to the exponential growth of the Hilbert space with increasing system size, the number of required measurements also scales exponentially. However, with the help of deep NNs, the density matrix reconstruction can be done with a polynomial number of measurements.

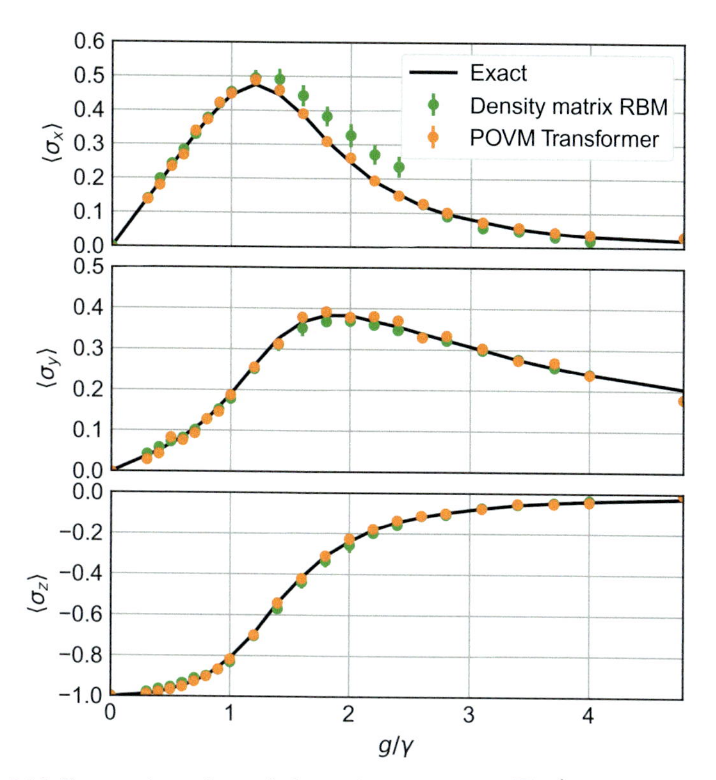

Figure 5.10 Expectation values of observables $\hat{\sigma}_k = 1/N \sum_i \hat{\sigma}_i^k$, $k \in \{x, y, z\}$ at the steady state of the open system described by the transverse-field Ising Hamiltonian and jump operators $\hat{J}_i = \sqrt{\gamma}\hat{\sigma}_i^-$ as a function of g/γ, with g the magnetic field strength. Results are shown for both the purified RBM and POVM approaches with a transformer network. Taken from [313].

Let us consider the task of reconstructing a wave function $|\psi\rangle$ from a limited number of snapshots $|\psi(s)|^2$ obtained by performing projective measurements in some basis spanned by $|s\rangle = |s_1, s_2, \ldots, s_N\rangle$, with s_i some local quantum numbers and N the size of the system. Then, the task, in the NQS language, is simply to minimize:

$$\min_{\theta} \operatorname{dist}(|\psi_\theta\rangle, |\psi\rangle) \tag{5.61}$$

with θ being some variational parameters, and $|\psi_\theta\rangle$ the variational state to optimize, parametrized by an NN. The architecture of this network is left unspecified here, and all architectures work provided training can be performed efficiently. Many distances can be considered, but we here focus on the Kullback–Leibler (KL) divergence (see Section 2.3) as was first presented in the work by Torlai *et al.* [345]. It is defined as:

$$\mathrm{D}_{\mathrm{KL}}(p||q) = \sum_{x \in \mathcal{P}} p(x) \log \frac{p(x)}{q(x)} \tag{5.62}$$

for two probability distributions p and q, defined on the same space $\mathcal{P}$. The application to quantum states is straightforward, as one can obtain probability distributions from the Born rule, that is, $p_\theta(s) = |\psi_\theta(s)|^2$, $q(s) = |\psi(s)|^2$. By taking x to be configurations s in some set S of snapshots, one can simply minimize:

$$\mathrm{D_{KL}}(\theta) = \sum_{s \in S} |\psi_\theta(s)|^2 \log \frac{|\psi_\theta(s)|^2}{|\psi(s)|^2}, \tag{5.63}$$

which concludes one possible approach.

However, recall that a quantum state is not simply a probability distribution. A probability distribution can *always* be defined from a quantum state but *not the reverse*. More explicitly, we want to reconstruct the full quantum state whose amplitudes are $\psi(s) = \sqrt{Q(s)}e^{i\phi(s)}$. By minimizing Eq. (5.63), *information about the phase, $\phi(s)$, is lost.* This difference is crucial and is at the heart of many issues in learning quantum states. As mentioned in Chapter 5, learning the phase of a frustrated quantum state is challenging [346]. The elegant solution to this problem is to *consider measurements performed in different bases.* Indeed, the form of the quantum state in a different basis involves the interference between amplitudes in different bases. Hence, matching the probability distribution defined by snapshots in different measurement bases leads to the correct quantum state as long as the bases contain enough information about the quantum state. Mathematically, one can simply replace Eq. (5.63) by:

$$\mathrm{D_{KL}}(\theta) = \sum_{B} \sum_{s \in S_B} |\psi_\theta^B(s)|^2 \log \frac{|\psi_\theta^B(s)|^2}{|\psi^B(s)|^2}, \tag{5.64}$$

where S_B is the set of snapshots of the quantum state in basis B, and $\psi^B(s) = \langle s| \hat{U}_B |\psi\rangle$ with $\hat{U}_B$ a unitary operator. Then, gradients are found as usual, either with automatic differentiation or analytically with simple models such as RBMs.

In Fig. 5.11, various observables are shown for a synthetic state and a reconstructed state. The synthetic state approximates the ground state of the Heisenberg model in a triangular lattice, that authors of the corresponding work generated with TN simulations [91]. The reconstructed state was obtained employing the ideas presented in this section with an RNN architecture (for more details, see the Introduction or Chapter 5 and Section 7.2), using a POVM representation of quantum states (for more details, see Section 5.3.6). Note that the approach has been extended to reconstruct mixed states [347], although additional care must be taken to avoid issues related to positivity of the reconstructed density matrix, similar to what was presented above.

Experimentally, one can implement this strategy by applying rotations with, for instance, laser pulses, and then measure the system repeatedly. In [348], authors demonstrate the first state reconstruction from experimental data from a programmable array of Rb atoms, using an RBM architecture. Here snapshots of the wave function in the $\hat{\sigma}^z$ basis are obtained through site-resolved fluorescence imaging. A challenge that arises when using real data is that noise is introduced, which comes from measurement errors, leading to a set of snapshots $|\psi(s)|^2$ that imperfectly match the state of the system. This is taken care of in this work by adding a noise layer to the NN, with which the snapshots are transformed to filter out the noise during training. At the expense of increasing the total number of parameters in the network, this is

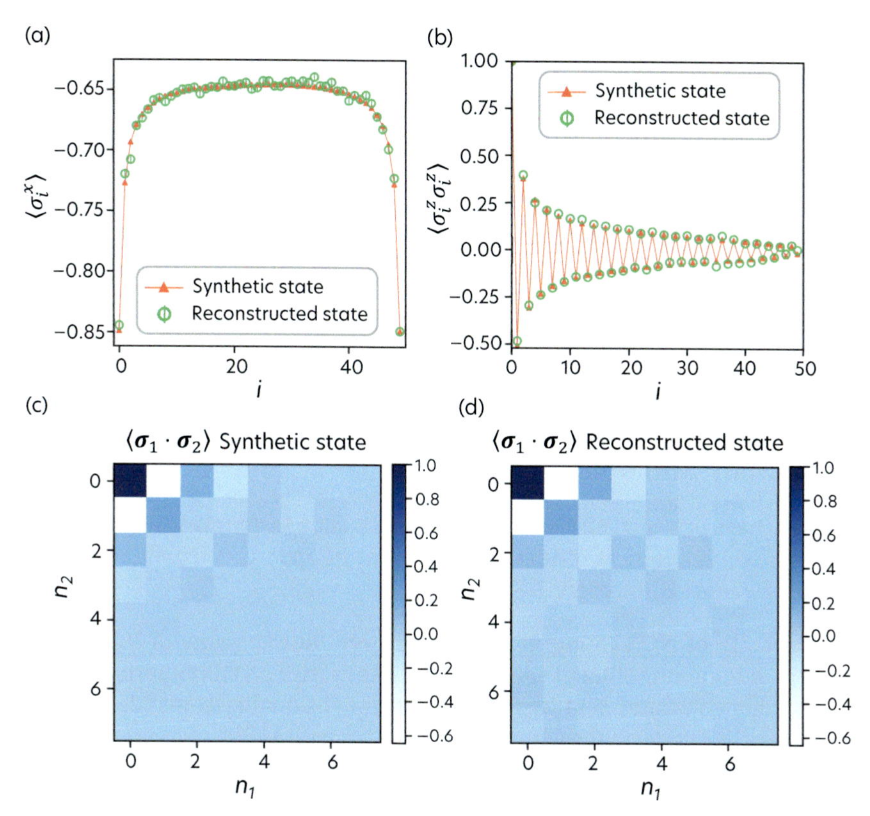

Figure 5.11 Various observables corresponding to the ground state of the Heisenberg model on a triangular lattice with $N = 50$ spins. (a) Average magnetization along x for each spin i. (b) Spin–spin correlation function between the spin at site 1 with spins at site i. (c) and (d) Average spin–spin correlation between the first and the ith spins. One can see that all observables are reproduced with a very high precision. Adapted from [91] with permission from Springer Nature.

a quick and easy strategy to deal with experimental noise, enabling high-fidelity state reconstruction. Since quantum-state tomography with NQS has been proposed, substantial efforts have been made to implement it in real-world experiments. For more details on the experimental challenges of such proposals, see for instance, [349–353]. The underlying principle behind these approaches is that with a polynomial number of bases B and a polynomial number of snapshots, one should be able to reconstruct states belonging to a certain class (not fully random states, for instance, which contain almost no structure). As underlined previously, this class is not exactly known and is the subject of current research. To draw an analogy with images, images are not fully random; they contain a lot of hidden structure, that can be learned by a properly designed and trained NN. The hope is that the same is true for quantum states, and investigating the limits of such techniques could also help us understand in more detail their hidden structure, beyond what has been found with entanglement properties through the study of TNs.

Finally, we mention the randomized measurement techniques that allow predicting selected properties of spin-1/2 quantum systems without reconstructing the full quantum state, the so-called shadow tomography (see [240, 354–360]). Shadow tomography allows estimation of the expectation value of the given observable based on data collected during repetitive measurements prepared in a randomly chosen basis of each spin separately. However, the number of required measurements scales exponentially with the locality of the operator averages to be reconstructed, albeit with a runtime that is typically better than the naive direct measurement of the observables from the data. In contrast to these approaches, in this section we have instead considered the task of training a low-dimensional representation of the full wave function from a limited number of measurements.

5.4 Outlook and open problems

We hope to have provided enough material to stimulate further research in the growing field of NQSs. Here is a non-exhaustive list of open problems and challenges related to the above discussion:

- **Capacity of NQS**. Some works have proven the capability of NQSs to represent volume-law entanglement, which means they could outperform TNSs for strongly correlated and two- and three-dimensional systems [261, 270, 271]. Others have proven the equivalence of RBMs with MPSs, meaning that the former cannot represent more states than the latter [262]. Even though general theorems have been found, knowledge about specific architectures is still rare, and understanding which architectures perform better on which problems is a crucial point. In addition, proving representativity does not mean that the models can be efficiently trained, thus understanding how the training of NQS models works is key.

- **Long-time dynamics**. Long-time dynamics remains a relatively untouched area for NQSs, due to stability issues of stochastic reconfiguration [288]. However, progress has been made thanks to regularization techniques [267]. Some works proposed infidelity minimization [287, 361], which enables going beyond stochastic reconfiguration for regimes where its performance is poor. In [362], a systematic bias that appears when performing time evolution was explained, which should stimulate progress in long-time dynamics, where ample results on large lattices are still lacking with NQS.

- **Open quantum systems**. No general method of encoding a density matrix into an arbitrary NN has been found yet; one is either forced to use an RBM, which has known limitations, or one can use a POVM approach, which may fail due to nonpositive density matrices.

- **Frustrated systems**. Finding the ground state of frustrated systems with an NQS approach has proved to be challenging [346], and understanding exactly how one can improve the optimization of the procedure to learn the phase (which has a nontrivial sign structure) is of particular interest.

- **Simulation of quantum circuits.** Few results have been obtained with networks other than RBMs, and investigating how different circuits affect the accuracy of the chosen ansatz can lead to results in two ways: understanding the complexity of a given circuit and the limitations of the chosen ansatz.
- **Quantum-state tomography.** Quantum-state tomography based on NNs is still in its infancy. So far, it has only been explored numerically on toy models and small experimental settings where traditional quantum-state tomography is still feasible. It is likely that its real benefits may emerge in the context of estimation of difficult quantities in quantum simulation. In this setting, the complexity of estimation arises because even simple quantities, such as energy and other correlation functions, can have high variance. This implies that some of these quantities have a sample complexity, which can grow quickly with the size of the system.
- **Quantum resources certification.** The generation of quantum resources can be performed dynamically by means of the one-axis twisting protocol [363, 364]. One-axis twisting can be implemented with ultracold atoms in optical lattices to generate many-body entanglement and many-body Bell correlations [365–370]. The challenge for this technique is to verify the quantum resources generated in many-qubit systems, which can be done with the help of DL [344].
- **Extension to continuous Hilbert spaces and bosonic systems.** For now most techniques and works have focused on systems with discrete degrees of freedom (such as spins). Extensions to continuous Hilbert spaces have been addressed, for example, in the context of quantum chemistry [293–295] and nuclear matter [371]. Efficient encodings for bosonic Hilbert spaces would also be of particular interest for photonic systems, for example, which are usually treated with mean-field-like approaches.
- **Applications in quantum information.** As mentioned previously, NQS have been used to simulate quantum circuits. They have also been applied to quantum codes [372] for quantum error correction and quantum communication. In this paper, the authors demonstrate that efficient quantum codes can be learned by NQS according to which noise channels a physical system is subject to. NQS have not yet been widely used for quantum information, and we expect them to be useful tools for this field in the coming years.

Further reading

- Carleo, G. & Troyer, M. (2017). *Solving the quantum many-body problem with artificial neural networks.* The original paper by Carleo and Troyer that introduced NQS [246].
- Becca F. and Sorella, S. (2017). *Quantum Monte Carlo Approaches for Correlated Systems.* A comprehensive book that includes details on quantum Monte-Carlo methods, and variational states [290].

- Vicentini, F. *et al.* (2021). *NetKet 3: Machine learning toolbox for many-body quantum systems.* The paper accompanying the open-source library NetKet 3, which contains an extensive discussion of how to implement several algorithms introduced in this chapter, as well as a collection of tutorials showing how to solve some benchmark problems with NQS [284, 373].
- Carrasquilla, J. & Torlai, G. (2021). *How to use neural networks to investigate quantum many-body physics.* A recent tutorial by Carrasquilla and Torlai that includes interesting applications and code snippets can help anyone who wants to start in the field [97].
- Carleo, G. (2017). Repository for example codes presented at the "Machine Learning and Many-Body Physics" workshop. Notes, exercises, and code produced for the 2017 Beijing workshop on Machine Learning and Many-Body Physics [374].

6 Reinforcement learning

So far, we have encountered multiple machine learning (ML) scenarios featuring supervised or unsupervised learning problems where we want to infer some labels, predict certain values, or find patterns in the data. In this chapter, we describe a different approach: *learning strategies.*

In the supervised learning framework, we can think of a student who learns from a teacher who knows the correct answers to all possible questions within a given domain. In this scheme, the student is limited by the teacher's knowledge and can never surpass it or address questions outside the teacher's expertise. To overcome this limitation, in reinforcement learning (RL), we remove the teacher and let the student try things out and learn from the resulting experience. We refer to the student as the *agent*, as it can actively take actions. Just like us humans, the agent learns from the interaction with an *environment*, understands the consequences of its actions, and finds strategies to achieve particular goals.

For instance, let us consider the case in which we teach an agent to play chess. A supervised learning approach would consist of training an ML model to reproduce the moves of recorded chess games by the best players in the world. In this setting, given a state of the game, that is, the position of the remaining pieces on the chessboard, the model predicts the move such reference players would make. However, this approach suffers from some major shortcomings. For example, there is no single optimal move for every situation, and the moves strongly depend on the game strategy adopted by the players. As a result, the agent may be unable to consistently execute a strategy through various actions. Additionally, the agent's performance is ultimately limited by the quality of the training data, meaning that it may be impossible to outperform the reference players. We refer to Section 6.6.2 for a related example.

Instead, we can let the agent play chess games, either against various opponents or even against itself, without providing any additional knowledge besides the rules. In that case, it develops its own understanding of the game and devises its own strategies. The resulting agent's potential is far superior to the previous one, as it is not limited by its teacher. Nevertheless, learning from experience may be challenging, provided that the quality of the actions is only assessed at the very end of the game when the outcome is decided: victory or loss.[1] Hence, the agent must develop a deep understanding of the long-term consequences of the actions based on the feedback from the sparse environment.

Framing problems as games to discover strategies has countless applications. In particular, control problems naturally fit this framework. However, we can design games to obtain any protocols or algorithms of interest, from new quantum experiments [375] to faster matrix multiplication or sorting algorithms [376, 377]. Here, we show how to tackle some paradigmatic problems in the field of quantum technologies with RL.

[1] In some cases, we may be tempted to add intermediate rewards, such as a bonus for taking out a piece from the opponent. However, in doing so, we effectively change the game and its goal, and, as a consequence, we might fail to find the optimal strategy of the original problem.

In this chapter, we introduce the field of RL. We start with an intuitive view on the concept of learning from experience and its mathematical foundations in Section 6.1. Then, we present two main approaches: value-based RL in Section 6.2 and policy gradient in Section 6.3. In Section 6.4, we combine the two paradigms, introducing actor-critic algorithms. Then, we provide an alternative approach to RL, projective simulation (PS), in Section 6.5. Finally, we present a series of application examples of RL in Section 6.6, featuring superhuman performance in games as well as various problems in quantum technologies.

6.1 Foundations of reinforcement learning

The general setting of any RL problem consists of two main elements: an *agent* and an *environment* that it interacts with, as illustrated in Fig. 6.1. The environment contains all the information defining the problem at hand, for example, the rules of a game, and it provides the agent with observations and feedback according to its *actions*. The environment defines the set of all possible *states*, $s \in \mathcal{S}$, which can range from an empty set, in the case of a stateless environment (see the first example in Section 6.6.1), to a multidimensional continuous space. For example, these could be all the possible configurations of a board game or all the possible combinations of joint angles in a robot.

The agent can observe (sometimes only partially) the state s of the environment, and it can choose an action a to perform, which may include the possibility of remaining idle. The action is chosen from the set of possible actions, $a \in \mathcal{A}$, which is defined by the environment and can be state-dependent. For instance, the action of pushing forward a pawn in chess is only possible if there is a free position in front of it. The actions may alter the state in which the environment is found, and they can have deterministic or stochastic outcomes. In the chess example, all the actions are deterministic. In contrast, in the case of a walking robot, the action to move forward may have different results: It can succeed in doing so, the robot may trip, or it may even remain idle with a certain probability due to a hurdle or malfunctioning. This information is encoded in the environment, and the agent may not have access to it.

Nevertheless, every time the agent performs an action, the environment provides it with an observation of the new state together with a feedback signal called *reward*, r. The reward can take any numerical value. It may depend on the previous state, the new state, and the action that was taken. The main purpose of the agent is to maximize the obtained rewards by the end of the task, and it is, therefore, the quantity that defines the objective task. Hence, the agent obtains higher rewards when accomplishing the objective task or progressing toward the goal, for example, winning a game, while it might receive penalties when performing harmful or bad actions, for example, losing a game.

The central objective of any RL problem is to learn the *optimal policy*, π^*, that maximizes the obtained rewards. A policy, π, dictates which actions to take given the observations and thereby defines the strategy followed by the agent.

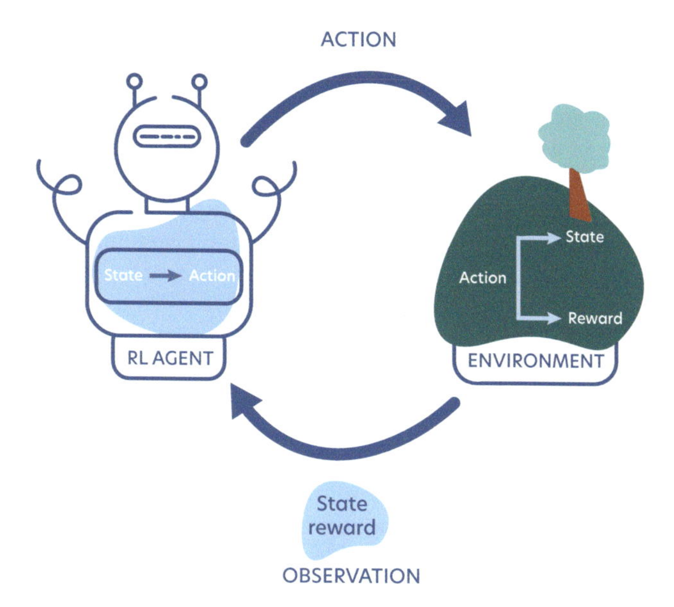

Figure 6.1 Overview of the basic RL setting. The agent receives an observation from the environment. Given the observation, it chooses the next action according to its policy. The environment determines the outcome of the action, and it returns an observation to the agent consisting of the new state and a potential reward.

In general, the policy can take any form. For example, it can be a table assigning the best possible action to every possible state or an ML model that, given a state, provides a probability distribution over all the possible actions. However, the learned policy is specific to the problem. We summarize the introduced key elements of the RL setting in Fig. 6.1.

Let us provide some insight on the main elements of the RL setting with a couple of examples. In the case of the chess game, the agent is one of the players. The environment models the game's rules, the opponent,[2] and its states that correspond to the piece positions on the board.[3] The state space contains all the possible board configurations that can be reached within a game, for example, excluding those where one of the kings is missing. The action space corresponds to all the possible legal moves that can be made at every turn. In this case, the agent does not obtain rewards until the game is resolved. At this point, the agent receives a positive or negative reward upon victory or defeat, respectively. In the case of a draw, the final reward could be zero or even negative. The goal is to learn the policy that yields the highest possible number of victories.

As a second example, we consider a robot trapped in a maze. The robot can only see its immediate surroundings and has to maneuver to reach a target location. In

[2]The opponent could be the same agent, which would play against itself, but each agent would perceive the other as part of their respective environment. This is known as *self-play*, and it helps explore new strategies faster.

[3]The state for chess can also contain extra information, such as whether castling is still possible. For the purpose of this example, and to keep it simpler, we restrict ourselves only to the piece positions here.

this case, the agent is the robot, and the environment models the maze, its walls, and the target location. The state is the current position of the agent plus its immediate surroundings, and the state space comprises all the reachable locations. The action space contains the moves in all possible directions, and the environment ensures that the agent does not cross the walls. Hence, moving into a wall would leave the agent in the same position and, therefore, would not modify the state. As a reward, we can provide the agent with a constant negative reward after every move to encourage it to take the least amount of steps toward the goal.

6.1.1 Delayed rewards

As we have previously introduced, the reward r is a key concept in RL. The agent learns to maximize the reward, and, therefore, it is the quantity that defines the problem. At a given discrete time t, the agent observes a state s_t and performs an action a_t according to its policy. Then, the environment presents the agent with a new state s_{t+1} and a reward r_{t+1}. Hence, r is time dependent, and it may depend on any of the other three quantities $r_t = r(s_{t-1}, a_{t-1}, s_t)$ (see Section 6.1.3 for further details).

So far, we have briefly talked about maximizing the rewards. For formalizing the RL objective, we need to introduce the notion of *delayed rewards*. They introduce the idea of "looking ahead" to the agent, allowing it to account for the future rewards obtained along a trajectory through the state space. However, we can penalize the rewards that are far into the future with a discount factor $\gamma \in [0, 1]$.

The discount factor weights the rewards according to their temporal separation. This way, immediate rewards have larger weights than those far into the future. The RL objective is to maximize the *discounted return*, defined as the weighted sum of future rewards

$$G_t = \sum_{k=0}^{T-t-1} \gamma^k r_{t+k+1}, \tag{6.1}$$

which accounts for the rewards obtained starting at time t until the final time T.[a]

[a] In RL, we typically consider finite trajectories. However, a discount factor $0 \leq \gamma \leq 1$ allows us to consider infinite trajectories $T = \infty$ with finite returns.

Notice that the return presents a recursive form that is essential for many RL algorithms

$$G_t = r_{t+1} + \gamma G_{t+1}. \tag{6.2}$$

This concept draws inspiration from human psychology, and it mimics our daily observation that far-term rewards, even if high, are less desired than near-term ones, for example, we favor procrastinating instead of reading this book. We can distinguish two limits: For a small discount factor, $\gamma \to 0$, the return becomes myopic, that is, immediate rewards predominate over any other possible future ones. On the

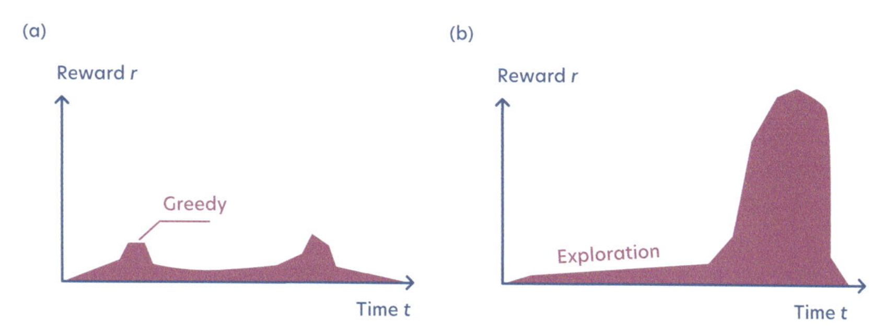

Figure 6.2 Impact of the discount factor, γ, in RL algorithms. (a) A myopic algorithm ($\gamma \to 0$) may settle for a greedy policy that leads to early immediate rewards, even if they are smaller than possible latter ones. (b) However, a long-term-oriented algorithm ($\gamma \to 1$) might sacrifice early rewards in favor of larger late ones.

other hand, large discount factors, $\gamma \to 1$, result in equal weights for early and late rewards, which encourage long-term-oriented strategies. This includes, in particular, the deliberate choice to perform a few seemingly suboptimal choices in the beginning that, however, result in a far greater final return. We depict the two cases in Fig. 6.2.

Hence, the discount factor strongly affects the resulting policy. In fact, it defines the RL task, as the agent aims to maximize the return, introduced in Eq. (6.1). Nevertheless, we often rely on trial-and-error methods to find the discount factor that best suits our needs.

6.1.2 Exploration and exploitation

In RL, we encounter a trade-off between exploration and exploitation. To maximize the return, the agent must *exploit* its knowledge about good strategies. However, the agent must *explore* other different actions to improve them or even discover better strategies in the future.

However, a learning algorithm cannot rely on exploration alone, as it would be reduced to a brute-force search algorithm. Conversely, in a case of pure exploitation, the agent would blindly commit to the first working strategy that it found, even if it was highly suboptimal. Hence, we need to find a balance between both regimes in which the agent can try several actions and progressively favor the best ones. This way, the exploration is conducted around the most promising areas of the state and action spaces, heavily reducing the amount of experience that the agent must gather to find the optimal policy.

A common strategy to balance exploration and exploitation is the so-called ε-greedy policy. In this case, the agent follows its policy to perform actions (exploits), and it may take a random action (explores) with probability $\varepsilon \in [0, 1]$ at any point. This approach encompasses both paradigms: For $\varepsilon = 1$, we have full exploration, whereas we have full exploitation for $\varepsilon = 0$. By tuning ε, we interpolate between both regimes. A common practice is to start with high ε, to enforce early exploration, and decrease it during the training process.

6.1.3 Markov decision processes

All RL problems are modeled by the same underlying mathematical structure: Markov decision processes (MDPs). They constitute a general framework to model environments with a notion of sequentiality between states. In such environments, the future is independent of the past, given the present. This is known as the *Markov property*.

> In essence, the *Markov property* means that the current state is a sufficient statistic containing all the required information relevant to the possible evolution of the environment. In particular, we do not have any memory effects from previously visited states. Formally, at any time step t,
>
> $$p(s_{t+1}|s_0, \dots, s_t) = p(s_{t+1}|s_t). \tag{6.3}$$

Mathematically, an MDP is a tuple $(\mathcal{S}, \mathcal{A}, p, G, \gamma)$, respectively, denoting the state space $\mathcal{S}$, the action space $\mathcal{A}$, the *dynamics* p, the set of total returns G, and the discount factor γ. In this formalism, the return G, together with the discount factor γ, determines the objective, and p describes the dynamics of the environment,

$$p(s', r|s, a) = p(s_{t+1} = s', r_{t+1} = r|s_t = s, a_t = a)\,, \tag{6.4}$$

which corresponds to the joint probability of observing a new state s' and obtaining a reward r by performing action a in state s. For fully deterministic environments, $p(s', r|s, a)$ is either zero or one.

From Eq. (6.4) we can derive all the relevant information about the environment. For instance, *state-transition probabilities* are a central quantity in many RL algorithms:

$$p(s'|s, a) = \sum_r p(s', r|s, a)\,. \tag{6.5}$$

Furthermore, it allows us to determine the reward functions. In Section 6.1.1, we briefly introduce the reward function $r(s, a, s')$. In the most general form, the reward is jointly determined with the state s', as shown in Eq. (6.4).[4] However, in many cases, we may need to consider the expected rewards for state–action pairs and state–action–next-state triplets:

$$r(s, a) = \sum_r \sum_{s' \in \mathcal{S}} r p(s', r|s, a)\,, \tag{6.6}$$

$$r(s, a, s') = \sum_r r \frac{p(s', r|s, a)}{p(s'|s, a)}\,. \tag{6.7}$$

In the iterative interaction between the agent and environment, the agent chooses the actions according to a *policy*. The policy is a mapping from states to the probability of performing each possible action

[4]In stochastic environments, the reward can be inherently sampled from a probability distribution. Consider the game of blackjack: With the same hand (state), the action of settling may have different rewards depending on the opponent's hand (environment). Hence, the reward is stochastic.

$$\pi(a|s) = p(a_t = a \,|s_t = s)\,. \tag{6.8}$$

In the limit of deterministic policies, $\pi(a|s)$ is one for a single action and zero for the rest.

The goal in RL is to modify the policy with the experience gathered from interacting with the environment to achieve the goal. This interaction generates *trajectories* of the form

$$s_0, a_0, r_1, s_1, a_1, r_2, s_2, a_2, \dots, s_T\,,$$

where all states, actions, and rewards are random variables. This way, the agent performs a trajectory through the state-action space $\tau = a_0, s_1, a_1, \dots, s_T$ with probability

$$p(\tau) = \prod_{t=0}^{T-1} p(s_{t+1}|s_t, a_t)\pi(a_t|s_t)\,, \tag{6.9}$$

starting from an initial state s_0. We denote the discounted return associated with the trajectory as $G(\tau) = \sum_{t=0}^{T-1} \gamma^t r_{t+1}$.

This entire formalism holds assuming the Markov property from Eq. (6.3), which implies that the environment is memoryless. However, we may encounter situations in which the environment has certain memory effects, such as games in which the execution of a sequence of actions yields an additional effect at the end. In these cases, we may recover the Markov property by considering an extended state space that already includes the memory. In return, this implies that even deterministic Markovian dynamics on the full state space can give rise to nondeterministic and non-Markovian dynamics on the smaller state space.[5]

6.1.4 Model-free versus model-based reinforcement learning

We can distinguish between two main paradigms in RL: model-free and model-based RL. In the first setting, the agent does not have any information about the underlying mechanisms of the environment, and it must purely learn by trial and error. In the second one, the agent either has access to a *model of the environment* or builds one from the gathered experience. Then, the agent can use this model to plan ahead, inferring the result of a sequence of actions before executing any of them, to choose the best possible ones.

Although we focus on model-free RL in the remainder of the chapter, we briefly elaborate on how to exploit the knowledge of a model. Building a model of the environment provides the agent with an enhanced understanding of the problem and can potentially help it face new situations. For example, in a case where an agent juggles a set of balls, if it has a good model of the laws of physics, it is much easier for it to learn to juggle a new set of balls with different shapes and weights.

These models can take various forms, but a general formulation is fully characterizable MDPs. This way, the model approximates the dynamics of the underlying MDP of the problem. In some situations, the true model is too complex to be grasped, and we may simply try to approximate the parts of the dynamics that are the most relevant to the problem. An example of a simple model would be an ML algorithm

[5] An analogous situation is encountered in the discussion of open quantum systems: Nonunitary dynamics in the subsystems arise despite a global unitary evolution of the system and its bath.

that predicts both the expected next state and the reward (s_{t+1}, r_{t+1}) given the current state and an action (s_t, a_t) at any time step t. Such a model allows us to predict the outcome of a series of future actions given the current state, and we can train it in a supervised way directly from the experience gathered by the agent.

In continuous-action spaces, the model provides a direct connection between the input action and the received reward, allowing us to employ backpropagation methods to maximize the return instead of mere sampling from the environment. See Section 7.1 for examples illustrating the process. In the case of discrete-state spaces, the model typically takes the form of a search tree that we can explore to our advantage. Models are especially convenient when the interaction cost with the environment is very high, such as realizing a physical or chemical experiment. In these cases, we try to augment our dataset of actual samples from the environment with artificial samples drawn from the model to minimize the total sampling costs.

However, we do not always have access to a model, or building one may not be in our interest. Building models is costly, especially in cases where we have limited knowledge about the environment, and they are only helpful when accurate. Furthermore, models are often tailored to specific problems. On the contrary, model-free RL algorithms come with the advantage that they are agnostic to the problem at hand and, thus, they are more versatile. Therefore, we focus on model-free RL for the rest of the chapter for pedagogical purposes, as they prove useful on the full range of RL tasks. In particular, we provide an introduction to value-based and policy-based RL in Sections 6.2.1 and 6.3, respectively.

6.1.5 Value functions and Bellman equations

As we have mentioned in Sections 6.1.1–6.1.4, the goal in RL is to find the optimal policy π^* that maximizes the return, introduced in Eq. (6.1). Such a clear objective allows us to define *value functions* that estimate how convenient it is for the agent to be in a given state or to perform a certain action to accomplish the task. For instance, consider the case in which we are looking for a treasure on a map. Being one step away from the treasure is, overall, much better than being ten steps away. However, not all actions in the close position are equally good, provided that one leads to the treasure, but the others move away from it. This is quantified by the expected future return that the agent may obtain, given the current conditions. However, given that the future rewards strongly depend on the actions that the agent will take, value functions are defined with respect to the policy.

The *state-value function*, $V_\pi(s)$, of a state s under the policy π is the expected return when starting at state s and following the policy π thereafter. We formally define it as

$$V_\pi(s) = \mathbb{E}[G_t | s_t = s, \pi] = \mathbb{E}\left[\sum_{k=0}^{T-t-1} \gamma^k r_{t+k+1} \middle| s_t = s, \pi \right]. \tag{6.10}$$

In a similar way, the *action-value function*, $Q_\pi(s, a)$, is the expected return when starting at state s, performing action a, and then following the policy π:

$$Q_\pi(s,a) = \mathbb{E}[G_t|s_t = s,\ a_t = a, \pi] = \\ = \mathbb{E}\left[\sum_{k=0}^{T-t-1} \gamma^k r_{t+k+1} \middle| s_t = s, a_t = a, \pi \right]. \tag{6.11}$$

The *advantage*, $A_\pi(s,a)$, is the additional expected return obtained by following an action a at state s, over the expected policy behavior:

$$A_\pi(s,a) = Q_\pi(s,a) - V_\pi(s). \tag{6.12}$$

The value functions fulfill a recursive relationship that is exploited by many RL algorithms, which stems from the recursive nature of the return Eq. (6.2). This allows us to write the state-value function $V_\pi(s)$ as a function of the next states

$$\begin{aligned} V_\pi(s) &= \mathbb{E}[G_t|s_t = s, \pi] = \mathbb{E}[r_{t+1} + \gamma G_{t+1}|s_t = s, \pi] \\ &= \sum_a \pi(a,s) \sum_{s',r} p(s',r|s,a) \left(r + \gamma\, \mathbb{E}[G_{t+1}|s_{t+1} = s', \pi]\right) \\ &= \sum_a \pi(a,s) \sum_{s',r} p(s',r|s,a) \left(r + \gamma V_\pi(s')\right) \\ &= \mathbb{E}[r_{t+1} + \gamma V_\pi(s_{t+1})|s_t = s, \pi]. \end{aligned} \tag{6.13}$$

We can do the analogous derivation for the action-value function $Q_\pi(s,a)$

$$\begin{aligned} Q_\pi(s,a) &= \mathbb{E}[G_t|s_t = s, a_t = a, \pi] = \mathbb{E}[r_{t+1} + \gamma G_{t+1}|s_t = s, a_t = a, \pi] \\ &= \sum_{s',r} p(s',r|s,a) \left(r + \gamma\, \mathbb{E}[G_{t+1}|s_{t+1} = s', \pi]\right) \\ &= \sum_{s',r} p(s',r|s,a) \left(r + \gamma V_\pi(s')\right) \\ &= \mathbb{E}[r_{t+1} + \gamma V_\pi(s_{t+1})|s_t = s, a_t = a, \pi], \end{aligned} \tag{6.14}$$

from which the relationship $V_\pi(s) = \sum_a \pi(a|s) Q_\pi(s,a)$ becomes evident. These are the *Bellman equations* for the value functions, and they lie at the core of RL as they define the relation between the value of a state s and its successors s', recursively capturing future information.

These concepts introduce the notion of partial ordering between policies. A policy π is better than another policy π' if it yields a higher return. Hence, $\pi > \pi'$ if and only if $V_\pi(s) > V_{\pi'}(s)\ \forall s \in \mathcal{S}$. Therefore, the optimal policy π^* is such that it is better than or equal to all the other possible policies.[6] Hence, the optimal policy maximizes the value function. Taking the Bellman equations, Eqs. (6.13) and (6.14), π^* is such that

[6]The ordering operator is not always defined between policies. Two policies π, π' cannot be ordered if and only if $\exists\ s, s' \in \mathcal{S}\ :\ V_\pi(s) > V_{\pi'}(s),\ V_\pi(s') < V_{\pi'}(s')$. However, for MDPs there always exist an optimal policy π^* s.t. $\pi^* \geq \pi\ \forall \pi$ [378].

$$\begin{aligned} V_{\pi^*}(s) &= \max_a \mathbb{E}[G_t|s_t = s, a_t = a, \pi^*] \\ &= \max_a \mathbb{E}[r_{t+1} + \gamma V_{\pi^*}(s_{t+1})|s_t = s, a_t = a, \pi^*] \\ &= \max_a Q_{\pi^*}(s,a)\,. \end{aligned} \tag{6.15}$$

Notice that in this new Bellman equation, there is a maximization over the first action, as opposed to the expectation over actions from Eq. (6.13). This is because the value of a state under the optimal policy must be equal to the expected return for the best action. In a similar way, we can find the Bellman equation for the action-value function $Q_\pi(s,a)$ for an optimal policy π^*. Together, they define the set of the *Bellman optimality equations*:

$$\begin{aligned} V_{\pi^*}(s) &= \max_a \sum_{s',r} p(s',r|s,a)\left[r + \gamma V_{\pi^*}(s')\right] \\ Q_{\pi^*}(s,a) &= \sum_{s',r} p(s',r|s,a)\left[r + \gamma \max_{a'} Q_{\pi^*}(s',a')\right]. \end{aligned} \tag{6.16}$$

These equations fulfill

$$\begin{aligned} Q_{\pi^*}(s,a) &= \max_\pi Q_\pi(s,a) \\ V_{\pi^*}(s) &= \max_\pi V_\pi(s) = \max_a Q_{\pi^*}(s,a)\,. \end{aligned} \tag{6.17}$$

We can define the optimal policy $\pi^*(a|s)$ and action a^* at a given state s as:

$$\begin{aligned} \pi^* &= \arg\max_\pi V_{\pi^*}(s) \\ a^* &= \arg\max_a Q_{\pi^*}(s,a)\,. \end{aligned} \tag{6.18}$$

The optimal policy π^* corresponds to the deterministic choice of the best action a^* for a given state s according to the optimal action-value function $Q_{\pi^*}(s,a)$. Due to the recursive nature of the value functions, a greedy action according to V_{π^*} or Q_{π^*} is optimal in the long term.

The Bellman optimality equations [Eq. (6.16)] are, indeed, a system of equations with one for every state. To solve them directly, we need to explicitly use $p(s',r|s,a)$.[7] If $p(s',r|s,a)$ is known, we know the underlying model of the system, and thus we deal with model-based RL, as discussed in Section 6.1.4. In a general model-free RL scenario, it is unknown and, as such, we need additional methods to solve them, such as the ones we introduce in Sections 6.2–6.4.

6.2 Value-based methods

In value-based RL, the goal is to obtain the optimal policy $\pi^*(a|s)$ by learning the optimal value functions, as in Eq. (6.18). This way, we start with an initial estimation

[7] Due to the maximization step in Eq. (6.16), this is a nonlinear optimization problem.

of the value function for every state, $V_\pi(s)$, or state–action pairs, $Q_\pi(s,a)$. Then, we progressively update them with the experience gathered by the agent following its policy.

Given that the value functions are defined with respect to a policy (recall Section 6.1.5), we need to define a fixed policy for this family of algorithms. A common choice is an ε-greedy policy, as introduced in Section 6.1.2, provided that the optimal policy is greedy with respect to the optimal value function. Hence, learning the value function for such policy provides us with the optimal one in the greedy limit.

One of the most straightforward and naive approaches to learn the value function would be to sample trajectories $\tau \sim p(\tau)$ (Eq. (6.9)) and then use the return G_t to update our value function estimation[8] for every visited state s_t along the way:

$$V_\pi(s_t) = V_\pi(s_t) + \eta(G_t - V_\pi(s_t)), \tag{6.19}$$

where η is a learning rate. We can do an analogous process for every visited state and action along the trajectory to learn $Q_\pi(s,a)$ instead.

However, with this approach we can only learn at the end of each trajectory, also known as *episodes*, which can be very inefficient in problems involving long episodes, or even infinite ones. On the contrary, temporal difference (TD) algorithms exploit the recursive nature of the value functions, Eqs. (6.13) and (6.14), to learn at every time step:

$$V_\pi(s_t) = V_\pi(s_t) + \eta\left(r_{t+1} + \gamma V_\pi(s_{t+1}) - V_\pi(s_t)\right). \tag{6.20}$$

Notice that, while $V_\pi(s_t)$ is an estimate, $V_\pi(s_{t+1})$ is also an estimate. This is known as a *bootstrapping* method, as the update is partially based on another estimate. Nevertheless, it is proven to converge to a unique solution. The term in brackets is known as *TD error*.

The algorithm implementing Eq. (6.20) is known as TD(0), which is a special case of the TD(λ) algorithms [379]. The analogous algorithm for the action-value function is known as SARSA [380, 381]:

$$Q_\pi(s,a) = Q_\pi(s,a) + \eta\left(r + \gamma Q_\pi(s',a') - Q_\pi(s,a)\right), \tag{6.21}$$

where we have recovered the notation s', a', and r to denote the next state, action, and reward. Replacing the term $Q_\pi(s',a')$ by an expectation over the next possible actions, such as $\sum_{a'} \pi(a'|s')Q_\pi(s',a')$, we obtain the expected SARSA algorithm [382]. If, instead, we take a maximization, as in Eq. (6.22) below, we obtain Q-learning [383], for which we provide a detailed introduction in Section 6.2.1.

6.2.1 Q-learning

Q-learning is one of the most widely used TD algorithms due to its desirable properties [383]. Most of the TD algorithms that we introduce in Section 6.2 learn the value functions for their given policies, mainly ε-greedy policies. These include exploratory random actions (recall Section 6.1.2) that have an impact on the learned

[8] The return is an unbiased estimator for the expectation $V_\pi(s_t) = \mathbb{E}[G_t|s_t,\pi]$ from Eq. (6.10). This is known as a sample update, as we only use a single sample to determine the expectation.

value functions. Therefore, the policy determines the result, and we must adjust ε during the training process to ensure their proper convergence toward the optimal value functions. However, Q-learning always learns the optimal action-value function regardless of the policy followed during the training.[9]

> The goal is to directly learn the optimal Q-values, $Q_{\pi^*}(s,a)$, hence the name Q-learning, to obtain $\pi^*(s|a)$ by performing greedy actions over them, as in Eq. (6.18).

We start by arbitrarily initializing our estimates $Q_\pi(s,a)\,\forall s\in\mathcal{S}, a\in\mathcal{A}$, which are typically stored in a table (see Section 6.2.3 for an implementation with NNs). Then, we sample trajectories $\tau\sim p(\tau)$ according to the policy to progressively update our estimates with the relation

$$Q_\pi(s,a) = Q_\pi(s,a) + \eta\left(r + \gamma\max_{a'} Q_\pi(s',a') - Q_\pi(s,a)\right). \tag{6.22}$$

We illustrate the process in Algorithm 9.

Algorithm 9 Q-learning

Require: learning rate η, maximum time T, policy parameter ε
  Initialize $Q(s,a)\,\forall s\in\mathcal{S}, a\in\mathcal{A}$
  while not converged **do**
    Initialize s_0
    for $t=0$ to $T-1$ **do**
      $\xi\leftarrow$ uniform$\in[0,1]$
      $a\leftarrow$ uniform a **if** $\xi\leq\varepsilon$ **else** $\arg\max_a Q_\pi(s,a)$ ▷ ε-greedy policy
      Move to next state s' and obtain reward r
      $Q(s,a)\leftarrow Q(s,a)+\eta\,(r+\gamma\max_{a'}Q(s',a')-Q(s,a))$.
    end for
  end while
  return $Q(s,a)$ ▷ Optimal action-value function for all states and actions

This method is guaranteed to converge to the optimal action-value function as long as all possible state–action pairs continue to be updated. This is a necessary condition for all the algorithms that converge to the optimal behavior, and it can become an issue for fully deterministic policies. However, with Q-learning, we can have an ε-greedy policy with $\varepsilon\neq 0$ that ensures that this condition is fulfilled.

The key element is that, while the policy determines which states and actions are visited by the agent, the Q-value update is performed over a greedy next action, as shown in Eq. (6.22). This way, the learned Q-values are those corresponding to the greedy policy over them, which is the one fulfilling the Bellman optimality equations Eq. (6.16).

[9] Q-learning is an off-policy algorithm, which means that the policy it learns (optimal $\pi^*(a|s)$) is different from the one it follows in the training episodes. Algorithms such as SARSA are on-policy, and learn the value function that corresponds to the policy with which they generate the training data.

6.2.2 Double Q-learning

Most of the TD algorithms suffer from a maximization bias that results in an overestimation of the Q-values, which can harm the performance. Especially in Q-learning, we encounter two maximizations: one in the ε-greedy policy and one in the greedy target policy (Eq. (6.22)). This way, we use a maximum overestimated value (see below) to update the maximum Q-value, which corresponds to the greedy action taken by the policy, potentially incurring into a significant positive bias for $Q_\pi(s,a)$.

The maximization over next possible actions in Eq. (6.22) is a sample estimate for the maximum expected value $\max_{a'} \mathbb{E}[Q_\pi(s',a')]$. However, it is a positively biased estimator, provided that the sample estimate actually corresponds to the expected maximum value $\mathbb{E}[\max_{a'} Q_\pi(s',a')]$ [384]. In [378], they provide a simple example to develop intuition on the matter: Suppose that the true Q-values for all actions in a state are zero and that our estimates $Q_\pi(s,a)$ are distributed around them taking positive and negative values. The maximum value is positive and, hence, it is an overestimation. The overestimation of the Q-values can prevent the algorithm from learning the optimal policy [385].

We overcome this issue with double Q-learning [386]. This way, instead of learning a single set of Q-values, we learn two: $Q_\pi^A(s,a)$ and $Q_\pi^B(s,a)$. However, to update one, we use the other to estimate the value of its corresponding next greedy action:

$$Q_\pi^A(s,a) = Q_\pi^A(s,a) - \eta\left(r + \gamma Q_\pi^B\left(s', \arg\max_{a'} Q_\pi^A(s',a')\right) - Q_\pi^A(s,a)\right), \qquad (6.23)$$

where A and B are interchangeable. This approach avoids using the same estimate to determine both the maximizing action and its value, yielding an unbiased estimate.

We learn both sets of values by randomly updating one at a time at every time step. The only additional difference with respect to standard Q-learning is that we take actions following an ε-greedy policy that combines the information of both $Q_\pi^A(s,a)$ and $Q_\pi^B(s,a)$, for example, using their sum or mean. With double Q-learning, we overcome a major limitation of Q-learning at the price of doubling the memory requirements.

6.2.3 Implementing Q-learning with a neural network

In Q-learning, as we have introduced it in Section 6.2.1, we store the Q-values, $Q_\pi(s,a)$, for every possible state–action pair. This approach allows us to find the exact optimal action-value function. However, it is only viable for small problems, as the memory requirement quickly becomes unfeasible for moderately large ones.

In these cases, we must rely on an efficient way to represent $Q_\pi(s,a)\;\forall s \in \mathcal{S}, a \in \mathcal{A}$. NNs are a prominent candidate to approximate the action-value function, as introduced in [28], with significantly less parameters than state–action pairs (recall Section 2.4.4). Using NNs to learn the Q-values is known as *deep Q-learning*, and the network is commonly referred to as deep Q-network (DQN). DQNs take a representation of state in the input layer $\phi(s)$ and have as many neurons as possible actions in the output layer, which encode $Q_\pi(s,a;\boldsymbol{\theta})\;\forall a \in \mathcal{A}$. Here, $\boldsymbol{\theta}$ denotes the set of learnable parameters of the neural network. This way, the DQN provides the Q-value of all possible actions given a state.

Nevertheless, DQNs may become highly unstable when directly applying Algorithm 9 with an update rule for the network parameters:

$$\theta = \theta + \eta \left(r + \gamma \max_{a'} Q_\pi(s', a'; \theta) - Q_\pi(s, a; \theta) \right) \nabla_\theta Q_\pi(s, a; \theta), \tag{6.24}$$

which is analogous to a regression problem in which we minimize the MSE loss (Eq. (2.1)) between the target, $r + \gamma \max_{a'} Q_\pi(s', a'; \theta)$, and the prediction, $Q_\pi(s, a; \theta)$, through gradient descent. The instabilities are mainly due to correlations in consecutive observations along the trajectories, correlations between target and prediction, and significant changes in the data distribution due to small variations in the parameters. The latter happen because the agent follows an ε-greedy policy, and small changes in the parameters may change the actions that have the maximum Q-value for the states, abruptly altering the course of the trajectories.[10] We overcome these limitations with *experience replay* [387], and introducing a *target network*.

With experience replay, instead of learning at every time step, we store the experience gathered along the episodes in a memory, which keeps the information of every transition (s, a, r, s'). Then, once the agent has gathered enough experience, it replays a randomly sampled batch of transitions in its memory to compute the loss and update the DQN parameters. This way, the agent alternates between episodes to gather experience and replaying it to perform the learning process. This technique removes the correlation between training samples and mitigates the sudden changes in data distribution. Furthermore, it allows the agent to reuse the experience to prevent forgetting and relearning.[11]

To remove the correlation between target and prediction, we consider a target network, which is a clone of the DQN that we update at a different rate. While we update the DQN parameters, θ, at every iteration, we only update the parameters of the target network, θ^-, copying θ every few iterations. Then, we use it to predict the target term $\max_{a'} Q_\pi(s', a'; \theta^-)$, hence the name of the network. This ensures that the prediction, $Q_\pi(s, a; \theta)$, and the target are uncorrelated.

Additionally, we can go a step further and use the target network for double Q-learning (see Section 6.2.2) to prevent the DQN from overestimating the action-value function, as introduced in [388]. Thus, the overall implementation consists of gathering experience by following an ε-greedy policy on the Q-values, $Q_\pi(s, a; \theta)$. Then, the agent replays randomly selected transitions from the experience to compute the MSE loss function between the target and the prediction but using a target network to perform double Q-learning:

[10] Consider the case of two separate paths that lead to different treasures. We initialize the Q-values arbitrarily, and the ε-greedy policy mainly takes the path with the highest one, while casually following the other with small probability ε. However, if the second one leads to a bigger treasure, its Q-value will eventually become the highest, and the data distribution will suddenly change to mainly sample this path and casually take the other.

[11] This is specially valuable when the experience is costly to obtain. For instance, if a robot receives severe damage, having a memory allows it to keep learning from the situation without receiving further injuries.

$$\mathcal{L} = \frac{1}{n}\sum_{i=1}^{n}\left(r_i + \gamma Q_\pi\left(s_i', \arg\max_{a'} Q_\pi(s_i', a'; \theta); \theta^-\right) - Q_\pi(s_i, a_i; \theta)\right)^2, \tag{6.25}$$

where i denotes the index in a batch of n randomly sampled transitions from the memory. Then, we perform a gradient descent step over the loss in Eq. (6.25) to update θ. Finally, every few iterations, we update the target network $\theta^- \leftarrow \theta$.

6.3 Policy gradient methods

The main goal of RL is to find the optimal policy $\pi^*(a|s)$ that maximizes the expected return for a given task. In policy gradient algorithms, we try to directly find the optimal policy by proposing a parametrized ansatz $\pi_\theta(a|s)$ and optimizing its parameters θ, similar to the variational wave functions from Chapter 5. Hence, finding the optimal policy $\pi^*(a|s)$ is equivalent to finding the optimal set of parameters θ^* that best approximates it $\pi_{\theta^*}(a|s) \approx \pi^*(a|s)$. This parametrization can take several forms, such as an NN, and controlling the shape of the policy may allow us to leverage prior knowledge about the task to obtain better results. Furthermore, the policies are stochastic, which have a natural exploratory character and the flexibility to also approximate deterministic policies.

To optimize the parameters, we use an objective function O_π that we aim to maximize. This can be any figure of performance, such as the state-value function V_π, the action-value function Q_π, or the return G. Having continuous parametrized policies, the objective function changes smoothly with changes in the parameters, which allows us to compute their derivatives. We approach the optimization by a gradient ascent method: We compute the gradient of the expectation value $\nabla_\theta\, \mathbb{E}[O_\pi|\pi_\theta]$ and perform a small update of the parameters θ. The expectation value is taken over the trajectories τ sampled according to the policy (recall Eq. (6.9)).

Directly evaluating the gradient is not straightforward because it depends on the stationary distribution of the states, to which we do not have access in model-free RL. Hence, it is difficult to estimate the effect of the policy update on the state distribution. However, the *policy gradient theorem* [389, 390] provides us with an analytical form for the gradient of the objective function that does not involve the derivative over the state distribution.

Policy gradient theorem: For any differentiable policy $\pi_\theta(a|s)$ and objective function O_π, the gradient of its expectation value $\nabla_\theta\, \mathbb{E}[O_\pi|\pi_\theta]$ can be expressed in terms of derivatives acting exclusively on the logarithmic policy $\nabla_\theta \log \pi_\theta(a|s)$. The term $\nabla_\theta \log \pi_\theta(a|s)$ is often referred to as the *score function*.

To get some additional intuition on the above theorem, let us consider an example with the total return $G(\tau)$ as objective function (see [378] for an extended proof with $V_\pi(s)$). Thus, we are interested in maximizing the expectation value $\mathbb{E}[G|\pi_\theta]$, which is performed over the trajectories $\tau \sim p_\theta(\tau)$. We restate Eq. (6.9) to explicitly show the parameter dependence

$$p_\theta(\tau) = \prod_{t=0}^{T-1} p(s_{t+1}|s_t, a_t)\pi_\theta(a_t|s_t). \tag{6.26}$$

Therefore, we can write the expectation as

$$\mathbb{E}[G|\pi_\theta] = \sum_\tau p_\theta(\tau)G(\tau). \tag{6.27}$$

To take the gradient, let us first recall the property of logarithmic derivatives $\nabla_\theta p_\theta = p_\theta \nabla_\theta \log p_\theta$, which we apply in the following derivation:

$$\begin{aligned}\nabla_\theta \mathbb{E}[G|\pi_\theta] &= \sum_\tau G(\tau)\nabla_\theta p_\theta(\tau) \\ &= \sum_\tau G(\tau)p_\theta(\tau)\nabla_\theta \log p_\theta(\tau).\end{aligned} \tag{6.28}$$

Then, from Eq. (6.26), we see that the only dependence on θ from $p_\theta(\tau)$ is in the policy. Therefore,

$$\nabla_\theta \log p_\theta(\tau) = \sum_{t=0}^{T-1} \nabla_\theta \log \pi_\theta(a_t|s_t), \tag{6.29}$$

which, combined with Eq. (6.28), we obtain the expression

$$\begin{aligned}\nabla_\theta \mathbb{E}[G|\pi_\theta] &= \sum_\tau p_\theta(\tau)G(\tau) \sum_{t=0}^{T-1} \nabla_\theta \log \pi_\theta(a_t|s_t) \\ &= \mathbb{E}\left[G(\tau) \sum_{t=0}^{T-1} \nabla_\theta \log \pi_\theta(a_t|s_t) \middle| \pi_\theta \right].\end{aligned} \tag{6.30}$$

The importance of the policy gradient theorem lies in the fact that it yields a closed form for the gradient as an expectation value. As a consequence, we can estimate it via Monte Carlo sampling over different trajectories τ. Furthermore, the gradient of the objective function is independent of the initial state s_0, as it does not depend on the policy.

6.3.1 REINFORCE

The REINFORCE algorithm [391] is one of the most commonly used policy gradient algorithms and it uses the return as objective $O_\pi = G(\tau)$.[12]

> The main principle of REINFORCE is to directly modify the policy to favor series of actions within the agent's experience that lead to a high return. This way, previously beneficial actions are more likely to happen the next time the agent interacts with the environment.

[12] In Section 6.1.5, we mention that the optimal policy maximizes $V_\pi(s)\ \forall s \in \mathcal{S}$. Taking $V_\pi(s)$ as objective, the gradient is $\nabla_\theta \mathbb{E}[V_\pi(s)|\pi_\theta] = \mathbb{E}[Q_\pi(s,a)\nabla_\theta \log \pi_\theta(a|s)|\pi_\theta]$ (see [378]). In REINFORCE, G_t acts as an unbiased estimator of $Q_\pi(a_t, s_t)$ to find the optimal policy, since $Q_\pi(a_t, s_t) = \mathbb{E}[G_t|s_t, a_t, \pi_\theta]$ from Eq. (6.11).

Formally, we solve the optimization problem $\theta^* = \arg\max_\theta \mathbb{E}[G|\pi_\theta]$. We find θ^* via an iterative update rule in which we compute the gradient $\nabla_\theta \mathbb{E}[G|\pi_\theta]$ and perform a gradient ascent step in its direction. In practice, we estimate it by sampling a batch of n trajectories $\tau \sim p_\theta(\tau)$, also known as *episodes*, to approximate the expectation value from Eq. (6.30). This way, at learning iteration k,

$$\Delta\theta_k \approx \frac{1}{n}\sum_{i=1}^{n} G(\tau_i) \sum_{t=0}^{T_i-1} \nabla_\theta \log \pi_\theta(a_t|s_t) \tag{6.31}$$

$$\theta_{k+1} = \theta_k + \eta\nabla\theta_k, \tag{6.32}$$

where η is the learning rate.[13] We illustrate the procedure in Algorithm 10.

Algorithm 10 REINFORCE

Require: learning rate η, number of trajectories n, maximum time T
Require: randomly initialized differentiable policy $\pi_\theta(a|s)$
  while not converged **do**
    for $i = 1$ to n **do**
      Initialize s_0
      for $t = 0$ to $T - 1$ **do**
        Take action $a_t \sim \pi_\theta(a_t|s_t)$ and store $\nabla_\theta \log \pi_\theta(a_t|s_t)$
        Move to next state s_{t+1} and store reward r_{t+1}
      end for
      $G^{(i)} \leftarrow \sum_t \gamma^t r_{t+1}$
      $z^{(i)} \leftarrow \sum_t \nabla_\theta \log \pi_\theta(a_t|s_t)$
    end for
    $\Delta\theta \leftarrow (1/n)\sum_i G^{(i)} z^{(i)}$
    $\theta \leftarrow \theta + \eta\Delta\theta$
  end while
  return θ ▷ Optimal policy parameters

However, the trajectory sampling introduces significant fluctuations to the expected quantities that result in large training variances, which is a general problem with any Monte Carlo-based approach. Some episodes may be quite successful, whereas some others could be a complete failure with very low returns. Such high variance results into unstable policy updates, which increase the convergence time toward the optimal policy. A common technique to tackle this issue is to introduce a *baseline* into the returns, which reduces the variance of the method without incurring any bias, and therefore *should always be used.*

For a better description of the baseline, let us first rewrite Eq. (6.30) in a more convenient way, and omitting the condition $\mathbb{E}[\cdot|\pi_\theta]$ for the rest of the chapter:

[13] In some cases, it is beneficial to compute the expectation of the gradient as a weighted sum over the trajectory returns. In this case, rather than dividing by n, we divide by $\sum_\tau G(\tau)$, which makes the update rule independent of the scale of the returns. This approach disregards trajectories with zero return, which do not contribute to the gradient and would dilute the information, yielding very small updates.

$$
\begin{aligned}
\nabla_\theta \mathbb{E}[G|\pi_\theta] &= \mathbb{E}\left[\left(\sum_{t'=0}^{T-1} \gamma^{t'} r_{t'+1}\right) \sum_{t=0}^{T-1} \nabla_\theta \log \pi_\theta(a_t|s_t)\right] \\
&= \mathbb{E}\left[\sum_{t'=0}^{T-1} \gamma^{t'} r_{t'+1} \sum_{t=0}^{t'} \nabla_\theta \log \pi_\theta(a_t|s_t)\right] \\
&= \mathbb{E}\left[\sum_{t=0}^{T-1} \nabla_\theta \log \pi_\theta(a_t|s_t) \sum_{t'=t}^{T-1} \gamma^{t'} r_{t'+1}\right] \\
&= \mathbb{E}\left[\sum_{t=0}^{T-1} \gamma^t G_t \nabla_\theta \log \pi_\theta(a_t|s_t)\right],
\end{aligned}
\tag{6.33}
$$

where in the first equation we write the explicit form of $G(\tau)$. In the second equation, we use the relation

$$
\begin{aligned}
\nabla_\theta \mathbb{E}[G|\pi_\theta] = \nabla_\theta \mathbb{E}\left[\sum_{t'=0}^{T-1} \gamma^{t'} r_{t'+1}\right] &= \sum_{t'=0}^{T-1} \nabla_\theta \mathbb{E}_{\tau_{t'}}\left[\gamma^{t'} r_{t'+1}\right] \\
&= \sum_{t'=0}^{T-1} \mathbb{E}_{\tau_{t'}}\left[\gamma^{t'} r_{t'+1} \sum_{t=0}^{t'} \nabla_\theta \log \pi_\theta(a_t|s_t)\right] \\
&= \mathbb{E}\left[\sum_{t'=0}^{T-1} \gamma^{t'} r_{t'+1} \sum_{t=0}^{t'} \nabla_\theta \log \pi_\theta(a_t|s_t)\right],
\end{aligned}
\tag{6.34}
$$

where $\mathbb{E}_{\tau_{t'}}$ denotes expectation over trajectories up to time t'. Then, in the third line of Eq. (6.33), we rearrange the terms in the summations and we find the explicit form of G_t offset by a γ^t factor. In the final expression, it becomes clearer how past rewards in the trajectories do not contribute to the gradient of the policy from a given time onwards, which recovers the Markov property.

We can reduce the variance in the gradient by introducing a state-dependent baseline $b(s_t)$ in Eq. (6.33) such that

$$
\nabla_\theta \mathbb{E}[G|\pi_\theta] = \mathbb{E}\left[\sum_{t=0}^{T-1} \gamma^t \left(G_t - b(s_t)\right) \nabla_\theta \log \pi_\theta(a_t|s_t)\right]. \tag{6.35}
$$

Any baseline is appropriate as long as it does not depend on the actions. This way, we do not introduce any bias, given that

$$
\begin{aligned}
&\mathbb{E}\left[b(s_t)\nabla_\theta \log \pi_\theta(a_t|s_t)\right] = \mathbb{E}_{\tau_t}\left[b(s_t)\, \mathbb{E}_{\tau_{t:T}}\left[\nabla_\theta \log \pi_\theta(a_t|s_t)\right]\right] \\
&= \mathbb{E}_{\tau_t}\left[b(s_t) \sum_{a_t} \pi_\theta(a_t|s_t) \nabla_\theta \log \pi_\theta(a_t|s_t) \underbrace{\sum_{s_{t+1}} p(s_{t+1}|s_t, a_t)}_{1} \underbrace{\sum_{\tau_{t+1:T}} p_\theta(\tau_{t+1:T})}_{1}\right] \\
&= \mathbb{E}_{\tau_t}\left[b(s_t) \nabla_\theta \underbrace{\sum_{a_t} \pi_\theta(a_t|s_t)}_{1}\right] = \mathbb{E}_{\tau_t}\left[b(s_t) \cdot 0\right] = 0,
\end{aligned}
\tag{6.36}
$$

where $\tau_{t:T}$ indicates a trajectory from time t until the end T. We move from the second to the third line using the property of logarithmic derivatives, as in Eq. (6.28). Notice that the expectation remains unbiased even if the baseline depends on θ.

While the expectation is unaffected, the baseline can have a major impact in the variance.[14] Let us consider the case of a state-independent baseline. We can find the optimal baseline that minimizes the variance in the gradient for each parameter. To simplify the notation, let z_k and b_k be the kth components of the score function $z_k = \partial_{\theta_k} \log \pi_\theta(a|s)$ and a state-independent baseline vector, respectively. Hence, the goal is to minimize the variance of the term $(G_t - b_k)z_k$,[15] which is the argument of Eq. (6.35). Formally, we aim to find $b_k^* = \arg\min_{b_k} \mathrm{Var}\left[(G_t - b_k)z_k\right]$, that is such that $\partial_{b_k^*} \mathrm{Var}\left[(G_t - b_k)z_k\right] = 0$. Therefore,

$$\mathrm{Var}\left[(G_t - b_k)z_k\right] = \mathbb{E}[((G_t - b_k)z_k)^2] - \mathbb{E}[G_t z_k]^2 \tag{6.37}$$

$$\partial_{b_k} \mathrm{Var}\left[(G_t - b_k)z_k\right] = -2\,\mathbb{E}[(G_t - b_k)z_k^2] \tag{6.38}$$

$$b_k^* = \frac{\mathbb{E}[G_t z_k^2]}{\mathbb{E}[z_k^2]}, \tag{6.39}$$

where in the first equation we have used Eq. (6.36) to remove b_k in the second term.

There are several other valid baselines that we can consider, besides the state-independent example above, with which we may obtain better results. For instance, an estimation of the value function $\hat{V}_\pi(s_t) \approx \mathbb{E}[G_t|s_t]$ is a common state-dependent baseline. This can either be learned, either directly from G_t as we show in Section 6.4, or it can be estimated through sampling in self-critic schemes (see [392]). With such baseline, actions that lead to returns higher than expected with the current policy are reinforced, while those that lead to lower rewards are penalized. This is equivalent to weighting the score function by the advantage. Given that $\mathbb{E}[G_t|s_t, a_t] = \mathbb{E}[Q_\pi(s_t, a_t)|s_t, a_t]$, from Eq. (6.11), subtracting a baseline $b(s_t) = V_\pi(s_t)$, we obtain the expectation of the advantage (recall Eq. (6.12)). Hence, $\nabla_\theta \mathbb{E}[G|\pi_\theta] = \mathbb{E}\left[\sum_t \gamma^t A(s_t, a_t) \nabla_\theta \log \pi_\theta(a_t|s_t)\right]$. Directly estimating the advantage provides the least possible variance; see [393] for further reference on this matter.

Another common practice is to *whiten* the return. This consists of subtracting the mean of the return along all the time steps of a trajectory and dividing by its standard deviation $\bar{G}_t = (G_t - \langle G \rangle)/\sigma_G$. Since this is not exactly a baseline, this method *does* introduce a bias.

6.3.2 Implementing REINFORCE with a neural network

The parametrized policy π_θ is a central quantity in policy gradient methods, and it can take any form as long as it is differentiable with respect to its parameters. One of the most common approaches in discrete action spaces is to define action probabilities according to a *softmax* distribution:

[14] Recall that $\mathrm{Var}[x] = \mathbb{E}[x^2] - \mathbb{E}[x]^2$. Hence, adding a term with null expectation does not affect the second term, but it does have an impact on the first one $\mathrm{Var}[x - b] = \mathbb{E}[(x - b)^2] - \mathbb{E}[x - b]^2 = \mathbb{E}[(x - b)^2] - \mathbb{E}[x]^2$.

[15] In this case, we take the approximation $\mathrm{Var}\left[\sum_t X_t\right] \approx \sum_t \mathrm{Var}\left[X_t\right]$.

$$\pi_\theta(a|s) = \frac{e^{x(s,a)}}{\sum_{a' \in \mathcal{A}} e^{x(s,a')}}, \tag{6.40}$$

where $x(s, a)$ is the *action preference* for action a in state s.

The simplest way to define action preferences is through a set of linear parameters θ applied to a feature representation of the state and action $\phi(s, a)$, such that $x(s, a) = \theta^T \phi(s, a)$. However, this approach may lack the expressive power to approximate the optimal policy π^* in complex problems.

In these cases, we may need to use a deep NN to parametrize the action preferences. NNs are a natural generalization of the linear parameter approach that we can tune to increase the expressive power by, for example, increasing the number of hidden layers or their size. This way, the NN parametrizing the policy takes a state representation in the input layer $\phi(s)$, and has as many neurons as possible actions in the output layer, which encode $x(s, a)\ \forall a \in \mathcal{A}$. Applying a softmax activation function in the output layer (see Eq. (2.37)), we obtain $\pi_\theta(a|s)\ \forall a \in \mathcal{A}$, as in Eq. (6.40).

The training process is analogous to training a supervised classifier on the experience gathered by the agent. Implementing REINFORCE with gradients from Eq. (6.35) is equivalent to performing gradient descent with a modified categorical cross-entropy loss (recall Eq. (2.3)):

$$\mathcal{L} = -\frac{1}{n} \sum_{i=1}^{n} \sum_{t=0}^{T-1} \gamma^t (G_{ti} - b(s_{ti})) \log \pi_\theta(a_{ti}|s_{ti}), \tag{6.41}$$

where i denotes the index in a batch of n trajectories. This way, the procedure is analogous to training an NN classifier in which the actions act as state labels. The main difference with supervised classification problems is that, given a state, we do not know the true probability distribution of the actions (true labels), as that would be the optimal policy. Instead, we assign the obtained return G_t as true label for the taken action a_t.[16] Intuitively, in classification problems, we aim to enhance the probability that the NN provides the right label, whereas here we reinforce the actions with high returns.

In many situations, actions can take a range of continuous values rather than a discrete set of categories. For instance, a robotic arm may rotate by a certain angle or we can tune various continuous parameters in an experimental setup. Sometimes, we can discretize the action space into small intervals at the cost of a loss in precision and an increasing amount of actions. Nevertheless, this may not always be possible depending on the problem requirements and the resulting number of actions.

In these cases, we model the stochastic continuous actions with a mean μ and a standard deviation σ, such that

$$a = \mu + \sigma\xi, \tag{6.42}$$

[16] The standard categorical cross entropy would be $\mathcal{L} = -\frac{1}{n}\sum_n \sum_k p(a_k) \log \pi_\theta(a_k|s)$, where $p(a_k)$ is the true probability distribution that we want to learn. In standard classification problems, this is typically 1 for the true label and 0 for the rest. Here, it corresponds to the optimal policy $p(a_k) = \pi^*(a_k|s)$. Since we do not have access to π^* (it is our goal!), we use the return G_t for the chosen action in its place, as π^* would favor actions with high returns. This effectively removes the expectation over actions, and we make the sum over time explicit in Eq. (6.41).

where ξ is a random normal variable with unit variance. Analogously to the action preferences above, we can parametrize $\mu_\theta(s), \sigma_\theta(s)$ in various ways, ranging from a set of linear parameters, for example, $\mu_\theta(s) = \theta^T \phi(s)$, to an NN with two output neurons that determine both $\mu_\theta(s)$ and $\sigma_\theta(s)$ for the given observation. Formally,

$$\pi_\theta(a|s) = \frac{1}{\sigma_\theta(s)\sqrt{2\pi}} \exp\left(-\frac{1}{2} \left(\frac{a - \mu_\theta(s)}{\sigma_\theta(s)} \right)^2 \right). \tag{6.43}$$

In many cases, as the learning advances, and the agent becomes better at taking the right actions (choosing $\mu_\theta(s)$), the deviations decrease and we obtain a quasi-deterministic policy.

6.4 Actor-critic methods

In Section 6.2, we introduce value-based RL, featuring the Q-learning algorithm in Section 6.2.1. These methods excel at dealing with discrete state–action spaces, and their TD character makes them data efficient and allows them to tackle continuing tasks (infinite episodes). However, they experience difficulties to deal with large state–action spaces and can't deal with their continuous version. Furthermore, they are bound to implement deterministic policies, while many problems present stochastic optimal policies. Finally, small changes in the value functions can cause large variations in the policy, which may cause instabilities in learning.

On the other hand, we introduce policy gradient methods in Section 6.3, featuring the REINFORCE algorithm in Section 6.3.1. These algorithms overcome the aforementioned limitations of value-based methods, provided that they can deal with continuous (infinite) state–action spaces, and they are based on continuous stochastic policies, which ensure smooth changes in the policy throughout the learning process, and can become deterministic when needed. However, the learning happens at the end of the episodes, once we know the return, which is an issue for long trajectories or continuing tasks.

> Actor-critic algorithms combine value-based and policy-based methods to obtain the best of both approaches. We can understand actor-critic methods as the TD version of policy gradient, with which we retain all its advantages and overcome its major limitation. It features two main elements: the *actor*, a parametrized policy that dictates the decisions, and the *critic*, a model that evaluates them.

The presence of the critic allows the agent to immediately learn from each action without waiting for the outcome at the end of the episode. Evaluating the policy mainly consists of learning its value functions, which allows the critic to assess whether the actions are more or less favorable. In Section 6.3.1, we introduce the state-value function, $V_\pi(s)$, as the optimal baseline to reduce the variance in policy gradient. Although, in this case, we only look at $V_\pi(s)$ of the initial state in the transitions, which does not allow us to evaluate the actions.[17]

However, we show that, with such baseline, we can compute the gradient in terms of the advantage $A(s,a)$, introduced in Eq. (6.12). The explicit form of the advantage sets the foundation for actor-critic methods [394–396]:

$$A(s_t, a_t) = \mathbb{E}[r_{t+1} + \gamma V_\pi(s_{t+1}) - V_\pi(s_t)], \tag{6.44}$$

which is derived from Eqs. (6.12) and (6.14). This expression lies at the core of TD algorithms, as it corresponds to the TD error from Eq. (6.20).

In Eq. (6.44), we use $V_\pi(s)$ to evaluate both the initial and final states of a given transition, thus constituting a critic of the action. This allows the agent to learn from every time step in REINFORCE, processing states, actions, and rewards as they occur, like the TD algorithms from Section 6.2. Nevertheless, this advantage comes as the cost of learning two models: the policy $\pi_\theta(a|s)$ and the state-value function $V_\pi(s;\boldsymbol{w})$, which are usually parametrized with NNs with parameters $\boldsymbol{\theta}$ and $\boldsymbol{w}$, respectively. The NN parametrizing the state-value function takes a feature representation of the state, $\phi(s)$, in the input layer, and has a single output neuron encoding $V_\pi(s;\boldsymbol{w})$. The policy parametrization is the same as in Section 6.3.2. We train both models simultaneously by following Algorithm 11.

Algorithm 11 Actor-critic

Require: learning rates η_θ, η_w, maximum time T
Require: randomly initialized differentiable policy $\pi_\theta(s|a)$
Require: randomly initialized differentiable state-value function $V_\pi(s;\boldsymbol{w})$
  while not converged **do**
    Initialize s_0
    for $t = 0$ to $T-1$ **do**
      Take action $a \sim \pi_\theta(a|s)$
      Move to next state s' and obtain reward r
      $A \leftarrow r + \gamma V_\pi(s';\boldsymbol{w}) - V_\pi(s;\boldsymbol{w})$
      $\boldsymbol{\theta} \leftarrow \boldsymbol{\theta} + \eta_\theta \gamma^t A \nabla_\theta \log \pi_\theta(a|s)$ ▷ Update actor
      $\boldsymbol{w} \leftarrow \boldsymbol{w} + \eta_w A \nabla_w V(s;\boldsymbol{w})$ ▷ Update critic
    end for
  end while
  return $\boldsymbol{\theta}$, $\boldsymbol{w}$ ▷ Optimal actor and critic parameters

We train the actor with the methods from Section 6.3 and the critic using the principles from Section 6.2. Hence, all the methods in both sections apply to this algorithm. The parameter updates in Algorithm 11 come from performing gradient ascent with Eq. (6.35) on the actor, and an analogous update rule to Eq. (6.22) for the critic, using $V_\pi(s)$ instead of $Q_\pi(s,a)$. The process is equivalent to performing

[17] To determine the quality of an action, we need to compare the initial and final positions. In a game, an action that escapes from the brink of a loss toward a less disadvantageous position may be more valuable than one that moves from an already favorable position to a slightly better one, despite the latter providing a higher final state-value function.

gradient descent on the losses $\mathcal{L}_\theta = \frac{1}{n}\sum_n \sum_t \gamma^t A(s_t, a_t; \boldsymbol{w}) \log \pi_\theta(a_t|s_t)$ and $\mathcal{L}_{\boldsymbol{w}} = \frac{1}{n}\sum_n A(s, a; \boldsymbol{w})^2$, respectively, in which we omit the index for the sum over n samples. They are based on the same principles as the ones in Eqs. (6.25) and (6.41).

This method is often referred to as advantage actor-critic (A2C). It has been further enhanced using asynchronous actors, giving rise to the asynchronous advantage actor-critic (A3C) algorithm [397]. Other improvements rely on implementing more advanced optimization techniques, such as the natural gradient [398], as in natural policy gradient [399], natural actor-critic [400, 401], or more involved parameter updates such as trust-region [402, 403] or proximal policy optimization algorithms [404].

6.5 Projective simulation

In recent years, there have been introduced novel approaches to RL that explore techniques beyond the prototypical value-based and policy gradient methods that we introduce in Sections 6.2 and 6.3. Among those, projective simulation (PS) [405] is of particular interest for the physics community, due to its numerous applications in the field.

Projective simulation considers an agent based on an episodic and compositional memory (ECM), a mathematical object capable of storing the information about visited states and actions, and drawing connections between them. This way, the ECM is continuously updated as the agent gathers experience, and it ultimately determines the policy at any given state, as we show below. Usually, the ECM is represented as a directed weighted graph, as shown in Fig. 6.3(a). The nodes, defined here as *clips*, represent either visited states, actions, or hidden information learned by the agent. As the agent explores, clips corresponding to new visited states are added to the graph. Similarly, an agent may create additional ones to accommodate new actions, for example, the combination of two actions, or hidden information. The edges are weighted, and every new node is initialized with uniform edge weights. The weights determine the transition probability between clips, and they are updated as the agent gathers rewards.

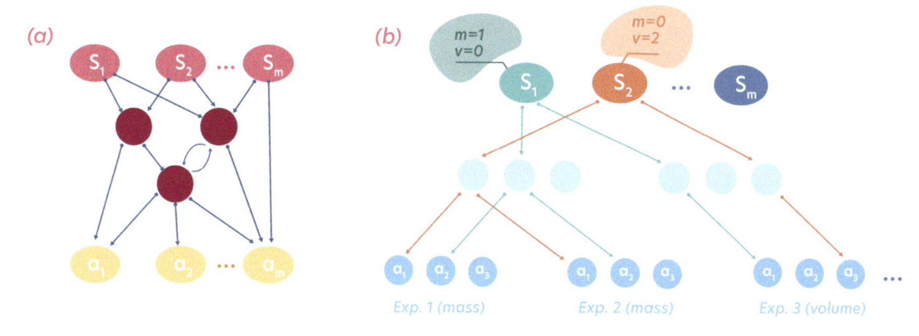

Figure 6.3 Schematic representation of the episodic and compositional memory (ECM) of various PS agents. (a) A multilayer ECM with m state nodes (pink), three hidden nodes (red), and m action nodes (yellow). The connectivity of the graph can also be set as a (b) three-layer ECM, used to demonstrate the possibility of feature extraction by the PS model. See the main text for details.

As we have previously mentioned, the ECM defines the policy of the PS agent. In the vanilla version of PS, given an observed state, the agent performs a weighted random walk through the ECM starting on the corresponding state clip. The walk ends as soon as it lands in an action node, and the corresponding action is chosen. The probability to jump from one clip, c_i, to another, c_j, can be any normalized function of the edge weights $h(c_i, c_j)$, such as

$$P(c_i, c_j) = \frac{h(c_i, c_j)}{\sum_{j \in \mathcal{I}} h(c_i, c_j)}, \tag{6.45}$$

where $\mathcal{I}$ is the set of edges of c_i. Other transition functions have also been introduced, such as softmax transitions, which allow us to have arbitrary h-values.

Following the previous scheme, training a PS agent consists of updating the ECM by adding new nodes, and learning the edge weights. The goal is that, for every state clip, the path through the ECM leads to the correct action with high probability. Thus, the training can then be reduced to the update of the h-values at every time step via

$$h(c_i, c_j) \leftarrow h(c_i, c_j) + \gamma(h(c_i, c_j) - 1) + r, \tag{6.46}$$

where c_i and c_j represent the clips traversed during the random walk through the ECM, γ is a damping parameter, and r is the reward given by the environment after performing the chosen action.

With this update rule, for every agent's decision, that is, every time it performs a walk from a state node to an action node, all h-values of the visited edges are updated. In this way, the h-values along the walk are always damped by a factor γ, and, in the case that they led to a rewarded action, they also increase their value by a factor r.

In many practical scenarios, rewards are obtained at the end of a long series of actions, for example, performing various steps in a grid-world to reach a target. Hence, it is important to *backpropagate* such reward through the sequence of all the actions that led to it. For instance, in TD algorithms, this is achieved by considering the expected value of future states to perform the updates, as we introduce in Section 6.2. To accommodate such property, we can generalize the update rule from Eq. (6.46) by introducing the concept of an *edge glow*: Every time an edge is traversed, it starts to glow decaying with time. This feature allows the agent to update all the edges in the ECM involved in the decisions to describe a trajectory $\tau = a_0, s_1, a_1, ...$[18] which led to a certain reward. The update rule can then be rewritten as

$$h(c_i, c_j) \leftarrow h(c_i, c_j) - \gamma(h(c_i, c_j) - 1) + g(c_i, c_j) r, \tag{6.47}$$

where g is the glow value.

Each time a certain edge is visited, its corresponding glow value is set to 1. Then, at every step, all the glow values are dampened via

$$g(c_i, c_j) \leftarrow g(c_i, c_j)(1 - \eta), \tag{6.48}$$

[18] Be careful to not confuse the trajectories through the ECM with the trajectories through the state and action spaces. Given a state s_t, the PS agent chooses the action a_t by performing a trajectory through the ECM that starts on the corresponding s_t node until it reaches an action node. Then, the corresponding action a_t is performed to move toward the next state s_{t+1}.

effectively decreases all of them with a rate η. This means that edges that have been recently visited and led to a reward $r \neq 0$ are strengthened, while those visited earlier on received a lesser update, analogous to TD algorithms. We refer to [406, 407] for an in-depth and practical description of the usage of the PS models.

The presented approach to PS is a tabular method, similarly to Q-learning from Section 6.2.1, as the agent's deliberation is saved in the adjacency matrix of the ECM, namely, the h-matrix. As commented previously, tabular methods have strong limitations when dealing with large action and state spaces. Nontabular approaches for PS have been proposed [408]. In that case, a neural network (and more precisely, an energy-based model) is trained to output the h-value for a certain state-action pair, analogously to how DQNs are used to predict Q-values, as we introduce in Section 6.2.3.

An important feature of the PS model is its transparency and potential interpretability power, in contrast to other approaches such as Q-learning. In the latter, the Q-values encode the expected reward received from an action–state tuple. As the policy relies on performing the action with largest Q-value, there is little to no room for interpretability, aside from such maximization. Conversely, PS constructs a visible graph encoding the probabilities to hop between nodes, which may represent both direct information from the RL task, that is, actions and states, but also hidden information extracted by the agent. For instance, as we describe in Section 6.6.6, the authors of [375] were able to interpret the hidden structure of the ECM, related in that example to different optical devices. Interestingly, the PS agent was able to create useful optical gadgets composed of multiple devices by composing actions together into new joint nodes (see [405]). Nonetheless, when working in the so-called two-layer PS (one layer of nodes for the states and one for the actions), PS reduces to a very similar model to Q-learning. Indeed, recent works have extensively compared both approaches [409]. However, we can introduce further *hidden* nodes to build deeper PS models, as shown in Fig. 6.3(a).

There have been multiple efforts to build such deep PS architectures and to show that they are indeed able to extract relevant hidden features from the environment or the task at hand [410, 411]. An enlightening example is shown in [411], which we schematically reproduce in Fig. 6.3(b). In this work, an agent is given a set of objects with different physical properties, such as mass, charge, and volume. For simplicity, these quantities can take only one of three values: 0, 1, or 2. The agent has access to different experiments, which measure each of these quantities separately. The states are then different objects with certain properties, for example, in Fig. 6.3(b), S_2 is an object of mass 0 and volume 2, obviously in arbitrary units or categories. On the other hand, the actions are the predictions over the various experiments. For instance, a_1 corresponds to the prediction that the object has the lowest value measured by experiment one (related in this case to mass), a_2 to an intermediate value of that same experiment, and so on. The authors show that the PS agent would assign the hidden nodes to meaningful features of the problem. In particular, each hidden node would represent a particular value of a physical quantity, as shown in Fig. 6.3(b). Such an interesting feature is not only a valuable sign of the interpretability of the PS model but also was shown to increase its generalization performance.

6.6 Examples and applications

In this section, we showcase a series of prominent applications of RL. Between all the examples, we find instances of each RL paradigm that we discuss in the previous sections. We start with two toy examples to settle the theoretical foundations of policy gradient, as they have analytical solutions. Then, we briefly comment on some of the most famous examples of RL: Atari video games and Go. Finally, we highlight a few applications of RL to quantum physics, more precisely, in the context of future quantum technologies such as quantum circuits, error correction, and certification.

6.6.1 Toy examples

Let us illustrate the REINFORCE algorithm, from Section 6.3.1, by solving a couple of toy examples. These simple scenarios allow us to solve all the equations analytically to lay down the foundations and become familiar with the basic concepts.

The random walker. Consider an agent that can move along a one-dimensional path with only two actions: move up or down. Every time the agent goes up, it receives a positive reward $r_t = +1$, and every time it goes down, it receives a negative reward $r_t = -1$. Considering the undiscounted case, $\gamma = 1$, the return of a trajectory of T steps is the final position $G(\tau) = x_T$. We can also express it in terms of the number of times the agent has taken the actions to go up or down $G(\tau) = n_{\text{up}} - n_{\text{down}} = 2n_{\text{up}} - T$. Clearly, the optimal policy is to always go uphill regardless of the current position.

In such a simple scenario, there is no notion of a state for the agent. Therefore, the policy only depends on the action. Furthermore, since there are only two possible actions, we can define the parametrized policy for one, for example, $\pi_\theta(\text{up}) \in [0, 1]$, and take the other as $\pi_\theta(\text{down}) = 1 - \pi_\theta(\text{up})$. Let us consider the parametrized sigmoid policy

$$\pi_\theta(\text{up}) = \frac{1}{1 + e^{-\theta}}, \; \pi_\theta(\text{down}) = \frac{1}{1 + e^{\theta}}, \tag{6.49}$$

which determine the probability to move upwards or downwards, respectively, in terms of the single parameter θ. Their score functions are

$$\nabla_\theta \log \pi_\theta(\text{up}) = \pi_\theta(\text{down}), \; \nabla_\theta \log \pi_\theta(\text{down}) = -\pi_\theta(\text{up}). \tag{6.50}$$

With Eqs. (6.49) and (6.50), we can compute the parameter update rule from Eq. (6.31) analytically. We can express each of its terms as a function of $\pi_\theta(\text{up})$:

$$\begin{aligned}
\mathbb{E}\left[G(\tau) \sum_{t=0}^{T-1} \nabla_\theta \log \pi_\theta(a_t)\right] &= \mathbb{E}\left[(n_{\text{up}} - n_{\text{down}})\left(n_{\text{up}}\pi_\theta(\text{down}) - n_{\text{down}}\pi_\theta(\text{up})\right)\right] \\
&= \mathbb{E}\left[(2n_{\text{up}} - T)\left(n_{\text{up}} - T\pi_\theta(\text{up})\right)\right] \\
&= 2\,\mathbb{E}\left[(n_{\text{up}} - T/2)\left(n_{\text{up}} - \langle n_{\text{up}}\rangle_{\pi_\theta}\right)\right] \\
&= 2\text{Var}\left[n_{\text{up}}\right] = 2T\pi_\theta(\text{up})(1 - \pi_\theta(\text{up})), \qquad (6.51)
\end{aligned}$$

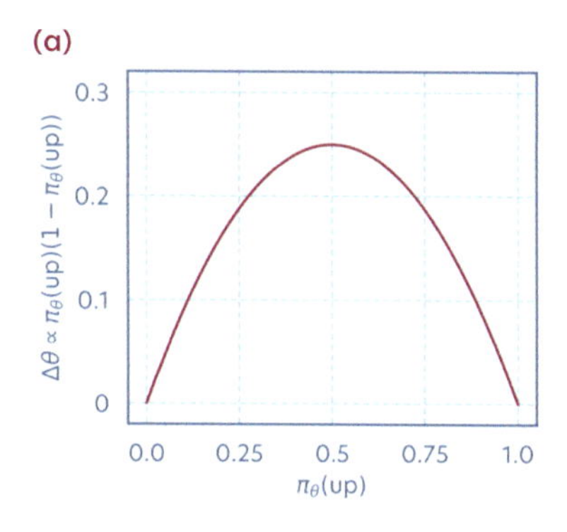

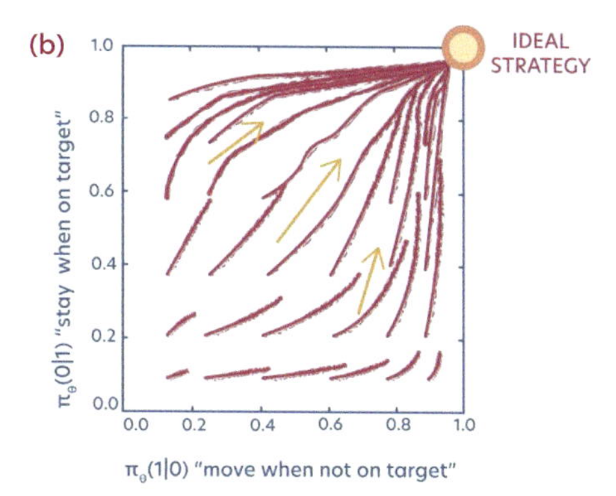

Figure 6.4 Walkers and RL. (a) Parameter update from Eq. (6.52) for the random walker. (b) Evolution of various policies trained on the walker with target.

where we have taken $T\pi_\theta(\text{up})$ as the expected number of upwards moves $\langle n_{\text{up}} \rangle_{\pi_\theta}$. With this, we are able to reach a closed analytical form for the parameter update rule in this simplistic scenario, which is not the usual case in RL. This allows us to understand the way that actions are reinforced. For instance, the term $n_{\text{up}} - \langle n_{\text{up}} \rangle_{\pi_\theta}$ reinforces actions that lead toward higher upwards moves than expected following the policy, and it penalizes those that lead to fewer.

The parameter update is a quadratic function on the policy, such that

$$\Delta\theta \propto \pi_\theta(\text{up})(1 - \pi_\theta(\text{up})), \tag{6.52}$$

which is minimal either close to the optimal policy $\pi_\theta(\text{up}) \simeq 1$ or far from it $\pi_\theta(\text{up}) \simeq 0$, as shown in Fig. 6.4(a). We can understand this in a very intuitive way: If the agent is already prioritizing the action to move upwards, it has very little to learn from there on. Conversely, if it barely takes this action, it cannot learn that it is the right choice. Hence, the agent learns the most whenever it takes both actions at a similar rate. This also reflects the importance of the initialization. If we initialize the policy to $\pi_{\theta_0}(\text{up}) \simeq 0$, the agent takes much longer to converge to the optimal policy than with $\pi_{\theta_0}(\text{up}) \simeq 0.5$.

The walker with target. Consider now a slightly more complex situation in which the agent moves along a one-dimensional path and has to stop at a target location. In this case, the two actions are to move forward, or to stay. The agent receives a reward every time step it stays at the target location. In this example, the optimal policy is to move forward until the agent reaches the target and then stop.

In contrast to the previous example, the agent is no longer blind, and the policy does depend on the state. Notice that, even though the agent moves in space, the actual position is completely irrelevant to the problem, and the only important information is whether the agent is in the right position or not. Therefore, we encode this information with Boolean indicators, assigning $s = 1$ when the agent is at the target location, and $s = 0$ elsewhere. Hence, despite the agent moving in real space, it only

Table 6.1 Optimal policy $\pi^*(a|s)$ for the walker with target example.

	Stay $a = 0$	Move $a = 1$
Out of target $s = 0$	0	1
On the target $s = 1$	1	0

navigates in a two-state MDP.[19] Then, we denote the actions "stay" and "move" with $a = 0$ and $a = 1$, respectively. This way, the optimal policy always takes the action to move when not in target and to stay when in target. We illustrate the optimal policy in Table 6.1. Additionally, we illustrate the convergence of various policies to the optimal one with REINFORCE in Fig. 6.4(b).

6.6.2 Go and Atari games

Games are one of the most natural applications for RL, and they serve as a benchmark for the state-of-the-art methods. Most games involve long-term strategies, and early actions may lead to completely different outcomes, even in short time scales. Furthermore, many games involve vast state spaces, or even infinite ones. Overall, they pose a great challenge that has motivated some of the greatest advances in the field.

The first applications of AI to games were board games. The first superhuman performance was demonstrated in chess when, in 1997, a knowledge-based system *Deep Blue* [412] beat Garry Kasparov, the highest-rated chess player in the world at the time. A more recent breakthrough has been achieving superhuman performance in the game of Go [30]. Go is a Chinese board game which is over 3 000 years old. Two players take turns to place stones on the board. The goal is to conquer as much space as possible, either by strategically surrounding empty spaces or capturing the opponent's stones by surrounding them. Once all stones are allocated, the player with the largest captured territory wins. Even with this simple set of rules, there are 10^{172} possible board configurations, making this game order of magnitudes more complex than chess [413].

The computer program developed by DeepMind, AlphaGo [30], combines a technique called Monte Carlo tree search [414] with deep NNs. With this approach, the goal is to progressively build a search tree of the state space that grows as the agent gathers experience. In the tree, each edge contains the learned action-value function $Q(s, a)$, which partially determines the policy, similar to Q-learning from Section 6.2.1. However, since the state space is virtually infinite, they implement two NNs that guide the search through the regions outside of the tree: a parametrized policy that guides the exploration and a parametrized value function that predicts the probability to win from each state. See [30] for a detailed explanation.

[19] We emphasize that, when we frame a problem as an RL instance, we only need to model and encode the information that is relevant to the problem. Hence, the resulting state and action spaces do not need to correspond directly to those in the "real world." The simpler the MDP, the easier it is be for the agent.

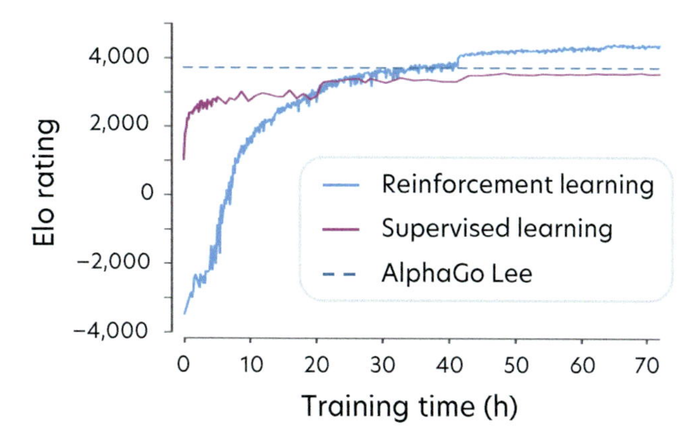

Figure 6.5 Performance comparison between AlphaGo (initial supervised learning) [30] and AlphaGo Zero (pure RL) [415] at the game of Go. Initially, AlphaGo has an advantage thanks to the initial supervised training. However, it limits its capabilities and is quickly outperformed by AlphaGo Zero. The horizontal dashed line corresponds to the Elo rating of the AlphaGo version that defeated Lee Sedol, the winner of 18 international titles, in March 2016, being a reference point for the supervised/pure RL performance. Taken from [415] with permission from Springer Nature.

Initially, they train the policy network by supervised learning, taking example moves from expert games. This provides them with an early advantage with respect to starting tabula rasa to build the search tree from already functional strategies. However, they then proceed to train the whole pipeline through *self-play*, that is, playing against itself, further refining the policy via policy gradient, as shown in Section 6.3. This model defeated the world champion of Go in 2015.

This approach has been improved by removing the initial supervised training over expert human games and purely training through self-play from scratch. This algorithm is known as AlphaGo Zero [415]. This new version defeated the previous one by a hundred games to zero. In Fig. 6.5, we see the performance of AlphaGo and AlphaGo Zero with training time in terms of Elo rating.[20] Initially, AlphaGo has a substantial advantage thanks to the previous supervised learning phase. However, this pretraining ultimately limits its capabilities, and AlphaGo Zero outperforms it in just a few hours of training. Furthermore, while these algorithms are generally tailored to the specific game, more general and recent approaches, defeated the previous benchmarks in chess, shogi, and Go at the same time [416].

Another exciting avenue for RL in games are video games. One of the first applications were Atari games, achieving superhuman performance with deep Q-learning [28], as we explain in Section 6.2.3. In this case, the state space is also infinite and the agent receives the screen pixels as input, together with the current score. However, the action space is limited by the game controller, which is very

[20]The Elo rating system, named after its creator Arpad Elo, is a method to calculate the relative skill level of players in zero-sum games. After every game, the winning player takes points from the losing one. The difference in rating between players determines the total number of points gained or lost after a game. If the higher-rated player wins, only a few rating points are taken from the lower-rated player. However, in the opposite case, the lower-rated player takes many points from the higher-rated one.

convenient for Q-learning. This approach achieved superhuman performance in 49 different games with the same algorithm.

Some other recent outstanding results in video games include competitive performance in StarCraft II [29], Dota 2 [417], and Minecraft [418]. Furthermore, advances in model-free RL have motivated the research on planning with model-based algorithms, with which some of the benchmarks that we introduce above have been bested [419]. This approach does not even require the explicit encoding of the game rules, as it builds a model of them while playing.

6.6.3 Quantum feedback control

Quantum control is a research direction in quantum technologies that aims to improve the initialization and stabilization of a desired quantum state. Deep RL algorithms have already been successfully employed in a wide range of applications for quantum feedback control [421–424]. In general, the quantum system is controlled by an RL agent with a feedback loop with some measurements periodically performed on the system. In this way, the agent drives the control scheme based on the measurement results. In [422], the authors consider a single-mode quantum cavity. The cavity mode is leaking, and this signal can be measured. The goal is to adjust an external drive amplitude of a beam entering the cavity to create and stabilize a cavity quantum state with a single photon as depicted in Fig. 6.6.

The agent observes the measured electric field that leaked from the cavity, which is the state of the environment. Given the observation, the agent can set the value of

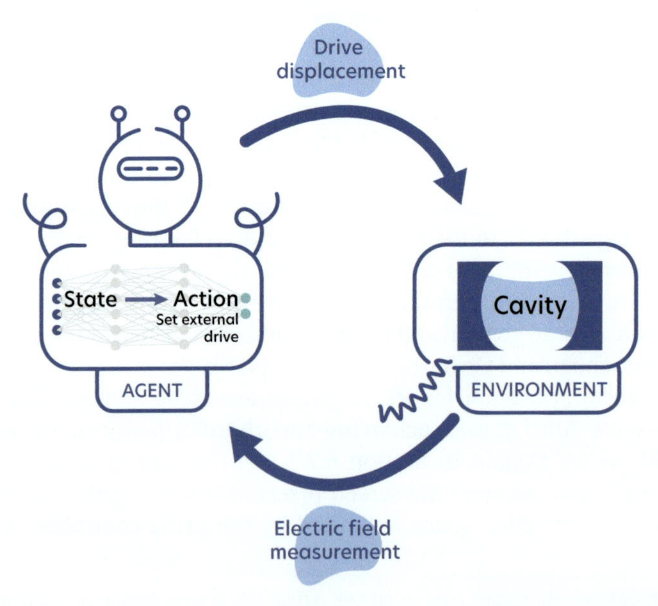

Figure 6.6 Driven single-mode microcavity as an RL environment proposed in [420]. The mode decays from the cavity. Its measurement serves as the observation for the agent represented by an NN. The network converts a measurement trace into probabilities for all the available actions, which give feedback for the displacement drive of the cavity.

the driving laser amplitude. The system evolves under the set parameters for a short period of time. Then, the leaked electric field is measured again and the process is repeated until a time limit is reached.

The agent is trained with a policy gradient approach, as introduced in Section 6.3. During training, the agent eventually finds a strategy to balance the electric field leakage with the proper drive, ending up in the stabilized target cavity state.

6.6.4 Quantum circuit optimization

Quantum computing based on quantum gates requires designing a quantum circuit for a specific quantum algorithm. However, there can be many different sequences of quantum gates implementing the same algorithm. Additionally, due to the fact that quantum gates have nonperfect fidelity, the more gate operations are performed, the more errors appear during the algorithm execution. As such, quantum circuits should be designed in the most optimal way, implementing the least possible number of quantum gates. This is especially important for NISQ devices, which currently allow for >100 qubits [425] but at the same time do not allow for high-level logical quantum error correction.[21] Quantum circuit optimization utilizes the fact that there exist certain sets of transformation rules that allow us to replace sequences of quantum gates by others that yield the same output. For example, these transformations could involve swapping the position of two gates, or moving one gate to a different position relative to another. Furthermore, some sequences of gates can be shortened by merging gates without changing the output.

We can naturally formulate quantum circuit optimization as an RL problem [426]. In the resulting framework, depicted in Fig. 6.7, the environment holds the quantum circuit, containing information about the different gates, such as their error rates. The agent can observe a representation of the quantum circuit, which corresponds to the state, and it can decide to perform a transformation to the circuit from a set of possible transformation rules. The environment can evaluate the resulting circuit after the transformation and provide the agent with a reward. The reward can account for various aspects, such as the reduction in the total gate count, the reduction in depth (the time needed for the circuit to run), or the combination of both. Additionally, the reward function can also depend on a decoherence estimate for the whole circuit, based on the decoherence that happens on each the gates.

This way, the resulting circuit optimization is an autonomous process that can account for specific information about the hardware when choosing the actions, for example, some gates involve longer execution times, or a given qubit may be prone to further errors than others. In the future, quantum compilers will be able to optimize circuits tailored to the hardware specifications and native gate implementation.

6.6.5 Quantum error correction

Whenever we perform any kind of computation, we have to ensure that it is performed flawlessly. In both classical and quantum computation, we need mechanisms to mitigate any possible effect of errors occurring during computations. Whereas classical

[21] In fact, we show how to employ RL methods to tackle quantum error correction in Section 6.6.5.

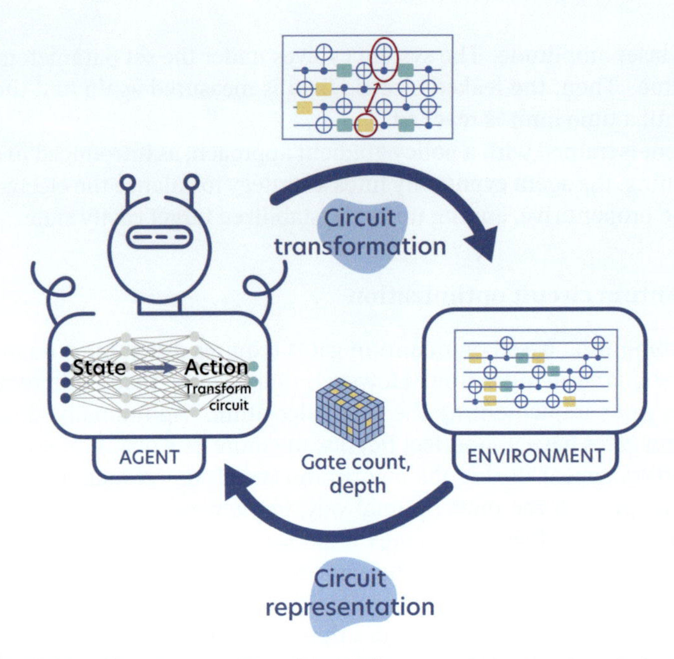

Figure 6.7 Schematic representation of the RL framework for circuit optimization proposed in [426]. The agent observes a representation of a quantum circuit given by the environment. Then, it can choose to perform a modification to the circuit. The environment calculates a reward depending on the gate count (or another metric) of the resulting circuit, and it provides the agent with the new circuit and the reward.

error correction methods have long been established, the current quantum error correction schemes come with a daunting overhead in the number of qubits. Moreover, classical correction schemes cannot be transferred directly to the quantum case, since we can neither simply copy arbitrary quantum states (known as the no-cloning theorem [427]) nor measure the quantum computer's state arbitrarily to find possible errors, as we would erase the state's superposition. Some error correction implementations tackle these challenges using RL methods. Here, we discuss two different approaches.

Error correction with qubit interaction. The first one proposes a suitable error correction scheme from scratch, simply interacting with a collection of qubits [422], as sketched in Fig. 6.8. This approach treats the actual hardware as a black box, and therefore it is versatile regarding the hardware's constraints, as it does not require any prior knowledge about the task. In this setting, the goal is to preserve an arbitrary single-qubit state, $|\phi(0)\rangle = \alpha\,|0\rangle + \beta\,|1\rangle$, over time. For this, the agent can choose to apply gates from a given set, or to perform measurements on auxiliary qubits. This way, any hardware limitation can readily be incorporated by a suitable choice of the available gates, which conforms the action space. Then, we can measure the performance in terms of the fidelity $F = |\langle\phi(T)|\phi(0)\rangle| \in [0, 1]$ after some arbitrary, but fixed time T.

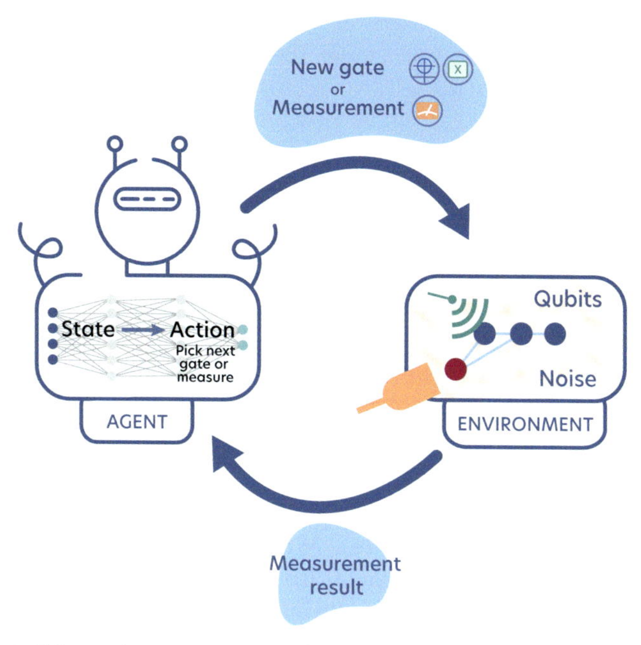

Figure 6.8 Schematic representation of the RL-based error correction framework proposed in [422]. The agent can choose the next gate or measurement to be applied to an ensemble of a few, possibly error-affected, qubits to protect a single target qubit.

However, a naive RL approach is bound to fail when we only consider the fidelity as the reward. Almost all possible circuit transformations reduce the fidelity, thus making random strategies worse than remaining idle. This happens even when considering the fidelity after each new gate or measurement, as the optimal scheme initially decreases the fidelity, and applies a recovery sequence to restore it afterward. Hence, the chance of finding the right gate sequence to protect the state vanishes for large times T. To overcome these challenges, the authors in [422] propose a two-stage learning scheme and a more convenient reward function.

The two-stage learning consist of training two models. First, we train an RL agent that has access to enhanced information with respect to what it is available in an actual device, such as a full description of the multi-qubit state. This also allows us to use a more convenient reward function: the *recoverable quantum information*

$$r_t = \frac{1}{2} \min_{\vec{n}} \|\hat{\rho}_{\vec{n}}(t) - \hat{\rho}_{-\vec{n}}(t)\|_1 , \tag{6.53}$$

where $\vec{n}$ denotes the vector in the Bloch sphere corresponding to the initial state. This reward uses the idea that $\vec{n}$ and $-\vec{n}$ are orthogonal to provide a reward at every time step that guides the agent toward the optimal gate sequence. See [422] for details.

Then, we train the second model using the previous one as a teacher. The second model only has access to the information available in a real device, such as the gates it applies, and the occasional measurement outcomes. Instead of using RL, we train it in a supervised way to mimic the behavior of the first one. The process is analogous to training a supervised classifier in which the labels are the actions of the first model.

The overall process of two-stage learning is way faster and much less computationally demanding than directly solving the original problem. The main limitation is that the teacher model requires a full state description of the multi-qubit system, which limits the application to just a few qubits, and requires a well-characterized noise map of the device that might not be known, in practice.

Error correction with stabilizer codes. Whereas the previous approach aims at discovering the best error correction scheme from direct qubit interaction, the second one implements a quantum code to represent logical qubits [428–430]. In the example we consider here [431], the authors use stabilizer codes to achieve error correction via redundancy. To understand the process, let us build some basic intuition about the stabilizer formalism. Consider a precursory code to correct arbitrary single bit flips of the physical qubits, $|0\rangle \leftrightarrow |1\rangle$ with the encoding

$$|0_L\rangle = \frac{1}{\sqrt{2}}(|000\rangle + |111\rangle)$$

for a single logical qubit state $|0_L\rangle$ in terms of three physical qubits. We can jointly measure subsets of qubits without changing the state with *stabilizer operations*. In this case, we can apply the operations Z_1Z_2 and Z_2Z_3 without altering the qubit state: $Z_1Z_2\,|0_L\rangle = |0_L\rangle = Z_2Z_3\,|0_L\rangle$. Moreover, we can use these operators to detect bit-flip errors on one of the physical qubits, as the stabilizer operators are designed to not alter the erroneous state either.[22] The stabilizer measurements have an outcome of ± 1, and applying them successively we can identify whether any qubit suffered an error to proceed with the correction. The series of outcomes is known as the *syndrome*, and, in practice, these can also have errors.

In this example, we can deal with single bit-flip errors but not with phase errors represented by Z_i operators. For the error correction of arbitrary single-qubit errors, we need five physical qubits with four stabilizer operations [430]. The amount of qubit overhead grows quickly with the number of qubit error classes to cover. The stabilizer code in [431] is a surface code to protect a single logical qubit against arbitrary errors affecting up to d qubits while using, at most, d^2 physical ones.

To properly perform error correction, we need a combination of accuracy, scalability, and speed to detect and correct errors. We can formulate this as an RL

[22]In practice, we would first devise a set of stabilizer operators, and then, we would define the logical 0 and 1 states as the simultaneous eigenstates of all of the stabilizers.

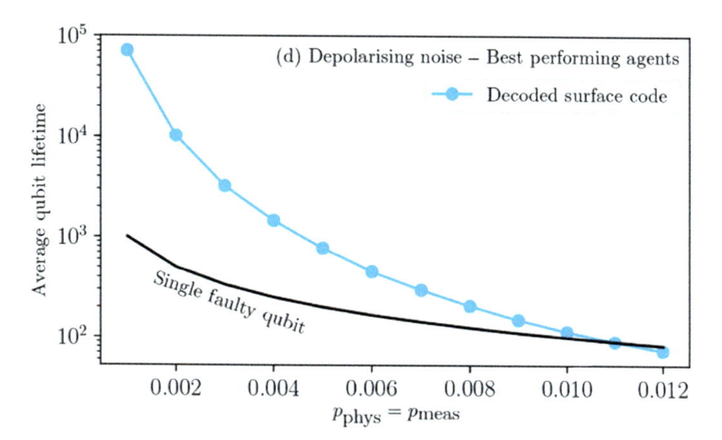

Figure 6.9 Average lifetime of the logical qubit encoded with the surface code. The physical qubits are affected by depolarizing noise with parameter p_{phys} that drastically decreases the unprotected qubit lifetime (black line). With agents trained on various noise levels, p_{phys}, we can dramatically increase the qubit lifetime (blue dots). Taken from [431] under the CC BY 4.0 DEED license.

task [431–434] implementing the full toolbox introduced in this chapter. In the setting from [431], the environment tracks the underlying quantum state, accounting for possible stochastic errors on the physical qubits in the form of depolarizing and bit-flip noise. The agent can choose to perform single-qubit X-, Y-, or Z- rotations,[23] or to perform syndrome measurements. Then, the environment provides the agent with the (possibly faulty) measurement outcome, and a reward, from which the agent can decide the new set of actions to perform. The environment employs a *referee decoder* that checks whether the multi-qubit state after the agent's actions leads to the same logical qubit state. If it is the case, the reward is positive, otherwise, it is negative and the episode terminates.

The authors of [431] consider an agent based on a DQN which they train with Q-learning algorithm, as we explain Section 6.2.3. After training, the average lifetime of the encoded logical qubit can be extended drastically as shown in Fig. 6.9.
Furthermore, they implement an alternative genetic algorithm within the described framework that results into significantly smaller ML models, which are better suited to run in actual devices.

6.6.6 Quantum experiment design

The design of new experiments is key for the development of the quantum sciences. The more complex the applications become, the harder it is to find suitable setups to

[23] Given that XZ = iY, we can reduce the action space in certain cases.

test our ideas. In the context of quantum physics, this can be illustrated in an optical experiment, where we combine different components such that the final quantum state has certain desired properties. For instance, finding the appropriate set of components to create multipartite entanglement in high dimensions is a nontrivial task and usually relies on sophisticated previous knowledge of the states and involved mathematical approaches [435]. Nonetheless, such states are of great importance in applications of quantum information and computation, and hence they are highly coveted.

In [375], the authors propose an autonomous approach to build experiments with RL, using the PS algorithm that we introduce in Section 6.5. The goal is to create high-dimensional many-particle entangled states, based on the orbital angular momentum of light. To do so, the agent has access to a set of optical elements, and the actions consist on placing one of such components in the optical table. The states are the different configurations of optical components in the table. After each placement, the environment analyzes the resulting quantum state generated by the setup. If it corresponds to the desired quantum state, it provides the agent with a reward and the episode ends. If not, the agent continues placing more elements. It is important to note that, due to the presence of noise in optical setups, the more elements, the harder it becomes to correctly find the target quantum state. Hence, the agent is given a maximal number of elements to reach its goal, after which the episode ends and the table resets.

From a technical point of view, the agent has a two-layer ECM: one representing the table configurations (states) and one representing the optimal components (actions). An interesting feature of PS is *action composition*: The agent can create new composite actions from simpler ones that were found useful in previous episodes. In the current context, if the agent finds a particular profitable action sequence leading to a reward, the actions can be added combined as a new single one in the ECM, hence allowing the agent to access rewarded experiments in a single decision step. This way, the agent can distill combinations of components that lead to well-known setups, such as optical interferometers, as well as completely novel ones, such as a nonlocal version of the Mach–Zehnder interferometer.

Hence, we can divide the general task of generating quantum states in two: finding the simplest optical configuration leading to the target state and finding as many experiments as possible that produce it. The former is crucial in terms of practical applications of quantum technologies, as shorter experiments are less noisy, and usually easier to implement. The latter allows us to explore to the full extent all possible solutions to the problem, which may lead to the discovery of new approaches to create the desired quantum states.

The automated design of quantum-optical experiments has also been tackled with non-RL approaches [436–438]. We describe them in more detail in Section 7.3.4.

6.6.7 Building optimal relaxations

In physics we often encounter optimization tasks that we cannot solve in a reasonable amount of time. In these cases, we rely on approximate methods to obtain solutions that are as close as possible to the exact one. There are two paradigmatic approaches: variational and relaxation methods. In the former, we parametrize a family of solu-

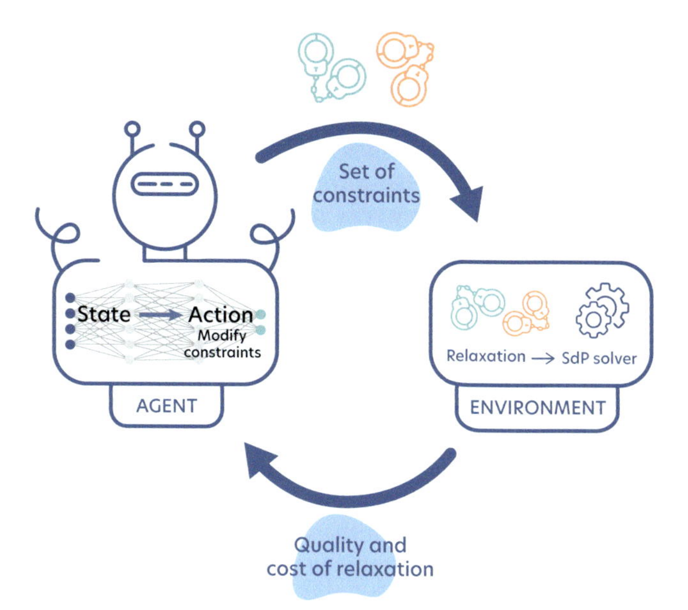

Figure 6.10 Schematic representation of the RL framework to find optimal relaxations. The agent can modify the set of active constraints of a problem with its actions. These constraints go into the environment, which solves the constrained optimization problem. Then, the agent observes a reward that depends on both the result of the problem and the computational cost incurred by the environment. Given this observation, it can decide to further modify the constraints.

tions with the hope that it contains the exact one, such as the variational quantum states introduced in Chapter 5. In the latter, we build a relaxed (easier) version of the problem to provide the optimization process with desirable properties, such as convexity.

Relaxation methods are broadly used in quantum physics, and they lie at the core of quantum information processing. One of the most paradigmatic examples in entanglement theory is the relaxation from the set of separable states to those that are positive under partial transposition (PPT) [439]. Determining whether a state belongs to the first class is hard, whereas it is straightforward to check the membership to the second one. This greatly simplifies the problem of determining whether a state is entangled: We simply need to check it is not PPT. However, while all the product states are PPT, there are some entangled states which also belong to this class, thus resulting into an outer bound to the set of separable states.

Just like with variational methods, we often encounter a trade-off between the computational cost that we can incur and the accuracy of the method. Hence, given a limited computational budget, it is crucial to find the relaxation that best approximates the optimal solution. Nevertheless, there is no clear way to know such an optimal relaxation beforehand. The most common practice relies on exploiting specific knowledge of the given problem, such as symmetries, to build hand-crafted relaxations which, in general, are suboptimal. However, we can combine RL with semidefinite programming to systematically build optimal relaxations [440].

A natural way to build relaxations is to remove or relax constraints of the optimization problem at hand. In the proposed RL framework, presented schematically in Fig. 6.10, the states encode the active constraints of the problem, and the agent can loosen or strengthen them with its actions. The environment acts as a black box that provides the agent with the associated reward to the action and the new state, that is, the new set of constraints. The rewards are engineered to guide the agent toward the optimal relaxation, evaluating both the quality and the cost associated with the current one.

The RL agent is completely agnostic to the problem. Therefore, the method can be applied in a wide variety of relevant problems in physics and optimization, such as entanglement witnessing, optimizing outer approximations to the quantum set of correlations, or finding better sum-of-squares representations of multivariate polynomials, to name a few. In [440], the authors show two applications: finding the ground-state energy of quantum many-body Hamiltonians and building energy-based entanglement witnesses. They can infer properties of the system from the resulting optimal relaxations, such as changes in the ground state, and, even more, they can explore the phase diagram in an autonomous way using transfer learning.

6.7 Outlook and open problems

In this chapter, we have introduced the field of RL and its main paradigms, featuring value-based RL (Section 6.2), policy gradient methods (Section 6.3), and actor-critic algorithms (Section 6.4). Additionally, we have explored other methods that present an alternative approach to RL, such as the PS algorithm (Section 6.5). These lay down the conceptual foundations to understand a whole plethora of other advanced RL techniques while already being competitive, as we have shown in Section 6.6.

In the context of quantum technologies, RL has been widely applied to quantum control problems and, especially, in quantum simulation. With the current boom in quantum computation, many problems involving state preparation, error correction, or controlling and preparing qubits have a natural mapping to the RL framework [441–448]. Furthermore, RL serves as an optimization tool for large problems with a clear structure, with applications as varied as quantum circuit optimization, the design of experimental setups, or the construction of relaxations in quantum information processing problems.

Similar to unsupervised learning, RL is an appealing technique for autonomous scientific discovery, as it does not require explicit fully characterized learning instances. However, while we can identify some previously known strategies in the resulting RL applications, as in the Section 6.6.4 example, there is still the need to develop further analysis techniques to fully understand the nature and rationale behind some of the most prominent results.

A big concern in the field of RL algorithms is data efficiency, which is crucial in applications involving costly experiments or simulations. In this regard, the field of RL can greatly benefit from the latest advances in physics, such as devising optimal exploration strategies for the most challenging problems, or leveraging the latest advances in quantum technologies to enhance RL, as we show in Section 8.2.7.

Further reading

- Sutton, S. R. & Barto, A. G. (2018). Reinforcement Learning: An Introduction. This textbook provides a comprehensive review on RL [378]. Specifically, Chapters 7 and 12 expand the TD concept, and Chapter 13 contains a full complementary derivation of policy gradient, actor-critic, and their application to continuing problems (infinite time).
- Marquardt, F. (2021). *Machine learning and quantum devices.* SciPost Phys. Lect. Notes 29. An introduction to RL for physicists [420].
- Silver, D. *et al.* (2014). *Deterministic policy gradient algorithms.* PMLR, 387–395 [449]. We have introduced policy gradient methods in Section 6.3 with stochastic policies. Here, the authors introduce policy gradient with deterministic policies and its corresponding implementation in actor-critic algorithms.
- Some of the current state-of-the-art algorithms, such as the ones we mention at the end of Section 6.4, feature additional terms in the objective function, usually in the form of an entropy or a Kullback–Leibler divergence. This results in more robust algorithms, and it is tightly close to the formulation of RL as probabilistic inference. We recommend reading [450] for a tutorial, [451] for a prominent algorithm, and [452] for another algorithm, featuring a great overview of the field. The latter proved its performance in the experimental control of a nuclear fusion reactor [453].
- Some of the most prominent applications of RL in quantum technologies are quantum control and error correction. To dive deeper into the quantum control field, we recommend reading [454] for an alternative (model-free) scheme to the one presented in Section 6.6.3. The authors present an experimentally friendly RL framework readily applicable to superconducting circuits and trapped ion platforms. On the quantum error correction side, we recommend reading [455] for a pioneering work demonstrating a fully stabilized and error-corrected logical qubit in a superconducting quantum device. The authors significantly extend the coherence time of the logical qubit using an error correction scheme trained with RL.

7 Deep learning for quantum sciences: Selected topics

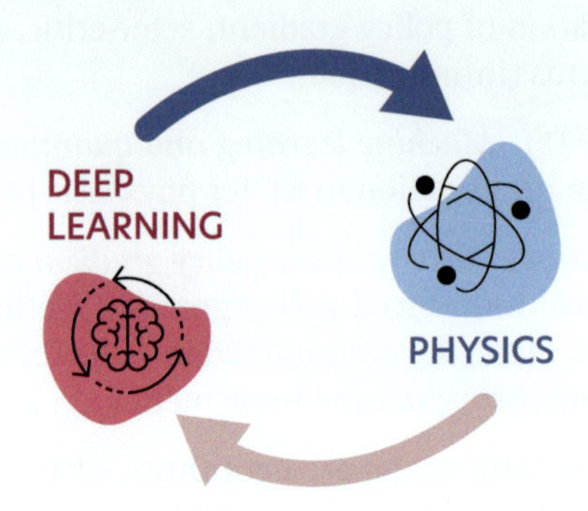

Figure 7.1 There exists a dual relationship between machine learning (ML) and physics. In this chapter, we focus on the more popular direction, where techniques from ML, in particular deep learning (DL), are used to solve problems in physics.

So far, this book has focused on four broad fields at the intersection of quantum sciences and ML: phase classification with unsupervised and supervised ML methods in Chapter 3, use of kernel methods especially in quantum chemistry in Chapter 4, representation of quantum states with ML models in Chapter 5, and use of RL in quantum sciences in Chapter 6. We have presented each of these ideas in detail after a (hopefully) exhaustive introduction. As such, Chapters 3–6 have highlighted a plethora of ML applications in quantum sciences. However, they obviously do not constitute a complete overview of the field.

To fill these gaps, this chapter and Chapter 8 aim at addressing more specialized topics located at the intersection of ML and quantum sciences. This chapter, in particular, discusses further how ML can be used to solve problems in quantum sciences (see Fig. 7.1). We start by explaining the concept of differentiable programming and its use cases in quantum sciences in Section 7.1. Section 7.2 describes generative models and how they can tackle density estimation problems in quantum physics. Finally, we describe selected ML applications for experimental setups in Section 7.3.

7.1 Differentiable programming

Differentiable programming (∂P) represents a fundamental shift in software development that emerged from DL [456]. In "standard" programming, each instruction is explicitly specified in the code, that is, one specifies a point in the *program space* with some desirable behavior (see Fig. 7.2). In ∂P, computer programs are instead composed of parametrized elements of code which can be adjusted. The programmer specifies the desired behavior of the program via a loss function. The space of programs is then searched for a suitable program by tuning the code parameters to minimize the given loss function using derivative information. An example of this

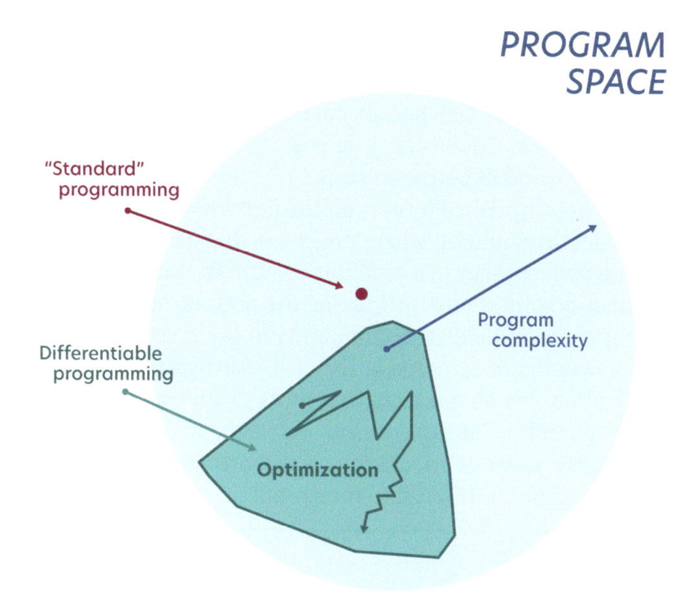

Figure 7.2 Illustration of the difference between "standard" programming and differentiable programming. In some cases, the complexity of programs found by ∂P exceeds human capabilities. Inspired by [456].

which we have continually encountered in this book is the use of backpropagation to efficiently tune the parameters of an NN to solve a given task, such as classifying different phases of matter or representing the ground-state wave function of a quantum many-body system.

In most real-world problems, collecting data in the form of instances in which a given task has been correctly solved is easier than writing a program that solves the task. Under these circumstances ∂P shines because it allows for the program that solves the task to be *learned* from data. This approach can be extremely powerful as demonstrated by the success of programs generated through deep learning. Indeed, as we have extensively discussed in this book, there are nowadays many instances where ∂P has led to algorithms that easily outperform humans, such as in AlphaGo [30].

Differentiable programming also has multiple other advantages compared to conventional programming. One aspect regards the possibility to develop customized optimization strategies. The typical instruction set of NNs consists of matrix multiplication, vector addition, and element-wise application of nonlinearities: Such a set is limited and much smaller, compared to the instruction set associated with the entire class of standard computer programs. This can allow for computational speedups through the design of hardware that is optimized for the limited instruction set underlying ∂P. Graphics processing units (GPUs) and tensor processing units (TPUs) are examples of such application-specific hardware. More recently, neuromorphic computing has emerged as a new paradigm that promises faster and more energy-efficient computation for machine intelligence through hardware systems that mimic the neuronal and synaptic computations of the brain [457, 458].

Differentiable programming also allows for more flexible programming: Consider the situation where you had standard code that performs a certain task and someone wanted you to make it twice as fast, possibly at the expense of its accuracy. This would be a highly nontrivial task. However, it is easy to incorporate such constraints by means of a cost function and hyperparameters in ∂P. For example, given that one uses an NN this could be accomplished by cutting the network's size in half and retraining it. Moreover, consider the situation where programs that were first optimized or coded individually are merged together in a modular fashion to create a new larger program. Then, ∂P offers an easy solution for optimizing the performance of this new program: simple fine-tuning of the individual components in the given configuration through optimization. The benefits of ∂P come at the cost of program interpretability. At the end of the optimization, we obtain code that works well but is very hard to read for a human and understand in intuitive terms. As such, we typically are left with the choice between a fairly accurate model that is understandable in human terms and a more accurate model that is difficult to interpret.[1]

In ∂P, arbitrary computer program structures can be differentiated in an automatic fashion. Importantly, this allows for NNs to be embedded into a plethora of existing scientific simulations and computations because the gradients required for training the NNs can be computed efficiently. In particular, one can differentiate through the NN, as well as surrounding nonparametrized/nontrainable parts of the program.

Recently, widespread interest in ∂P has arisen in the area of scientific computing [459]. Examples of algorithms that have been written in a fully differentiable way are Fourier transforms, eigenvalue solvers, singular value decompositions, or ordinary differential equations (ODEs) [460, 461]. As such, one is able to differentiate through domain-specific computational processes to solve inverse problems, such as learning or control tasks: Tensor networks [462–464], molecular dynamics [465, 466], quantum chemistry [467–476], quantum optimal control [477–488], or quantum circuits [489, 490] have all been formulated in a fully differentiable manner. We discuss several examples in detail in Section 7.1.2.

Notably, ∂P enables scientific ML which combines the best of two worlds: In general, black-box ML approaches are flexible but require a large amount of data to be trained successfully. The amount of required data can be reduced by incorporating our scientific knowledge of the structure of a problem into the program. The training of the parametrized program part is then enabled via ∂P. This allows for the learning task to be simplified because only the parts of the model that are "missing" need to be learned.

Perhaps the biggest feat of ∂P is the ability to compute gradients of loss functions with respect to the NN parameters (see Section 2.5). Recall that we require these gradients for NN training when using gradient-based optimizers (as is typically done). Crucially, the computation is efficient, precise, and occurs in an automated fashion. In particular, it allows for arbitrary NN architectures to be differentiated automatically without implementation overhead. Compare this to the tedious computation of

[1] Again, interpretability appears as a central issue (see Section 3.5).

analytical gradients, which needs to be performed again given different NN architectures.

However, ∂P is not restricted to the computation of gradients with respect to NN parameters for NN training. It enables the automatic computation of gradients and higher-order derivatives of arbitrary program variables. These can, for example, be tunable parameters of a Hamiltonian whose ground state we are interested in. Being able to differentiate through the eigensolver, we can tune the Hamiltonian's parameters via a derivative-based optimizer such that its ground state satisfies desired properties (as specified by a loss function), see Section 7.1.2 for details. This is an example of an *inverse problem* which can be solved efficiently through ∂P. However, the applicability of ∂P goes beyond solving optimization tasks. Gradients and higher-order derivatives contain highly valuable information on the relationship between model parameters and outputs which can, for example, facilitate the interpretation of phase classification methods [164] (see Section 3.5.3) or help to characterize variational quantum circuits [491].

7.1.1 Automatic differentiation

Differentiable programming allows us to compute the gradients and higher-order derivatives of arbitrary computer programs.

In general, methods for the computation of derivatives in computer programs can be classified into four categories [492]: (1) manually working out derivatives and coding them, (2) numerical differentiation using finite difference approximations, (3) symbolic differentiation using expression manipulation,[a] and (4) automatic differentiation (AD) which is the workhorse behind ∂P.

[a]This is done by computer algebra systems such as `Mathematica`, `Maxima`, or `Maple`.

Let us briefly discuss these different approaches.

Manual differentiation is time consuming and prone to errors. *Numerical differentiation* is quite simple to implement. Its most basic form is based on the limit definition of a derivative: Given a multivariate function $f : \mathbb{R}^m \to \mathbb{R}$, the components of its gradient $\nabla f = (\frac{\partial f}{\partial x_1}, \ldots, \frac{\partial f}{\partial x_m})$ can be approximated as

$$\left.\frac{\partial f}{\partial x_i}\right|_{\boldsymbol{x}} \approx \frac{f(\boldsymbol{x} + h\boldsymbol{e}_i) - f(\boldsymbol{x})}{h}, \tag{7.1}$$

where $\boldsymbol{e}_i \in \mathbb{R}^m$ is the ith unit vector and h is a small step size. Approximating ∇f in such a fashion requires $\mathcal{O}(m)$ evaluations of f. This is the main reason why numerical differentiation is not useful in ML where the number of trainable parameters m can be as large as millions or billions. Also note that for the gradient approximation to be somewhat accurate, the step size h needs to be carefully chosen: While the truncation error of the approximation in Eq. (7.1) can be made arbitrarily small as $h \to 0$, eventually round-off errors due to floating-point arithmetic dominate the calculation.[2]

[2]In computing, floating-point numbers are typically represented approximately through a fixed number of significant digits that are scaled through an exponent in some fixed basis $a \times b^c$, where a, b, and c

Symbolic differentiation is the automated manipulation of mathematical expressions for obtaining explicit derivative expressions, for example, by using simple derivative rules such as the product rule

$$\frac{d}{dx}(f(x)g(x)) = \frac{df(x)}{dx}g(x) + f(x)\frac{dg(x)}{dx}. \tag{7.2}$$

Symbolic expressions have the benefit of being interpretable and allow for analytical treatments of problems. However, symbolic derivatives generated through symbolic differentiation typically do not allow for efficient calculation of derivative values. This is because they can quickly get substantially larger than the expression whose derivative they represent. Consider a function of the form $h(x) = f(x)g(x)$ and its derivative, which can be evaluated by the product rule in Eq. (7.2). Note that $f(x)$ and $\frac{df(x)}{dx}$, for example, appear separately in such an expression. A naive calculation of the derivative according to Eq. (7.2) thus involves duplicate computations of any expressions that appear both in $f(x)$ and $\frac{df(x)}{dx}$. Moreover, manual and symbolic methods require the underlying function to be defined in a closed-form expression. As such, they cannot easily deal with programs that involve conditional branches, loops, or recursions. That means, for symbolic differentiation to be efficient, there must exist a convenient symbolic expression for computing the derivative under consideration.

When we are concerned with the accurate numerical evaluation of derivatives and not their symbolic form, it is possible to significantly simplify computations by storing the values of intermediate sub-expressions in memory. This is the basic idea behind *automatic differentiation (AD)*. AD provides numerical values of derivatives (as opposed to symbolic expressions), and it does so by using symbolic rules of differentiation (but keeping track of derivative values, as opposed to the entire symbolic expression). As such, it may be viewed as an *intermediate between numerical and symbolic differentiation*. AD makes use of the fact that every computer program, no matter how complicated it may look, simply executes a sequence of elementary arithmetic operations (e.g., additions or multiplications) and elementary functions (e.g., exp or sin). We refer to the sequence of elementary operations that a computer program applies to its input values to compute its output values as *evaluation trace* [493]. The derivative of every computer program can therefore be computed in an automated fashion through repeated application of the chain rule. As such, the number of arithmetic operations required to compute the derivative is of the same order as for the original program. Moreover, this results in derivatives that are accurate up to machine precision. In the following, we illustrate how AD is done in practice.

Forward-mode AD. Conceptually, AD in so-called forward-mode is the simplest type. Consider the evaluation trace of the function

$$f(x_1, x_2) = \ln(x_1) + \cos(x_2) - x_1x_2 \tag{7.3}$$

given in Table 7.1(left). The associated computation graph is shown in Fig. 7.3, where the computation of a function f is decomposed into variables v_i. We follow the standard notation used in [494], where v_{1-i}, $i = 1, \dots, n$ are the input variables,

are all integers. Because of the limited number of representable numbers, round-off errors can occur when performing computations.

Table 7.1 Workflow for computation of derivatives in forward-mode AD given the function $f(x_1, x_2) = \ln(x_1) + \cos(x_2) - x_1 x_2$. Left: Forward evaluation trace for the choice of initial inputs $(x_1, x_2) = (2, 1)$. Right: Forward derivative trace resulting in the computation of $\frac{\partial f}{\partial x_1}$ at $(x_1, x_2) = (2, 1)$.

$v_{-1} = x_1$	$= 2$	$\dot{v}_{-1} = \dot{x}_1$	$= 1$
$v_0 = x_2$	$= 1$	$\dot{v}_0 = \dot{x}_2$	$= 0$
$v_1 = \ln v_{-1}$	$= \ln 2$	$\dot{v}_1 = \dot{v}_{-1}/v_{-1}$	$= 1/2$
$v_2 = v_{-1}v_0$	$= 2$	$\dot{v}_2 = \dot{v}_{-1}v_0 + \dot{v}_{-1}\dot{v}_0$	$= 1$
$v_3 = \cos v_0$	$= \cos 1$	$\dot{v}_3 = -\dot{v}_0 \sin v_0$	$= 0$
$v_4 = v_1 - v_2$	$= -1.307$	$\dot{v}_4 = \dot{v}_1 - \dot{v}_2$	$= -1/2$
$v_5 = v_3 + v_4$	$= -0.767$	$\dot{v}_5 = \dot{v}_3 + \dot{v}_4$	$= -1/2$
$v_5 = y$	$= -0.767$	$\dot{v}_5 = \dot{y}$	$= -1/2$

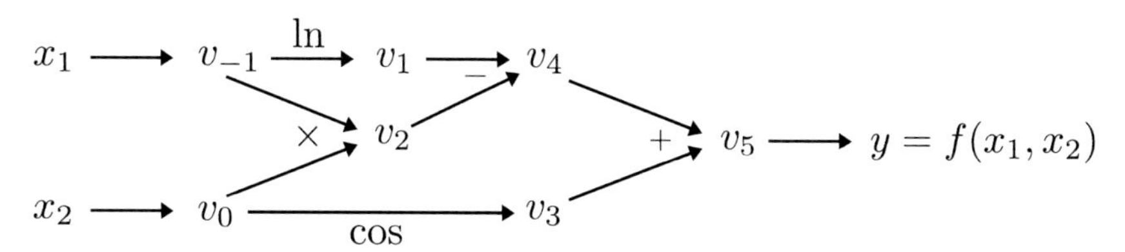

Figure 7.3 Computation graph associated with the forward evaluation trace of $f(x_1, x_2) = \ln(x_1) + \cos(x_2) - x_1 x_2$.

v_i, $i = 1, \dots, l$ are intermediate variables, and v_{l+i}, $i = 1, \dots, m$ are output variables. For computing the derivative of f with respect to x_1, we start by associating with each variable v_i a derivative

$$\dot{v}_i = \frac{\partial v_i}{\partial x_1}. \tag{7.4}$$

Applying the chain rule to each elementary operation in the evaluation trace, we generate the corresponding derivative trace, given in Table 7.1(right). In forward-mode AD, the desired derivative $\dot{v}_5 = \frac{\partial y}{\partial x_1}$ (where y is the output variable) is obtained by computing the intermediate variables v_i in sync with their corresponding derivatives $\dot{v}_i$.

This can be generalized to the computation of the full Jacobian of a function $f : \mathbb{R}^m \to \mathbb{R}^n$ with m input variables x_i and n output variables y_j. In this case, each forward pass of AD is initialized by setting $\dot{x}_i = 1$ for a single variable x_i and zero for the rest. That is, we choose $\dot{\boldsymbol{x}} = \boldsymbol{e}_i$ where $\boldsymbol{e}_i$ is the ith unit vector. The forward pass with given input values $\boldsymbol{x} = \boldsymbol{a}$ then computes

$$\dot{y}_j = \left.\frac{\partial y_j}{\partial x_i}\right|_{\boldsymbol{x}=\boldsymbol{a}} \quad \text{for } j = 1, \dots, n. \tag{7.5}$$

This corresponds to the ith column of the Jacobian matrix

$$J_f = \left[\begin{array}{ccc} \frac{\partial y_1}{\partial x_1} & \cdots & \frac{\partial y_1}{\partial x_m} \\ \vdots & \ddots & \vdots \\ \frac{\partial y_n}{\partial x_1} & \cdots & \frac{\partial y_n}{\partial x_m}. \end{array}\right]\Bigg|_{x=a} . \tag{7.6}$$

Thus, the full Jacobian can be computed in m forward passes, that is, m evaluations of the function f. As such, forward-mode AD is efficient if $m \ll n$. In the other limit, so-called reverse-mode AD is preferred which we discuss shortly.

In practice, forward-mode AD is implemented by augmenting the algebra of real numbers and introducing a new arithmetic: To every number, one associates an additional component with the derivative of a function computed at that particular value. We call this composite number a *dual number*

$$\text{dual}(v) = v + \dot{v}\epsilon, \tag{7.7}$$

where $\epsilon \neq 0$ is a number such that $\epsilon^2 = 0$. The extension of all arithmetic operators to dual numbers allows for the dual number algebra to be defined. Observe, for example, that

$$f(\text{dual}(v)) = f(v) + \dot{f}(v)\dot{v}\epsilon, \tag{7.8}$$

where we obtain the function value in the first part and the corresponding derivative $\dot{f}(v)\dot{v}$ in the ϵ part.[3] This follows from expanding the function in its Taylor series and noting that terms $\mathcal{O}(\epsilon^2)$ vanish due to the property that $\epsilon^2 = 0$. Equation (7.8) resembles the computation of the derivative using the chain rule.

Reverse-mode AD. As the name suggests, in reverse-mode AD the derivatives are propagated backward from a given output.[4] This is in contrast to forward-mode AD where we saw that the derivatives are propagated forward in sync with the function evaluation. Reverse-mode AD is done by complementing each intermediate variable v_i with a so-called adjoint

$$\bar{v}_i = \frac{\partial y_j}{\partial v_i}, \tag{7.9}$$

where y_j is the output variable with respect to which we desire to compute derivatives. In reverse mode AD, derivatives are computed in the second phase of a two-phase process. In the first phase, the original function code is run forward: Intermediate variables v_i are populated and their dependencies in the computational graph are tracked through a bookkeeping procedure. In the second phase, derivatives are calculated by propagating adjoints $\bar{v}_i$ in reverse, that is, from the outputs to the inputs. This

[3]Under the hood dual numbers are typically handled through so-called operator overloading, that is, overloading all functions to work appropriately on the new algebra.

[4]Historically, reverse-mode AD can be traced back to the master thesis of Seppo Linnainmaa in 1970 [495] in which he described explicit, efficient error backpropagation in arbitrary, discrete, possibly sparsely connected, NN-like networks [496].

Table 7.2 Workflow for computation of derivatives in reverse-mode AD given the function $f(x_1, x_2) = \ln(x_1) + \cos(x_2) - x_1 x_2$. Left: Forward evaluation trace for the choice of initial inputs $(x_1, x_2) = (2, 1)$. Right: Reverse (adjoint) derivative trace resulting in the computation of $\frac{\partial f}{\partial x_1}$ and $\frac{\partial f}{\partial x_2}$ at $(x_1, x_2) = (2, 1)$.

Forward trace		Reverse trace	
$v_{-1} = x_1$	$= 2$	$\bar{v}_5 = \bar{y}$	$= 1$
$v_0 = x_2$	$= 1$		
$v_1 = \ln v_{-1}$	$= \ln 2$	$\bar{v}_4 = \frac{\partial v_5}{\partial v_4}\bar{v}_5 = \frac{\partial v_3 + v_4}{\partial v_4}\bar{v}_5$	$= 1$
$v_2 = v_{-1}v_0$	$= 2$	$\bar{v}_3 = \frac{\partial v_5}{\partial v_3}\bar{v}_5$	$= 1$
$v_3 = \cos v_0$	$= \cos 1$	$\bar{v}_2 = \frac{\partial v_4}{\partial v_2}\bar{v}_4$	$= -1$
$v_4 = v_1 - v_2$	$= -1.307$	$\bar{v}_1 = \frac{\partial v_4}{\partial v_1}\bar{v}_4$	$= 1$
$v_5 = v_3 + v_4$	$= -0.767$	$\bar{v}_0 = \frac{\partial v_2}{\partial v_0}\bar{v}_2 + \frac{\partial v_3}{\partial v_0}\bar{v}_3$	$= -2.841$
		$\bar{v}_{-1} = \frac{\partial v_1}{\partial v_{-1}}\bar{v}_1 + \frac{\partial v_2}{\partial v_{-1}}\bar{v}_2$	$= -1/2$
$v_5 = y$	$= -0.767$	$\bar{v}_0 = \bar{x}_2$	$= -2.841$
		$\bar{v}_{-1} = \bar{x}_1$	$= -1/2$

is illustrated in Table 7.2 for the function given in Eq. (7.3), where the reverse pass is started with $\bar{v}_5 = \bar{y} = \frac{\partial y}{\partial y} = 1$. As a result, we obtain both $\bar{x}_1 = \frac{\partial y}{\partial x_1}$ and $\bar{x}_2 = \frac{\partial y}{\partial x_2}$ in a single reverse pass.

This example illustrates the complementary nature of the reverse mode compared to the forward mode: The reverse mode is cheaper to evaluate than the forward mode for functions with a large number of inputs, that is, where $m \gg n$ with $f : \mathbb{R}^m \to \mathbb{R}^n$. As we just saw, in the extreme case of $f : \mathbb{R}^m \to \mathbb{R}$, only one application of the reverse mode is sufficient to compute the full gradient compared with the m passes of the forward mode. The typical case encountered in ML applications corresponds to the evaluation of the derivatives of a loss function $y_j = \mathcal{L} : \mathbb{R}^m \to \mathbb{R}$ with respect to m trainable parameters, where m is typically large. As such, reverse-mode AD is the preferred method for computing gradients automatically as it is computationally more efficient compared to forward-mode AD.[5] In the context of ML, reverse-mode AD applied to NNs is typically referred to as backpropagation; see Section 2.5. It is the working horse behind NN training as it allows for efficient computation of the gradients for arbitrary NN-based architectures in an automated fashion. In the following, we discuss several ways how reverse mode AD is implemented in practice.

[5] Note that forward-mode and reverse-mode AD are just two (extremal) ways of applying chain rules. Finding the optimal way to traverse the chain rule to compute a Jacobian for a given function (i.e., the choice that results in the smallest number of arithmetic operations) is known as the optimal Jacobian accumulation problem and is NP complete.

Static graph AD. A basic implementation of reverse-mode AD makes use of static computation graphs. This choice is natural, given that we chose to illustrate reverse-mode AD using computation graphs [497]. TensorFlow is an example of a platform that uses this approach. Here, the user must define variables and operations in a graph-based language. Subsequent executions of the computation graph allow for the program to be differentiated in a straightforward manner. However, this requires all existing programs to be rewritten as a static computation graph which is inconvenient.

Tracing-based AD. This can be circumvented by building computation graphs *dynamically* at runtime which is achieved by "tracing" all the operations encountered in the forward pass given a particular input [497]. Dynamic computation graphs are the basis of many reverse-mode AD implementations in Julia (Tracker.jl, ReverseDiff.jl, or Autograd.jl) or Python (PyTorch, TensorFlow Eager, and Autograd [JAX]). The fact that it is simple to implement makes this approach widely adopted in practice. An issue of such tracing-based implementations is that each trace is value dependent, meaning that each run of a program (with different inputs) can build a new trace. Moreover, these traces can be much larger than the code itself, for example because loops are completely unraveled.

Source-to-source AD. In source-to-source AD, one overcomes these issues by generating source code for the backward pass that is able to handle all input values [497]. In particular branches, loops, and recursions are not explicitly unrolled. The right branch in the reverse passes through recall of the intermediates values used in the forward pass. It turns out that the implementation of a source-to-source AD system poses many requirements on the underlying language.[6] Source-to-source AD is used in programming languages such as Julia (Zygote.jl).[7]

High-level adjoint rules. The advantages of reverse-mode AD in ML applications come at the cost of increased storage requirements which (in the worst case) is proportional to the number of operations in the evaluated function. This is because the values of the intermediate variables populated during the forward pass need to be stored when using reverse-mode AD, whereas they can be directly used for the derivative computation within forward-mode AD. Improving storage requirements in reverse-mode AD implementations is an active research area. In general, reverse-mode AD can be made more efficient by deriving adjoint rules at a higher level. Consider, for example, the case where your program involves solving a nonlinear problem $f(\boldsymbol{x}, \boldsymbol{p}) = 0$ with an iterative method, such as Newton's method. A naive application of the reverse-mode AD system results in a backward pass through all iteration steps. Not only is this computationally expensive but also requires storing the values of all intermediate iteration steps. Instead of unrolling the entire computation, one can analytically derive an appropriate adjoint rule which can be used to compute the derivatives in reverse-mode via a separate linear equation. In particular, it only

[6] In particular, it should possess a strong internal graph structure.

[7] TensorFlow considered building a source-to-source AD based on the Swift language. Older AD systems for Fortran were also source-to-source.

requires knowledge of the final solution $\boldsymbol{x}$ of the nonlinear problem as opposed to values in the intermediate iterations. For further details, see [239, 460]. Other examples for which adjoint rules can be derived also include ODEs [461, 484, 497, 498] and eigenvalue solvers [460]. In the case of ODEs, for example, the adjoint rule involves solving a second, augmented ODE backward in time.

7.1.2 Application to quantum physics problems

In this section, we illustrate the application of ∂P to problems from quantum physics through two simple examples.

Inverse Schrödinger problem. As a concrete example of an inverse-design problem, we consider the time-independent Schrödinger equation in one dimension

$$\left[-\frac{1}{2}\frac{d}{dx^2} + V(x)\right]\Psi(x) = E\Psi(x), \tag{7.10}$$

where we set $\hbar = m = 1$ [460]. Typically, we are given a potential $V(x)$ and solve for the corresponding eigenfunctions Ψ and energies E. Here, we consider the inverse problem. Given a particular wave function $\Psi(x)$ we want to construct a potential $V(x)$ with a ground-state wave function $\Psi_0(x)$ that closely matches $\Psi(x)$. We restrict ourselves to the domain $x \in [-1, 1]$ and define the following mean-squared error (MSE) function

$$\mathcal{L} = \int_{-1}^{1} |\Psi(x) - \Psi_0(x)|^2 dx. \tag{7.11}$$

The potential $V(x)$ is discretized on a grid and each individual amplitude is tuned using gradient-based optimization methods to minimize Eq. (7.11). An implementation of this problem in JAX can be found at [499]. The results are illustrated in Fig. 7.4. Here, the gradient is calculated by the AD system underlying JAX and involves propagating the derivative through the eigenvalue solver. For more details on ∂P applied to inverse-design problems in quantum mechanics and adjoints for eigensolvers, see [500].

Quantum optimal control. Next, we consider a problem from quantum optimal control. We would like to find the time-dependent amplitudes $\{u_i(t)\}$ of the following Hamiltonian:

$$\hat{H} = \hat{H}_0 + \sum_{i=1}^{n_{\text{ctrl}}} u_i(t)\hat{H}_i, \tag{7.12}$$

such that the time evolution under $\hat{H}$ realize a CNOT gate. Here $\hat{H}_i$ with $i \in [1, n_{\text{ctrl}}]$ are time-independent Hamiltonians that can be tuned through the time-dependent amplitudes $\{u_i(t)\}$. We parametrize these amplitudes using Fourier series $u_i(t) = \sum_{j=1}^{n_{\text{basis}}} u_{ij} \sin(\pi jt/T)$, where we introduce n_{basis} as a cutoff enabling numerical evaluations. To find the gate which a given choice of $\{u_i(t)\}$ implements, we integrate the

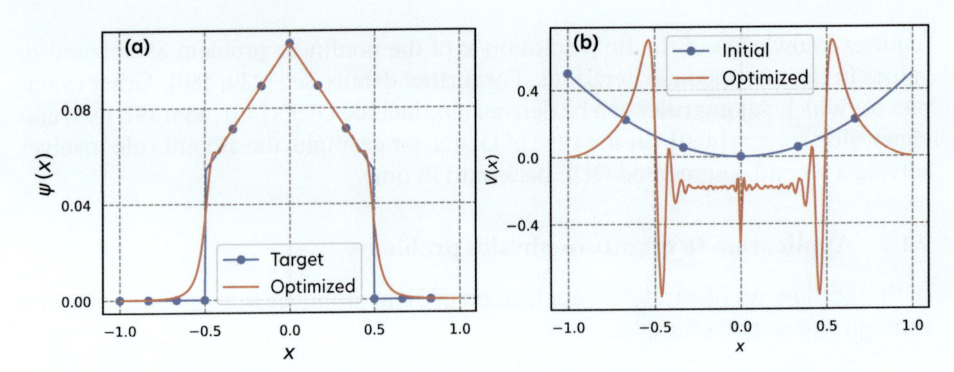

Figure 7.4 (a) Optimized ground-state wave function $\Psi_0(x)$ after $\approx 1\,000$ iterations of the L-BFGS algorithm with box constraints (L-BFGS-B) given the target wave function $\Psi(x) = 1 - |x|$ with $|x| < 0.5$. (b) Optimized potential $V(x)$ (rescaled by 1/300) and the initial harmonic potential. Figure reproduced from [499].

time-dependent Schrödinger equation from 0 to T under initial conditions $U(0) = \mathbb{1}$ with

$$\frac{dU}{dt} = -i\hat{H}(t)U, \tag{7.13}$$

where we set $\hbar = 1$. Ideally, $U(T)$ realizes a CNOT operation ($U_{\text{target}} = \text{CNOT}$). Hence, we setup our loss function as

$$\mathcal{L} = 1 - \frac{1}{d}|\text{tr}\,(U(T)^{\dagger}U_{\text{target}})|, \tag{7.14}$$

where d is the dimension of the associated Hilbert space. Note that when $U(T) = U_{\text{target}}$ we reach the global minimum of $\mathcal{L} = 0$. The coefficients $\{u_{ij}\}$ are tuned to minimize the loss function in Eq. (7.14) using gradient-based optimization methods. An implementation of this problem in JAX can be found at [501], where the gradient is calculated via AD in JAX and involves propagating the derivative through the ODE solver (Eq. (7.13)). For more details on differentiable programming applied to quantum optimal control, see [132, 478].

7.1.3 Outlook and open problems

In this chapter, we have introduced the novel programming paradigm that is ∂P. Most notably, ∂P enables NN training via the efficient, precise, and automated calculation of the corresponding gradients (see Section 2.5). Having the ability to differentiate arbitrary computer programs, in addition, allows for NNs to be seemingly incorporated in scientific workflows. By now there are many applications of ∂P in scientific computing, including quantum physics. However, the field is still in its infancy and many open problems remain to be tackled. For instance, finding efficient high-level adjoint rules for algorithms used in quantum physics problems, such as solvers for stochastic dynamics, is still a current topic of research [487]. Another example are chaotic systems for which standard AD methods can fail [502, 503]. The development of AD systems is also still an ongoing effort: In Enzyme.jl [504], for example, the idea is to perform reverse-mode AD on the portable, low-level intermediate representation of Julia which is language-agnostic. This allows for performance improvements due to low-level optimizations. NiLang.jl [505] on the other hand tries to build a reverse-mode AD system based on the paradigm of reversible programming. Running the program in reverse in the backward pass allows the overhead in memory in standard reverse-mode AD to be circumvented.

Further reading

- Baydin, A. G. *et al.* (2018). *Automatic differentiation in machine learning: A survey*. J. Mach. Learn. Res. 18(1), 5595–5637. Good overview on AD in the context of ML [492].
- Innes, M. *et al.* (2019). *Differentiable programming system to bridge machine learning and scientific computing*. arXiv:1907.07587. Discussion on the role of ∂P in scientific computing [459].
- Google Colab notebooks by Lei Wang [499, 501].

7.2 Generative models in many-body physics

Deep generative models are (mostly) neural network (NN) architectures designed to approximate the probability density underlying a system or a dataset we aim to describe. This task of constructing the probability density of a given problem is often called *density estimation* [506]. When such an underlying probability density is learned, one can then sample the obtained density and generate artificial new samples that are characteristic of the problem (hence the name "generative" models).

The difficulty of the density estimation task comes from the fact that underlying probability densities cannot be computed exactly in most cases. For a large class of problems, this issue can be attributed to the difficulty of calculating normalization constants (i.e., partition functions in statistical mechanics). As such, one often has to resort to different methods and techniques that approximate underlying probability density functions in which deep generative models are very useful.

Throughout this book, the reader encounters various examples of density estimation tasks in quantum physics. For example, finding a (variational) representation of a quantum-many body system wave function can be viewed as a density estimation task, see Chapter 5. More generally, in quantum physics, the concept of density estimation often appears in the context of reconstructing the many-body state from measurements performed on a quantum many-body system – a task known as *quantum state tomography*, described in more detail in Section 5.3.7. This is particularly challenging because only a reduced set of quantities, such as single-body density matrices or higher-order correlation functions, is experimentally accessible [507–509]. Moreover, the tomography experiment can be very demanding, and a single experimental run can take a long time (i.e., hours or days). One, therefore, has to face the challenge of estimating a high-dimensional probability density from a rather small number of measurements of a restricted set of observables.

In general, one can think of two different approaches to density estimation: *parametric* and *nonparametric* density estimation. In the parametric approach, one fixes a parametrized functional form for the approximated density. The free parameters are then tuned such that the trial density best matches the density of the system under consideration. This can be done by comparing the trial density against training data, which corresponds to observations drawn from the target density, or against an unnormalized target probability density (e.g., an unnormalized Boltzmann distribution). A simple example of a parametrized trial density would be a Gaussian, where its mean and variance can be adjusted accordingly. In parametric approaches, one reduces the problem of finding an appropriate density function to the problem of finding appropriate parameters. Clearly, the choice of the functional form of the density is crucial for the success of this approach, as it can substantially restrict the family of target densities that can be effectively approximated through the chosen ansatz.

By contrast, for nonparametric approaches, the structure of the trial density function is not set *a priori*.[8] Instead, a trial density function is directly constructed based on training data. The simplest example of a nonparametric approach to density estimation corresponds to building a histogram. Clearly, histogram binning comes with some drawbacks. For example, one must carefully choose the size and location of bins. Moreover, histograms are nondifferentiable functions. In general, nonparametric methods cannot leverage a priori information about the system at hand, contrary to parametric methods. While this makes nonparametric methods robust and applicable to general problems, they typically also require a generous number of samples to reach a suitable level of accuracy.

7.2.1 Training with or without data

In recent years, several new methods for density estimation have emerged, in particular from the interplay between ML and physics. These approaches are typically parametric in nature: the parameters $\boldsymbol{\theta}$ of a suitable variational (parametric) ansatz for the trial density $q_{\boldsymbol{\theta}}$ are optimized to match a target density p. The trial density $q_{\boldsymbol{\theta}}$ is sometimes referred to as *the model*. For now, let us assume it to have a generic form while we discuss the precise parametrizations of $q_{\boldsymbol{\theta}}$ in Section 7.2.2.

A natural way to match the trial and target densities is based on the maximum likelihood estimation (MLE) principle, where a likelihood function is maximized (or equivalently, a negative log-likelihood is minimized) to optimize the parameters $\boldsymbol{\theta}$. For a given dataset of independent observations $\mathcal{D} = \{\boldsymbol{x}_i\}_{i=1}^n$, the log-likelihood function of the model is defined as

$$\ell(\boldsymbol{\theta} \mid \mathcal{D}) = \log\left[\prod_{i=1}^{n} q_{\boldsymbol{\theta}}(\boldsymbol{x}_i)\right] = \sum_{i=1}^{n} \log q_{\boldsymbol{\theta}}(\boldsymbol{x}_i)\,, \qquad \boldsymbol{x} \in \mathcal{D} \sim p \tag{7.15}$$

where $q_{\boldsymbol{\theta}}(\boldsymbol{x})$ is the probability density of the model evaluated at each independent observation of the system $\boldsymbol{x}$, sampled from the true density p that is to be approximated. The set of observation $\mathcal{D}$ can, for example, refer to images of dogs, spin configurations $\boldsymbol{x}$ drawn from an Ising model at a fixed temperature, as well as a set

[8] The trial density function in nonparametric approaches to density estimation does not lack parameters entirely but rather is not fixed in advance, that is, is learned from scratch.

of molecule conformations from density functional theory calculations. The model trained on one such dataset can then be sampled to generate, respectively, different images of dogs (bearing similar features to dogs seen during the training), unseen Ising spin configurations, and new molecule conformations.

The MLE optimization requires training samples and is therefore often described as *learning from data*. However, in some cases, data may not be readily available, for example, because it is difficult or expensive to generate. For such cases, the data-driven procedure just described may not be applied. In physics, however, we sometimes have the advantage of knowing the closed form of the underlying density up to a normalization constant (a.k.a. the partition function in the Boltzmann distribution of a thermodynamic system). Then, we can use an alternative approach to density estimation, sometimes referred to as *variational inference*. This approach is based on encoding physical laws and prior knowledge of the physical system into a family of parameterized densities $\{q_\theta\}_\theta$ and choosing the best parameters by minimizing the so-called *reverse* Kullback–Leibler (KL) divergence [510], between the target density p and the variational ansatz q_θ,

$$\mathrm{KL}(q_\theta || p) = \int_\Omega q_\theta(\boldsymbol{x}) \log \frac{q_\theta(\boldsymbol{x})}{p(\boldsymbol{x})} d\boldsymbol{x}, \tag{7.16}$$

where Ω is the relevant integration domain. This quantity measures the discrepancy between the target density and the trial density.[9] Importantly, it can be interpreted as an expectation value with respect to the model q_θ such that its evaluation only requires $\{q_\theta\}_\theta$ to be easily sampled.[10] This learning objective for the parameters θ does not require either to know the normalization constant of p as it only amounts to a constant shift. We stress that the optimization through reverse KL divergence is equivalent to the variational free-energy principle in statistical mechanics [88, 89].

If the variational approach optimizes the reverse-KL divergence, the maximum likelihood MLE approach is equivalent to optimizing the so-called forward-KL divergence which reads

$$\mathrm{KL}(p || q_\theta) = \int_\Omega p(\boldsymbol{x}) \log \frac{p(\boldsymbol{x})}{q_\theta(\boldsymbol{x})} d\boldsymbol{x}. \tag{7.17}$$

Indeed, Eq. (7.17) can be interpreted as an expectation over p from which the previously introduced Eq. (7.15) estimates the parameter-dependent term. For a general overview of both data-driven and data-free methods, see [511, 512].

As a side remark, we note that another popular type of learning objective is adversarial training which we describe very briefly in Section 7.2.2. However, this is not extensively used in physics applications as it is unstable and requires a large amount of data.

The remainder of this chapter is structured as follows. In Section 7.2.2, we introduce different popular types of deep generative models. In Section 7.2, we put the emphasis on a family of models particularly useful in physics applications: normalizing flows (NFs) (Section 7.2.3).

[9] Note that this metric may lead to some pathological behavior, already mentioned in Section 2.3.

[10] Conversely to the maximum likelihood objective with requires i.i.d. draws from p.

7.2.2 Taxonomy of deep generative models

Deep generative models, also called generative neural samplers (GNSs), are ML models with NNs used to define probability distributions. They now come in a large variety, each with different properties. In Sections 5.2.1 and 5.2.2, we have already encountered restricted Boltzmann machines (RBMs) [513] and autoregressive neural networks (ARNNs) [514]. Therefore, we discuss them in this section only briefly. Moreover, we describe here variational autoencoders (VAEs) [515] and generative adversarial networks (GANs) [516, 517]. Finally, we dedicate the whole Section 7.2.3 to NFs [518–520]. Another type of generative models goes under the name of a diffusion model [521, 522]. While being very promising, their deployment in the field of quantum science is still very limited, and for this reason, we do not discuss them further. The interested reader can, however, find a comprehensive review in [522].

Choosing between the different types of generative models can reflect some prior knowledge of the system. For example, we can use VAEs or GANs to encode a density that can be represented by the low-dimensional manifold. Additionally, there is typically a trade-off between the capacity of the model and its tractability. This means that, often, more expressive models are at the same time harder to train and therefore difficult to manipulate. In particular, the choice of architecture determines whether the model comes with a direct sampling method or whether it will be inevitable to resort to the more intricate Markov chain Monte Carlo (MCMC) sampling algorithm to generate samples.[11] Moreover, in physics applications, it is highly desired to have access to a *tractable density*. Specifically, this is necessary to apply reweighting techniques such as neural importance sampling (NIS) [89, 523, 524] and neural Markov chain Monte Carlo (NMCMC) [89, 525, 526] as will be further discussed in Section 7.2.3.

> If a model gives access to a *tractable density*, it means it allows to compute the normalized modeled density $q_\theta(\boldsymbol{x})$ for any $\boldsymbol{x}$ in polynomial time in the system size N.

Energy-based models. Historically, energy-based models, such as binary Boltzmann machines and RBMs, were among the first proposed deep generative models. They were heavily inspired by statistical physics (see Section 5.2.1 for more details) as they parametrize the logarithm of the probability density directly, which is equivalent to parametrizing an energy in a Boltzmann distribution. Despite being flexible and very elegant in their formulation, these models are very hard to train because their normalization constant remains intractable. As a result, Boltzmann machines are learned from data through an *approximate* maximum likelihood principle [527, 528]. Moreover, sampling from RBMs is not as efficient as for other GNSs as there exists no direct sampling method. Therefore, while RBMs have been extensively used in the early stages of density estimation in quantum physics [513], they are not always the most suitable tool for density estimation.

[11] Even MCMC's performance is hindered by the correlation between samples (discussed widely in Section 5.2.2) and can severely fail in the case of multimodal distributions (where different blobs of probability are far apart in the configuration space).

Variational autoencoders (VAEs). When it comes to sampling efficiency, developments from the last decade introduced more sophisticated algorithms that significantly simplified sampling configurations from approximated densities. Two major examples are VAEs and GANs. While both algorithms push a low-dimensional random variable through an NN, VAE only pursues the maximization of a lower bound of the *intractable* likelihood [79, 80]. However, VAEs allow to draw new configurations very efficiently.

Generative adversarial networks (GANs). Instead GANs are trained following the so-called adversarial objective [516]. They consist of two NNs called a generator and a discriminator. The generator learns to generate new data, while the discriminator learns to distinguish between real and generated data. The generator and discriminator are trained together in an adversarial manner, where a generator tries to fool the discriminator. This type of game-theoretic optimization strategy may result in very unstable training routines. However, when successful, the joint optimization of those two agents makes the generator more and more capable of generating very realistic samples. This approach works well in practice and allows for easy sampling but, unfortunately, does not allow access to the likelihood of the model. Hence, GANs are not the most suitable for physics applications.

Autoregressive (AR) models. Finally, we turn our attention to deep generative models that give access to a tractable normalized density, namely, ARNNs and NFs. For reasons that will become apparent, we put particular emphasis on NFs and dedicate it to the whole Section 7.2.3. For now, let us discuss ARNNs. The main advantage of using an AR model is that it is easily trained through maximum likelihood or variational inference thanks to its closed-form likelihood and its uncorrelated and fast sampling [529]. The samples are uncorrelated thanks to the AR structure of the probability distribution, which allows for applying a direct sampling algorithm rather than using an MCMC. Direct sampling with AR models has been described in detail in Section 5.2.2, we remind the reader that it goes as follows: sample state $\boldsymbol{x}_1$ from $q_\theta(\boldsymbol{x}_1)$, then sample from $q_\theta(\boldsymbol{x}_2 \mid \boldsymbol{x}_1)$, and so on until $q_\theta(\boldsymbol{x}_N \mid \boldsymbol{x}_{N-1}, \dots, \boldsymbol{x}_1)$. Direct sampling from a probability distribution over an exponentially large configuration space (e.g., if the $\boldsymbol{x}_i$ are binary variables) is possible with AR models and makes them extremely powerful. The sampling can also be made faster as many samples can be processed in parallel in a single pass. Note that imposing an AR structure on a model constrains it: for instance, to build a deep CNN, one must introduce masked layers (to keep the AR structure), which alters the capacity of such models.

Recurrent neural networks (RNNs). A special type of AR models are RNNs introduced already in Section 2.4.6. Among RNNs, there is a plethora of sub-models, such as the Long-Short Term Memory (LSTM) and the Gated Recurrent Unit (GRU). The latter has been used in the context of ground state search [266]. In the context of statistical mechanics, an RNN model has been applied to complex problems such as spin glasses. These models are typically hard to sample from. Therefore, the direct sampling helped reach a high accuracy in a simulated annealing task [274].

While AR models crucially give access to tractable normalized densities, they are originally limited to discrete random variables and are affected by a relatively inefficient sampling process. That is, AR follows an *ancestral sampling* scheme where new samples, for example, new images, are built sequentially, constraining, for instance, new pixels on the image on previously sampled ones [530]. On the other hand, NFs also allow for exact likelihood density estimation, but they are also more flexible, can deal with continuous random variables, and allow for even faster, more efficient sampling. This is why we focus on this example in Section 7.2.3.

We conclude this section with a few remarks. This lightning overview suggests that the most interesting GNSs for physics applications are those giving direct access to exact (or approximate) normalized probability densities. An overview of deep generative models with their appealing properties and applications in physics can be found in [89]. We hereby list some of these GNSs and provide some references where they have been used in quantum physics: ARNNs [88–92], RNNs [266, 274], and latent variable models such as VAEs [531] and NFs [524, 525, 532].

7.2.3 An important example: normalizing flows

Normalizing flows share similar advantages to AR models though they deal with continuous random variables and allow for more efficient sampling (see [519, 520] for a review). Recently, many works, among which we mention [524, 525, 532], successfully used NFs for sampling configurations in lattice quantum field theory. While examples from this section focus on sampling methods for lattice field theories, the application of flows is indeed not limited to this specific domain. We refer to [533] for a broader overview of applications of deep generative models also in the context of statistical physics and chemistry.

Normalizing flows' construction. NFs rely on the simple idea of reparametrizing a variable of interest using a change of coordinates transformation. They generate new realizations of the variable as follows: First, a latent random variable is sampled from a tractable, known distribution (e.g., a normal or a uniform). The random variable transforms through a bijective mapping learned under specific constraints. Namely, the so-called trivializing map is learned so as to transform the base probability density of the base distribution into one that approximates a given target density. As detailed below, leveraging the change of variable formula, one can thus obtain a variational approximation of the target density, properly normalized, and easy to handle. In other words, NFs transform some Gaussian noise into real samples from a learned bijective mapping.

> In an NF, we define a diffeomorphic[12] transport map that redistributes the probability mass from an easy-to-treat base distribution to a more complicated target density.

[12] A function is diffeomorphic if it is differentiable and has a differentiable inverse.

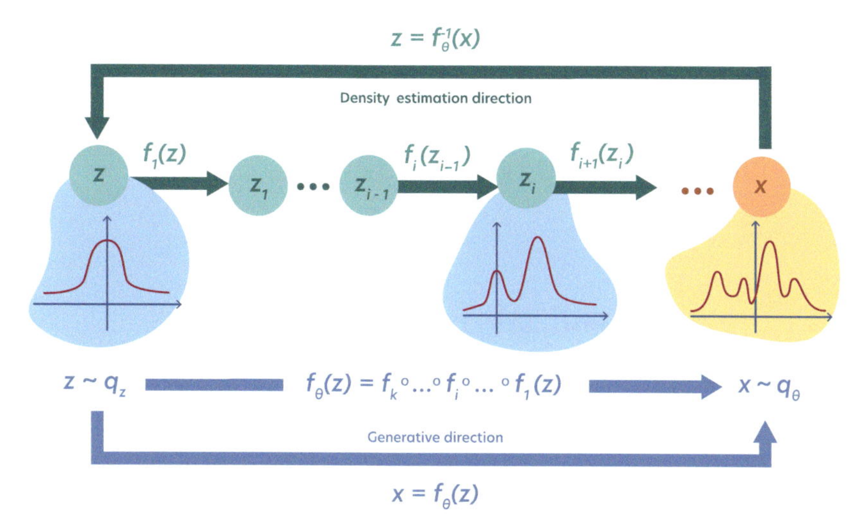

Figure 7.5 Sketch of an NF architecture. A sequence of bijective transformations are combined to construct a complicated nonlinear invertible transformation transporting the probability mass from a base distribution to a learned density. The learned density is tractable and properly normalized, which enables fast and efficient sampling. Additionally, through the learned density, new samples can be generated and the exact likelihood of given configurations can be computed.

We define a latent variable $z \in \mathbb{R}^D$ distributed according to a known, easy-to-treat distribution. A common choice is a standard normal $q_{\boldsymbol{z}} = \mathcal{N}(0_D, I_D) : \mathbb{R}^D \to \mathbb{R}^D$. We also define a parametrized transport map f_θ with the following properties:

- The transport map f_θ is a diffeomorphism.
- The inverse f_θ^{-1} is easy to compute.
- The determinant of the Jacobian $\boldsymbol{\nabla}_z f_\theta(\boldsymbol{z}) \in \mathbb{R}^{D \times D}$ is efficient to compute.

In the ML literature, many candidate transformations satisfying these requirements have been proposed. Among others, we mention the affine coupling transformations, such as the nonlinear independent component estimation (NICE) [534] and the real non-volume preserving (RealNVP) [535], transformations based on convolutions [536] and those base on splines [537]. Typically, the desired bijective transformation is the composition of many of these simple transformations (see sketch in Fig. 7.5), themselves parametrized through NNs carrying the set of trainable parameters $\boldsymbol{\theta}$:

$$f_\theta(\boldsymbol{z}) = \left(f_{\theta_k} \circ ... \circ f_{\theta_i} \circ ... \circ f_{\theta_1}\right)(\boldsymbol{z}), \tag{7.18}$$

where all intermediate transformations are invertible, differentiable, and are parametrized with their own subset of parameters $\boldsymbol{\theta}_i$. We again refer to Refs. [519, 520] for an up-to-date overview of these transformations.

The combination of the base density $q_{\boldsymbol{z}}$ and the map f_θ defines a new density $q_\theta(\boldsymbol{x})$ as the map "pushes forward" the base density $q_{\boldsymbol{z}}$. To build intuition, we focus on a one-dimensional problem. We start by noting that the probability mass for the random variable $z \in \mathbb{R}$ should be preserved through a change of coordinates $x = f(z) \in \mathbb{R}$. Therefore, one can write

$$|p_x(x)\,\mathrm{d}x| = |p_z(z)\,\mathrm{d}z|\,. \tag{7.19}$$

It then follows that

$$p_x(x) = p_z(f^{-1}(x))\left|\frac{\mathrm{d}z}{\mathrm{d}x}\right| = p_z(f^{-1}(x))\left|\frac{\mathrm{d}f(z)}{\mathrm{d}z}\right|^{-1}, \tag{7.20}$$

where we used the definition of our bijective transformation $x = f(z)$ and its inverse. More specifically, when the bijection is parametrized by $\boldsymbol{\theta}$, it follows that

$$q_\theta(x) = q_z\left(f_\theta^{-1}(x)\right)\left|\frac{\mathrm{d}f_\theta(z)}{\mathrm{d}z}\right|^{-1}. \tag{7.21}$$

Under our assumptions, f_θ is invertible by construction and q_z is the Gaussian distribution we chose as the reference density to sample a latent noise z. For a D-dimensional problem, $f_\theta : \mathbb{R}^D \to \mathbb{R}^D$ and the derivative with respect to $\boldsymbol{z}$ is replaced in Eq. (7.21) by the determinant of the Jacobian $\det\big(\boldsymbol{\nabla}_{\boldsymbol{z}} f_\theta(\boldsymbol{z})\big)$, which represents the volume transformation depicted in Fig. 7.6.

Once the trivializing map f_θ is trained, a new data point $\boldsymbol{x}^*$ can be generated by sampling $\boldsymbol{z}^* \sim q_{\boldsymbol{z}}$ in the latent space and transform it through the bijection to obtain a corresponding sample in the data space. Reversely, one can also take a given sample, plug it into Eq. (7.21) and obtain the exact likelihood of the sample. For making both learning and sampling processes smooth and efficient, the determinant of the Jacobian of the transformation needs to be tractable and efficient to compute as already stated in our assumptions. We further refer the reader to [520] for more details on the coupling transformation and the computation of the Jacobian.

Training. Depending on applications, the training of NF can be performed with or without data following the methods described in Section 7.2.1. When a dataset $\mathcal{D} = \{\boldsymbol{x}_i\}_{i=1}^n$ from the target distribution $p(\boldsymbol{x})$ is available, the model can be trained through maximum log-likelihood:

$$\log p(\mathcal{D}|\theta) = \sum_{i=1}^{n} \log q_\theta(\boldsymbol{x}_i). \tag{7.22}$$

which is equivalent to minimizing the forward-KL divergence Eq. (7.17).

However, as mentioned earlier, for physical systems it is often the case that an *unnormalized* target density is available while samples are not readily available. In these circumstances, one can train an NF by instead minimizing the *reverse*-KL divergence Eq. (7.16). This approach does not need data as it trains by self-sampling, meaning that during optimization, configurations are directly sampled from the variational density q_θ we seek to optimize. Specifically, given a target Boltzmann-like

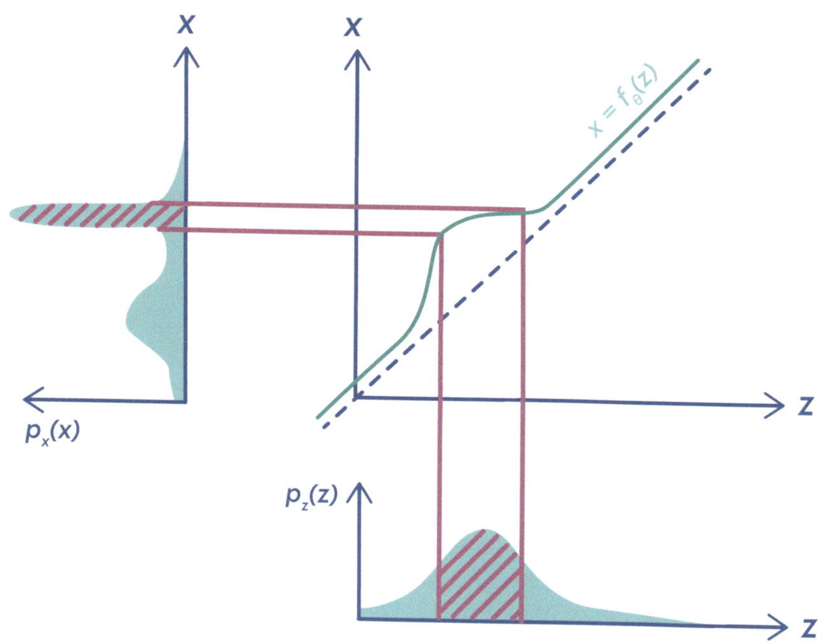

Figure 7.6 The change of variables formula Eq. (7.19) relies on the assumption that the total probability needs to be preserved. A base density p_z transforms according to the determinant of the Jacobian, which is responsible for redistributing the probability mass into a nontrivial one. In the one-dimensional case, the red lines identify the probability area in p_x and map it back to the corresponding area of the base density p_z through the bijection f_θ. If the determinant of the Jacobian is 1, the transformation is said to be volume-preserving.

target density, that is, $p(\boldsymbol{x}) = Z^{-1} \exp\{-H(\boldsymbol{x})\}$, the reverse KL objective can be rewritten as

$$\begin{aligned}
\mathrm{KL}(q_\theta || p) &= \int_\Omega q_\theta(\boldsymbol{x}) \ln\left(\frac{q_\theta(\boldsymbol{x})}{p(\boldsymbol{x})}\right) \mathrm{d}\boldsymbol{x} \\
&= \int_\Omega q_\theta(\boldsymbol{x}) \left(\ln q_\theta(\boldsymbol{x}) + H(\boldsymbol{x}) + \log Z\right) \mathrm{d}\boldsymbol{x},
\end{aligned} \tag{7.23}$$

where in the last step, we explicitly used the known form of the target density where $H(\boldsymbol{x})$ is the Hamiltonian and Z is the partition function. This latter now becomes just a constant shift and can be ignored for training purposes as it vanishes when computing the gradient during optimization. Using the analytic form for the model likelihood from Eq. (7.21), the KL divergence becomes

$$\mathrm{KL}(q_\theta || p) = \int_\Omega H(f_\theta(\boldsymbol{z})) - \log\left|\frac{\mathrm{d}f_\theta}{\mathrm{d}\boldsymbol{z}}\right|(\boldsymbol{z}) + \log q_z(\boldsymbol{z}). \tag{7.24}$$

Nevertheless, training models following the reverse-KL objective can lead to inaccurate outcomes when the target distribution is multimodal. Namely the model can miss out a mode entirely following a phenomenon known as *mode collapse*, which we briefly discuss in the Outlook of this section.

7.2.4 Applications

Once the model q_θ is trained to be a good approximation of the target density p, we can use the generative model, the flow in this case, to generate configurations for a physical system approximately following the Boltzmann distribution. This flow-generated configurations can be exploited to create an efficient Monte Carlo estimator of physical observables. For instance, NFs can be incorporated in MCMC sampling schemes as a proposal move to lower autocorrelation and effectively reach the ergodicity regime (see Refs. [89, 525, 526] among others). This strategy often goes under the name of NeuralMCMC sampling [89] or flowMCMC [538]. Flows can also be incorporated as powerful trial distributions in importance sampling schemes [89, 523, 539]. One further conceptual advantage of flow-based models, and generative models allowing for exact likelihood density estimation in general, is that they allow estimating quantities normally not accessible by standard methods. These include the partition function Z, the free energy, and other thermodynamic observables such as entropy and pressure. We refer to [89, 512, 524] for more details. We refer to the literature review of [540] for a wider overview of the many more applications in physics and quantum chemistry.

7.2.5 Outlook and open problems

In recent years, NFs demonstrated great potential and appealing conceptual properties, thus making them promising candidates for dealing with density estimation and sampling in the physical sciences. Nevertheless, many challenges need to be faced still.

First, a major drawback of flow-based methods is that they cannot easily scale to larger systems [541–543]. The promising results achieved so far by leveraging flow-based samplers, in particular in the context of lattice field theory, were obtained on relatively small lattices. In the limit of small lattice spacing and infinite volume limit, thus approaching the continuum, it is clear that the scaling of both standard methods and generative models is clearly going to be in favor of the former. However, there are strategies currently being explored to improve the scaling of these methods. One example is leveraging inductive biases, meaning exploiting prior physical knowledge, thus incorporating existing symmetries into the flow. This enables more effective training since the model does not have to learn physics from scratch. The idea of incorporating symmetries in the context of a flow-based sampler for lattice field theory has been successfully applied to U(1) [544] and SU(N) [545] lattice gauge theories. Moreover, [546, 547] give a complementary intuition on how to incorporate equivariance into an NF.

Another relevant challenge briefly mentioned at the end of Section "Training" indeed relates to the issue of mode collapse. This drawback, affecting systems trained by self-sampling, has been discussed in [511, 512, 526] and is currently an open problem not only in the domain of NFs in physics but also in the entire ML community [548]. Learning a multimodal distribution is often more challenging as it often prevents using the reverse-KL objective, prone to mode collapse. When this happens, the generative model may perfectly cover one (or more) modes of the target density yet completely neglect the others. It follows that when sampling from the learned variational distribution, for example, a flow, we don't have full support over the target density. Our ansatz thus badly approximates the target and leads to biased estimates [549]. This problem is very often found in the context of sampling and density estimation within physics and quantum chemistry applications. Mode collapse may be hard to detect in some scenarios and hence be very harmful when accurate estimates of observables are of interest. Some recent works tried to address this problem by combining NFs with initial knowledge of modes and *adaptive training methods* [526], *path-gradients* [550, 551], or *annealed importance sampling* [552–554]. While preliminary results are encouraging, this is still very much an open problem.

Further reading

- Wang, L. (2018). *Generative models for physicists* [555].
- Noe, F. *et al.* (2019). *Boltzmann generators.* Science, 365, eaaw1147 [523]. The seminal paper on Boltzmann Generators. Using normalizing flows (NFs) in the context of quantum chemistry.
- Köhler, J. *et al.* (2020). *Equivariant flows.* arXiv:2006.02425 [546], Satorras, V. G. *et al.* (2022). *E(n) equivariant normalizing flows.* arXiv:2105.09016 [547]. How to incorporate equivariances into normalizing flows (NFs).
- Nicoli, K. A. *et al.* (2021). *Machine learning of thermodynamic observables in the presence of mode collapse.* arXiv:2111.11303 [512]. Hackett, D. *et al.* (2021). *Flow-based sampling for multimodal distributions.* arXiv:2107.00734 [511]. Nicoli, K. A. *et al.* (2023). *Detecting and mitigating mode-collapse for flow-based sampling of lattice field theories.* Phys. Rev. D 108, 114501 [549]. Further discussions on the problem of mode collapse.
- Albergo, M. S. *et al.* (2021). *Flow-based sampling for fermionic lattice field theories.* Phys. Rev. D, 104, 114507 [556]. Sampling lattice field theory with fermions using normalizing flows (NFs).
- Abbott, R. *et al.* (2022). *Aspects of scaling and scalability for flow-based sampling of lattice QCD.* arXiv:2211.07541 [543]. Issues of scaling normalizing flows (NFs) to larger systems.

7.3 Machine learning for experiments

Quantum experiments pose tough technical challenges, and the task of optimizing their performance while interpreting the output data can seem daunting. It is informative to note that the output of quantum devices naturally generates large-scale data, which is the regime where ML thrives. In Chapter 3, we have already seen how ML can be used to detect phase transitions and although such efforts are much more challenging when dealing with experimental data, a few works managed to successfully address this real-world problem [106, 140, 557]. In Section 4.5, we have discussed the applications of Gaussian processes (GPs) and Bayesian optimization (BO) for inverse problems involving experimental data [194–196] and optimizing experiments [197, 205–207, 214–218]. In Section 5.3.7, we have also demonstrated how ML can boost quantum tomography [91, 345, 347, 348] with NQSs presented in Chapter 5. Another research direction pursued in the context of quantum experiments is the application of RL for quantum feedback control [421–424], quantum error correction [431–434], quantum circuit optimization [426], and experiment design [375], all described in Sections 6.6.3–6.6.6.

This section focuses on other ML approaches for experimental data. First, we acknowledge that there is an important niche of experimental physics that can be revolutionized by ML and that is the automation of (tedious) repetitive tasks. We show examples of successful realizations of this idea with actual experimental data in Section 7.3.1. In Section 7.3.2, we discuss the theoretical proposal of ML-based analysis of time-of-flight images, which is a standard measurement technique in ultracold-atom setups. Then, in Section 7.3.3, we describe a powerful scheme for quantum experiments, that is, learning the Hamiltonian governing the system from measurements. We conclude this section with Section 7.3.4 discussing the successes of the computer-guided design of experiments that does not include RL.

7.3.1 Automation of experimental setups

This section is devoted to novel ideas for the automation of physical experiments. Specifically, we present the automated identification of nanomaterial samples for quantum device technologies [558] and the automated tuning of double quantum dots [559] for quantum information devices. Both examples have one thing in common: at some point, a large amount of human labor becomes necessary that is tedious and repetitive but not trivial enough to be replaced by a simple knowledge-based algorithm.

Automated identification of nanomaterials. In the case of the preparation of nanomaterials for quantum devices, as detailed in [558], an important step is the selection of appropriate two-dimensional flakes from a wafer under a microscope. The flakes in question can differ depending on their desired use in the final device, but they all share their flat shape and approximate size due to the exfoliation-based technique with which they are prepared. Examples include hexagonal boron-nitride (hBN), graphite, and bilayer-graphene. Figure 7.7 depicts the scanning setup along with a typical image for hBN. The microscope in a typical setup can operate at different magnifications and can scan a 1 cm^2 wafer in roughly 3 min. In practice, however, this takes much longer, as the human operator has to slowly move the frame across the wafer and decide for each frame which of the depicted flakes are suitable for future device building. In short, the human operator classifies the flakes; a well-trained NN could do this as well. Hence, the design and training of a suitable NN architecture was at the core of the automation scheme developed in [558]. However, it is worth noting that the automation scheme did more than

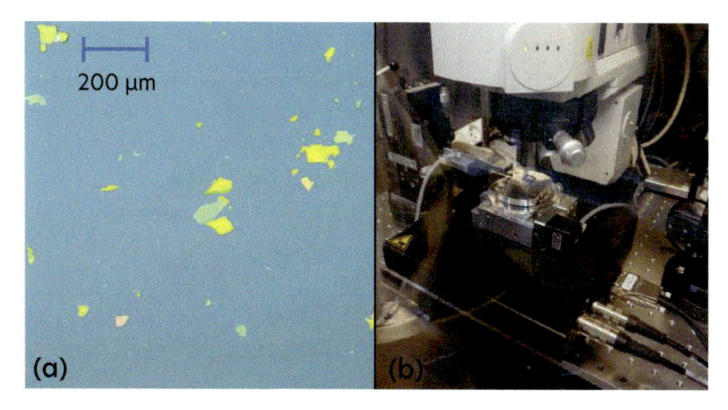

Figure 7.7 Experimental nanoflake setup. (a) A typical microscope image of hBN from [558]. (b) Typical microscope setup with waver already placed under the microscope. Photo credit: Klaus Ensslin Lab, ETH Zürich.

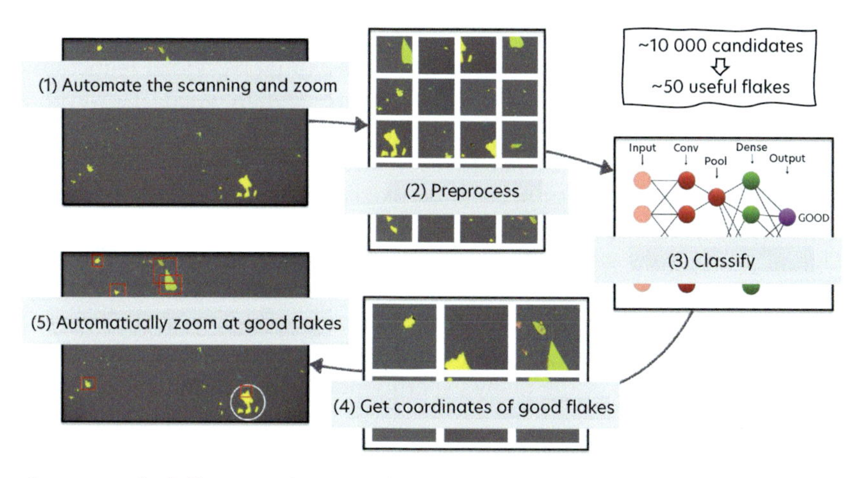

Figure 7.8 The full automation procedure consisted of scanning, *labeling*, preprocessing, *training*, classifying, and then collecting and presenting the good flakes. The steps marked with italics were done only prior to application. Figure credit: QMAI group at TU Delft.

just the classification task, as summarized in Fig. 7.8. For example, prior to even implementing anything network-related, they provided the experimental team with a program with a simple GUI for click-based flake labeling of pre-processed images to simplify the generation of an adequate dataset. This is noted here to truly reflect the additional steps that need to be taken into consideration when working with experimental setups.

In this work, more than one network was used to minimize classification errors: Three networks were used and applied consecutively, each consisting of four convolutional layers and one dense layer. The reason why this stacking of networks was necessary is the immense imbalance between good and bad flakes in the dataset. While a batch optimization procedure paired with data augmentation can usually account for this to some extent, here this was not sufficient: after passing new data that contained approximately 1 000 flakes, 10.8% of which were good, through a single trained network, the classified results yielded the 86% accuracy with 13% of false positives (bad flakes classified falsely as good) and 1% of false negatives (good flakes classified falsely as bad). Since there are so few truly good flakes, avoiding false negatives is of utmost importance.

An important and more practical aspect of this number and accuracy of leftover "good" (correctly or incorrectly classified) flakes is concerned with the additional human labor that would follow this classification result: After classifying, the automation scheme (compare Fig. 7.8) automatically zooms in on the good flakes after which a human operator steps in again. More specifically, this means that instead of manually scanning the probe, looking at the, for example, 1 000 flakes available and zooming in on the selected flakes to then decide whether they are good or not, in the automated scheme, the human operator only acts after the entire wafer is scanned and the microscope is shifted to the different locations of the good flakes on which it appropriately zooms in. In this way, the experimentalist only makes the final decision on which flakes to use; this represents a substantial workload reduction from 944 flakes to 224, approximately 44% of which are actually good. However, since there is still work involved, three networks instead of one are used to reduce the amount of flakes that need to be looked at even further. After passing the data through all 3 networks, only 150 flakes need to be looked at, approximately 57% of which are good.

One of the reasons why the accuracy is still comparatively low is due to the discrepancies in the classification choices among the humans who did the labeling beforehand. Given that it was already time consuming enough to label a large enough dataset even with the helpful GUI program written for that purpose, every flake was only labeled once by one of the many experimentalists participating in this project.

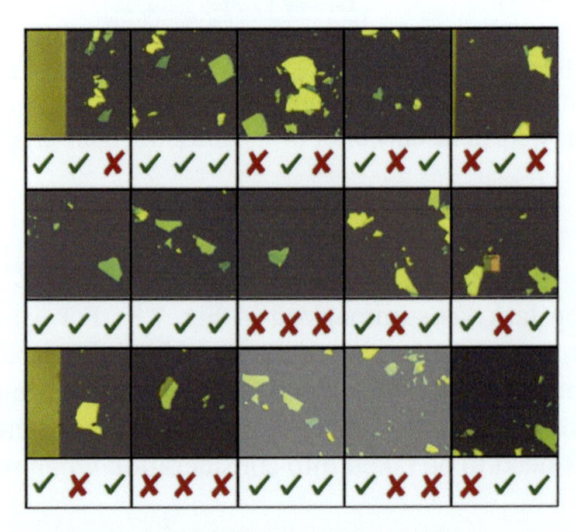

Figure 7.9 Differences of human judgment visualized. Selection of 15 frames that were to be judged on whether they contained good flakes or not along with the judgment results from three different human operators. Adapted from [558].

Figure 7.9 captures the differences in judgment of three different participants all looking at a selection of single shot frames. It is important to note that while here we were able to explicitly show one source of uncertainty and errors in the preparation of the data, usually we do not see it directly. We should therefore always acknowledge the possibility of their existence! Evidently, there is a degree of uncertainty and disagreement about whether the individual frames contain good flakes which introduced a bound on the model performance. This is reflected in the accuracy of the classification.

All in all, the developed automation procedure including and revolving around the NN is still clearly a success: This type of material control is a common step in the field of nanomaterial device development and the method generalizes satisfyingly to, for example, graphite and bilayer graphene (compare [558]). The implementation is available on GitHub [560].

Quantum dot tuning. The next example is the automated tuning of double quantum dots in quantum information technology research. A detailed discussion can be found in [559]. A quantum dot is a nanostructure that is confined so strongly in all three spatial dimensions that it is essentially zero dimensional. The confinement, similar to a particle-in-a-box scenario, leads to the emergence of quantum effects, that is, energy quantization (as opposed to having a continuous energy spectrum in larger structures) and thus discrete states.

In quantum information research, quantum dots are used to create qubits by putting together two dots as discrete states. There are various reasons why this technology is challenging in the context of universal quantum computation: on the one hand, there are difficulties associated with making two dots interact in a controlled manner, and, on the other hand, there are issues associated with reproducibility. Both concerns are related to preparation techniques, and the example discussed here offers a new ML-based remedy for the former.

Like in the previous example, the experimental procedure contains a tedious step that one can seek to automate. While in the flake example, this step revolved around human operators looking at images from a probe under a microscope, in this quantum dot setup the human operator looks at graphs created from changes in current measurements in the quantum device with changing applied voltages. When conducting a measurement like this, the quantity of interest is the occupation of the quantum dots, that is, the state of the quantum dot. The occupation can be changed by applying a voltage to the dot: as seen in Fig. 7.10(a), every one of the three quantum dots has a plunger gate (PG) associated with it that is used to tune the voltage. Underneath these dots, the current of a quantum point contact is measured.[13] As defined in Coulomb's law, the occupation of the dots, that is, the negative charge of the respective electrons, affects the electric current close to it. Hence, a change in the electron occupation causes a discrete change in the current flow, which corresponds to spikes in the conductance ($\partial I_{\mathrm{QPC}}/\partial V_{\mathrm{PG}}$) measurement. These spikes are the dark blue lines in the charge-stability diagram presented in Fig. 7.10(b). Figure 7.10(b) can then be interpreted as follows. In the bottom left corner, both quantum dots (QDs) are unoc-

[13]The quantum point contact in Fig. 7.10(a) is formed by the three gates at the bottom responsible for the measurement.

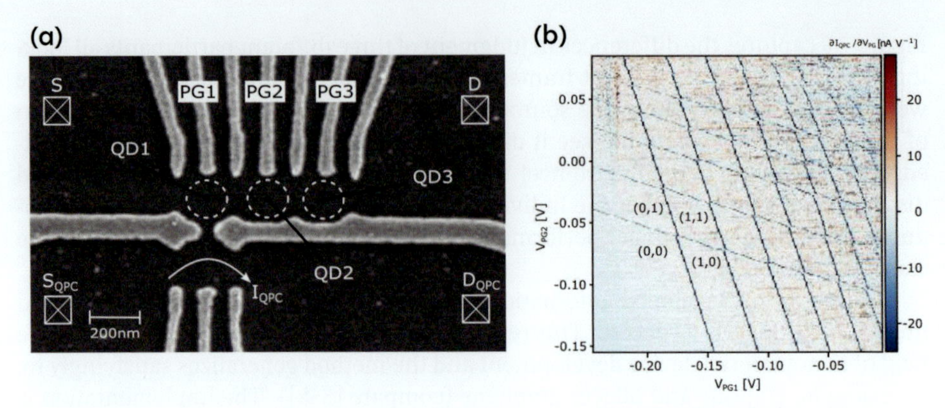

Figure 7.10 Experimental setup for quantum dots. The device is built and then tuned using measurements that are made possible by the quantum point contact built into it. (a) Scanning microscope image of an example device. Base material is a GaAs heterostructure with an electron gas embedded at the position where the three quantum dots (QD1, QD2, and QD3) are intended to be. It also contains a number of finger gates for confinement and measurement. The gates responsible for the measurement are the three at the bottom (quantum point contact). (b) The charge-stability diagram for a double quantum dot device that uses QD1 and QD2 in (a). Correspondingly, the changed voltages are the ones for PG1 (corresponds to QD1) and PG2 (corresponds to QD2). Taken from [559].

cupied but whenever a vertical (horizontal) line is crossed, an electron is added to QD1 (QD2).

The goal in device preparation and tuning is to prepare different discrete states by applying the adequate voltages that correspond to the correct current-spike-line framed, diamond-shaped area in the charge-stability diagram. To do this, the operator needs to know the charge-stability diagram. Thus, tuning a double quantum dot device, such as this one, requires measuring the entire charge-stability diagram, that is, performing many subsequent measurements where one voltage is kept constant and the other one is gradually changed. In this very time-consuming scenario, a classification-based ML scheme can be of help. For ML to bring a significant improvement, it is essential so that measuring the entire charge-stability diagram is not required for the input data. A suitable scheme should be able to produce the two PG voltages for a specified desired occupation state from any starting state (corresponding to a starting pair of voltages along with their current flow). However, without the charge-stability diagram, there is no way of knowing which occupational state the two starting voltages correspond to. For example, if the starting state had both voltages at $0\,V$, then a human operator with knowledge of Fig. 7.10(b) would know that this places the state somewhere in the top right of the diagram and would, by means of counting lines, be able to specify the state.

To avoid having to measure the entire diagram, the ML scheme uses an approach for which small, low-resolution excerpts of the diagram suffice:

1. **Finding the (0,0) state:** In a first step, one utilizes the fact that any state except the (0,0) state is framed by four lines in the diagram, whereas the (0,0) state only

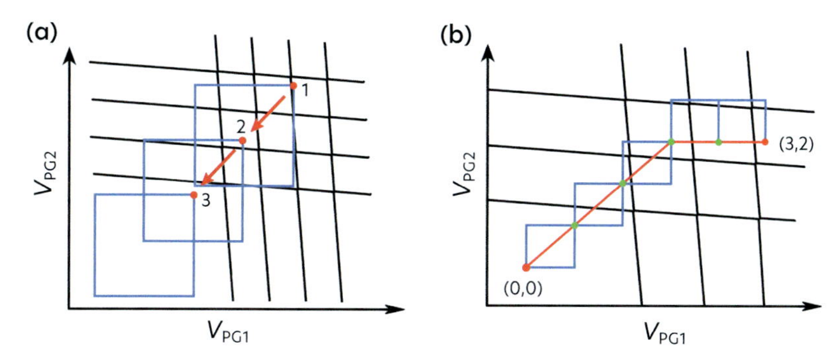

Figure 7.11 (a) Finding the (0,0) state using a first classification network: Starting with random initial voltages corresponds to point "1." In this example, the network had to go through three iterations of classification and frameshifting until (0,0), here marked by "3," was reached. (b) Finding the state of desired occupation (m,n): Depending on the different classification results, the frame can be shifted either only to higher PG1 voltages (in the case of one vertical line), only to higher PG2 voltages (in the case of one horizontal line) or diagonally, increasing both gate voltages (in the case of two lines). Adapted from [559].

has neighboring lines in the positive x- and y-direction. Therefore, a first classification network is trained to recognize if there are more lines to cross in the negative x- or y-direction. The output is true or false. If there are more lines, both plunger voltages are lowered by a set amount (as depicted in Fig. 7.11(a)), and the classification is performed again until there are no more lines to cross.

2. **Finding any desired, given state:** To get from the (0,0) state to any desired state (m,n), one has to cross *exactly* m vertical lines and n horizontal lines. Thus, a second network is now trained to more accurately classify which lines there are. This network uses smaller frames of a higher resolution that allow a more differentiated distinction between the cases where there are no lines, there is one vertical line, there is a horizontal line, and there are both in the considered frame (compare Fig. 7.11(b)). Just like in the first step, each classification is followed by a change in voltages and this two-step procedure is repeated until the desired state is reached.

For the training of the first network, 470 charge-stability diagrams were measured in fairly low resolution, whereas for the second network, 128 charge-stability diagrams were measured in higher resolution. The PG voltage ranges were varied for the different measurements to foster better generalization later on. In both cases, datasets were created by cutting out numerous random frames from the diagrams and labeling them with a script. Note that while full charge-stability diagrams were measured for the generation of the training dataset, the input for the eventual application of the network only needs small windows. Measurement of full charge-stability diagrams and use of many windows therein was just a convenient way to create a dataset.

When tested on the actual device, the success rates of the two loops were 90% (step 1 loop) and 63% (step 2 loop), which combines to an overall success rate of 57%. It is important to keep in mind that those individual success rates are not the equivalents of the accuracy rates of the two networks: Each loop calls the network multiple times, so errors are doomed to accumulate, and the second loop usually requires more calls to the network than the first loop because the frames are smaller; see panels (a) and (b) of Fig. 7.11. In fact, when tested separately and only a single time on a labeled dataset, the accuracy rates reached by the two networks were 98.9% and 96%. In the article, the authors stated that the primary error source was identified as a weak signal-to-noise ratio and improving on this would surely improve the scheme.

In conclusion, the integration of DNNs into a larger scheme can lead to an accumulation of errors, and this should be taken into account when planning the implementation of the automation routine. In general, the integration of ML approaches into broader automation schemes calls for different levels of network accuracy and, as was the case in the first example, some scenarios might even have limited network accuracy in general. It is important to take these things into account before implementation and gauge the benefits of automation versus the remaining workload.

Of course, the approaches presented are not the only ideas for automating experimental setups. For example, CNNs also help in detecting contamination by ice crystal diffraction in macromolecular diffraction data [561] or analyzing cryo-electron microscopy maps of proteins [562].

7.3.2 Machine-learning analysis of time-of-flight images

When it comes to analysis of the experimental data, we present one more example related to ultracold-atom experiments. In contrast to the two highly specialized applications to actual experimental data discussed so far in Section 7.3.1, we consider a proposal that is based on theoretical data but is readily extendable to experiments [563]. There is a number of theory-based, yet application-oriented, proposals that are currently being published, and discussing their differences should prove insightful. The focus of this discussion is on the feasibility of making the transition from theory to application.

For any such transition from theory to experiment within an ML scope, the following aspects should be examined:

- *Specificity vs. flexibility of the method:* As was discussed in the two earlier examples, when the ML model does not provide truly new insights into the physics of the model, the automation should instead yield a significant reduction of human labor. This can be achieved by designing a specific scheme for *one* scenario that requires a large expenditure of work or by designing a flexible scheme for a large number of scenarios of medium expenditure.
- *Similarity of theoretical and experimental results:* More often than not, theoretical models produce results that diverge quite significantly from their experimental counterparts. This can be due to experimental noise or limitations in the theoretical model. For a model that has been trained on theoretical results only, it is important to evaluate whether further preprocessing, such as the inclusion of artificial noise, could be sufficient to prepare the network architecture for an input of experimental data and/or how much the network needs to be retrained.

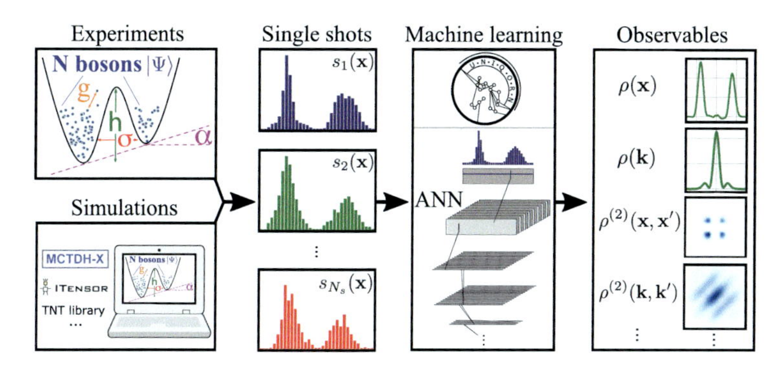

Figure 7.12 Implementing ML into the extraction of observables from measurements of ultracold atoms: Conventional experimental approaches are usually only able to extract the first- and second-order density matrix through averaging approaches, whereas the ML approach also extracts real-space observables from momentum-space measurements as well as correlation functions. Taken from [563].

Unlike the work done with quantum dots and flakes that utilized *specific schemes with high impact*, the scheme proposed now is very general and takes advantage of the *flexibility* of the probed system: ultracold atoms. Due to the high level of control available in such setups, ultracold atoms represent an exemplary quantum simulator for a large variety of few to many-body physics phenomena. It is worth noting that regardless of whether the considered experimental effect is a dynamic transition from superfluid to Mott-insulating states, the quantization of conductance through a quantum point contact, or simply the many-body nature of condensed versus fragmented states in a double-well potential, the standard output of experiments remains similar. It is namely a time-of-flight image.

In an experimental setup, an initially trapped cloud of ultracold atoms is allowed to expand, and time-of-flight imaging captures snapshots of the cloud. These single shots carry an amount of information as they can unambiguously be linked to a large variety of physical quantities and phases. Although experimentalists can usually only extract a few observables through averaging techniques, it is shown that an ML tool should be able to exploit the information contained in the data more accurately and access a larger selection of observables (compare Fig. 7.12). The ANN-based approach proposed by the authors exploits the shot-to-shot fluctuations to implicitly reconstruct the many-body state. This is promising for widespread application in experimental realizations.

When it comes to comparing theoretical predictions and experimental results, noise becomes an important factor. In ultracold atoms, Lode *et al.* (Fig. 7.12) proposed a method for an optimized observable readout from single-shot images of ultracold atoms, arguing that the similarity of theoretically simulated and experimentally detected single-shots is good enough that the addition of artificial Gaussian noise to the theoretical data during training should suffice. Figure 7.13 shows a comparison of simulated and experimental single shots of ultracold atoms at different points of a phase transition in an optical cavity upon increase in one of the external laser intensities. Although this example comes from a different framework, both publications

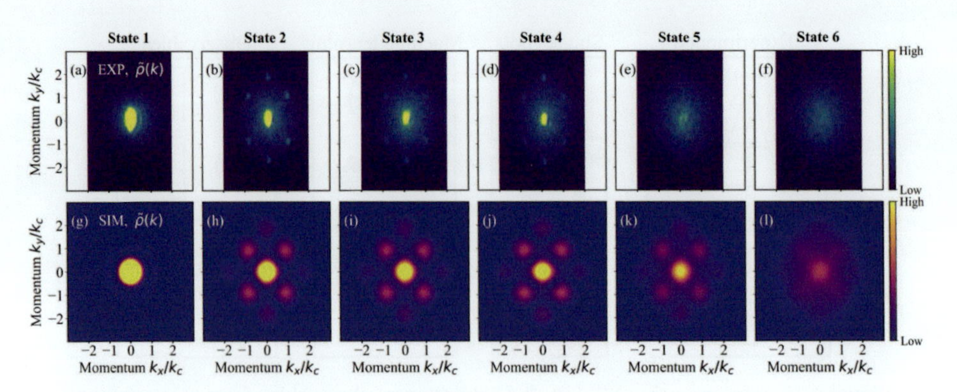

Figure 7.13 Comparison of experimentally measured (upper row) and simulated (bottom row) momentum-space density distributions of a system undergoing a phase transition from a superfluid self-organized phase (State 1) to a superfluid Mott insulating phase (State 6). Adapted from [564] under the CC BY 4.0 DEED license.

used the same simulation method for the single-shot generation [564, 565]. Noise is evidently present, but the agreement is satisfactory for the different stages of the phase transition. An alternative to adding artificial noise to theoretical data is attempting to subtract noise from experimental data, for example, by means of denoising autoencoders. The option chosen eventually naturally depends on the given experimental and theoretical data. In the case of single-shot images, denoising methods may not be ideal owing to the presence of quantum noise, inherent in many-body systems, which is difficult to discern from other noise sources and, therefore, selectively remove.

Overall, we have seen that ML techniques can help bridge the gaps between theoretical models and noisy or resource-constrained experimental realizations and measurements. These findings represent a solid groundwork demonstrating experimental quantum physics enhancement via ML and indicate a promising avenue toward the hybridization of ML and the quantum realm in the coming years.

7.3.3 Hamiltonian learning

The focus of this section is the verification of quantum simulators such as trapped ions, Rydberg atoms, superconducting qubits, or ultracold atoms in optical lattices [566–569].[14] These experimental setups are well understood and can be used to simulate more complex and challenging systems governed by the same Hamiltonians. We enter exciting times when quantum simulators start to be very complex and, in particular, not solvable with classical computers. For example, when working with quantum simulators with 50 qubits, we have to deal with enormous Hilbert spaces of the order of 10^{15}. Therefore, how can we know that these simulators are working as they should be if we cannot verify their results with classical computers? One possible

[14]Different experimental setups have different advantages and disadvantages for specific quantum simulation problems and Hamiltonians. A difference between quantum simulation and quantum computation is that quantum simulators are engineered for specific problems, and quantum computing is more versatile and capable of solving general problems.

solution to this problem is called *Hamiltonian learning* which is the main topic of this section. In particular, we discuss here the approach presented in [570].

> The main idea of Hamiltonian learning is to reconstruct the map from experimentally accessible measurements to the parameters of the underlying Hamiltonian.

The approach discussed in this section employs NNs to extract parameters governing the created quantum simulator. An exemplary procedure is as follows. We conduct numerical simulations and generate experimentally accessible data (e.g., real-space images) for the corresponding Hamiltonian whose parameters are known. Then, the NNs are trained via supervised learning to predict the parameters of these Hamiltonians. Then they can be tested on measurements generated with experimental quantum simulators, where the underlying Hamiltonian is not fully known. It is also possible to reverse the procedure: Given the defining parameters of the Hamiltonian of a quantum system, relevant characteristics of the system can be efficiently learned by an NN [571].

A very simple example to illustrate the process of Hamiltonian learning with NNs is a single spin system as shown in Fig. 7.14. First, we prepare an initial state of a known Hamiltonian, $\hat{H}_0$. In this case, it is an eigenstate of σ_z (spin "up"). Second, we perform a unitary evolution under an unknown Hamiltonian, $\hat{H}_1$, which leads to a precession of the spin around the axis of the Bloch sphere. We let the system evolve for some time t_{m}, after which we measure it. The process of preparing the initial state and letting the system evolve is repeated multiple times (potentially for different times t_{m}) to collect the dataset of measurements. We now want to learn from these measurements the unknown $\hat{H}_1$, that is, how fast the spin precesses around the sphere. In

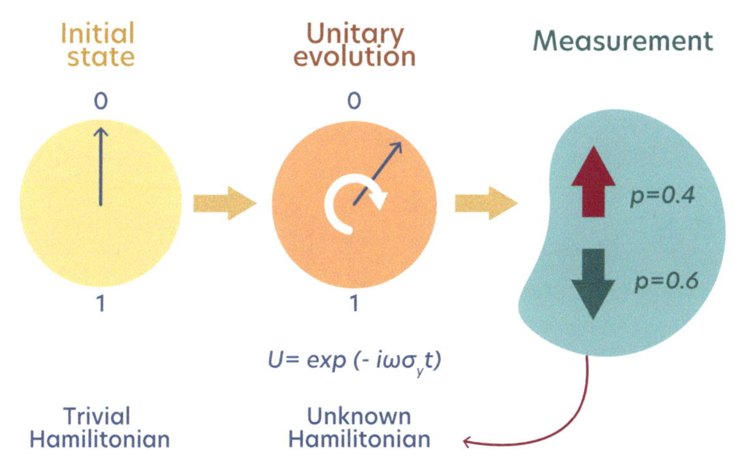

Figure 7.14 Illustration of the procedure of Hamiltonian learning of a one-spin system. The spin is prepared in an initial state and driven by an unknown Hamiltonian for some time t_m, after which the system is measured. The process is repeated multiple times (potentially for different times t_m) to collect the dataset of measurements, from which the rotation frequency can be learned.

this case, a sequence of measurements is required to obtain the oscillation frequency, ω. This procedure can be generalized to arbitrary known initial $\hat{H}_0$ and unknown $\hat{H}_1$, driving the unitary evolution of the system.

Now, we focus on another experimental setup of a quantum simulator consisting of neutral atoms in a harmonic potential in a system of 2×50 lattice sites, realizing the Bose–Hubbard Hamiltonian

$$\hat{H}_{\rm BH} = -\sum_{\langle i,j \rangle} J_{ij}\hat{a}_i^\dagger \hat{a}_j + \sum_i \frac{U_i}{2}\hat{a}_i^\dagger \hat{a}_i(\hat{a}_i^\dagger \hat{a}_i - 1) - \sum_i \mu_i \hat{a}_i^\dagger \hat{a}_i, \tag{7.25}$$

where J_{ij} describes the hopping between lattice sites i and j, U_i – the onsite energies, and μ_i is the chemical potential of the atoms in the optical lattice. If we consider only ten particles in this lattice, the corresponding Hilbert space is of dimension 10^{13} with 350 Hamiltonian parameters to estimate. This leads to two main issues: First, the wave function is too large, making it impossible to simulate this system, and second, it leads to a 350-dimensional optimization problem. Therefore, let us first consider a small system consisting of 4 atoms, as illustrated in Fig. 7.15(a), which reduces the number of Hamiltonian parameters to 25 and the Hilbert space size to 330. This eliminates the problem of the large Hilbert space and leaves us with the optimization problem.

Now, we want to create a mapping from the measurements to parameters of the Hamiltonian in Eq. (7.25).[15] To do so, supervised learning is used to train an NN and

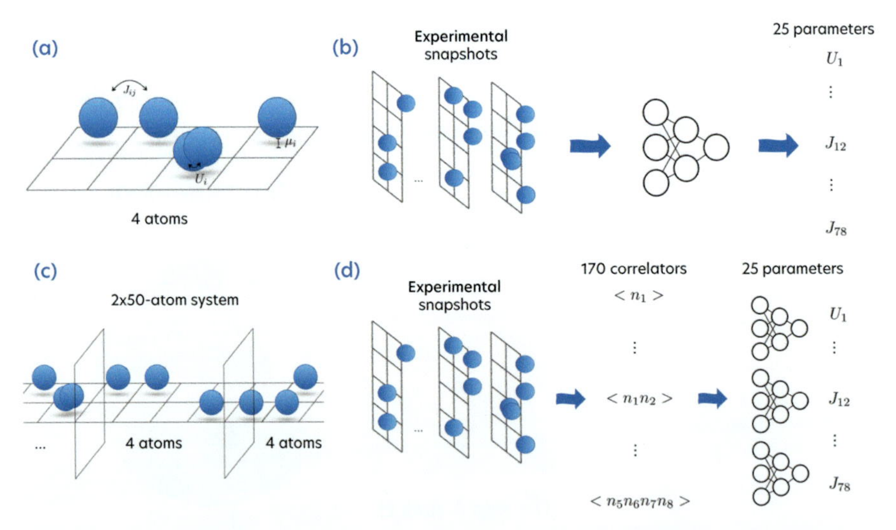

Figure 7.15 (a) Illustration of a four-atom system with 25 parameters. (b) Mapping from the experimental snapshots to the Hamiltonian parameters using one NN and supervised learning. (c) Extension to larger system sizes by "dividing" the lattice into subsystems with walls. (d) A more efficient way to map measurements to the parameters. Adapted from [570]. Additional credit: QMAI group at TU Delft.

[15]Here, exact simulations were used as "measurements" instead of experimental snapshots, and the input images are the real-space positions of the atoms in the optical lattice.

perform regression. The challenge in this setup is the scaling of the training data with the output size. To train a single NN to predict all parameters, as shown in Fig. 7.15 (b), several examples are required for all combinations of the 25 parameters, which is unfeasible for most applications due to the enormous size of the required training set. The solution to this problem is quite simple: Instead of using a single NN to predict all parameters, 25 NNs are trained to predict each parameter separately with continuous regression.

Moreover, the experimental snapshots may not be the best representation of the dataset. A more effective representation is to switch from experimental snapshot batches to the correlators of the specific Hamiltonian. In this example, density correlators are used, which enable a way more efficient way to train the NN by reducing the input dimension. This approach is shown in Fig. 7.15(d). After successful training, the NN achieves around 0.1% error rates for experimental parameters with 2 500 snapshots. Using Bayesian inference as a benchmark, the NN approach outperforms the Bayesian results for small datasets of 2 500 snapshots. However, for large datasets of about 20 000 samples, both approaches achieve the same accuracy in the predictions of the parameters.

So far, we have only considered small system sizes of four atoms, which can be solved exactly with classical computers. In the following, we present a scheme to scale to larger system sizes of this specific Hamiltonian. In this experimental setup, it is possible to modulate the lattice and create walls to separate the chain of 50 lattice sites into 4-site units (see Fig. 7.15(c)). In this system, the Hamiltonian parameters are local, and only the terms of $\hat{H}$ that are unaffected by the boundary have to be learned. They are called the "effective parameters" (see Fig. 7.16). Now, the boundary is shifted by one lattice site at a time, and 2 500 shots are measured for each position. Once the system is shifted up to the point of translational invariance, all parameters were at least in one configuration unaffected by the boundary wall and were successfully learned by the NN.

As mentioned above, this procedure is very specific to this system and cannot easily be generalized to different systems. The field of Hamiltonian learning is still

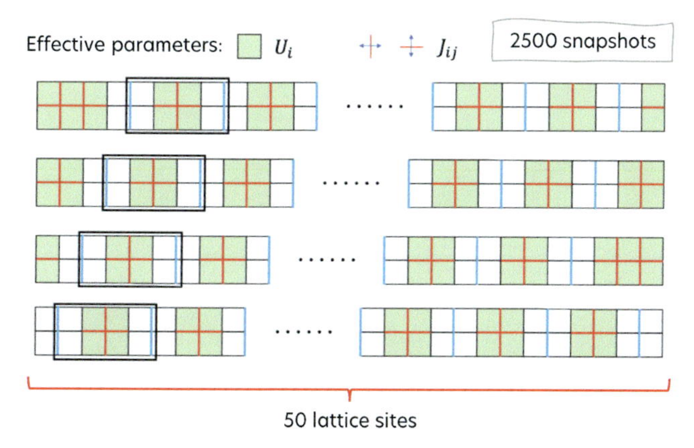

Figure 7.16 Scaling scheme from four lattice sites to 50 for this specific system and Hamiltonian. For each of the four wall configurations, 2 500 snapshots were taken to train the NN and learn all parameters. Adapted from [570].

in its early stages, and general schemes for large systems and complex Hamiltonians have yet to be developed [572]. However, it is a promising approach for the important task of validating if quantum simulators work correctly, which becomes increasingly important with the increasing size and applicability of these simulators, which might have the possibility to go beyond classical computation.

7.3.4 Automated design of experiments

Among the proposed ML applications for experiments, we have already discussed how NNs can be used to speed up, optimize, and verify the setups, as well as to analyze the generated data. One further application is the AI-guided design of experiments that may one day arguably revolutionize science.

When it comes to designing new experiments, most of the efforts have focused so far on quantum optics [375, 436–438, 573, 574]. The design of such an experiment consists of combining different optical laboratory components, for example, beam splitters, mirrors, and crystals, so that the final quantum state has specific desired properties. For example, we may be interested in obtaining a quantum state with a high-dimensional multipartite entanglement (i.e., between multiple particles), which is of great importance in applications of quantum information and computation [435]. While a trained physicist can design an experimental setup to create a quantum state with nontrivial properties, this task can be very challenging and heavily relies on trial and error.

In Section 6.6.6, we have already presented an example [375] of an autonomous approach to building quantum-optical experiments with reinforcement learning (RL), using the projective simulation (PS) algorithm that we introduced in Section 6.5. Interestingly, there is another AI-guided approach for designing optical experiments that has already allowed for a dozen new experiments in several laboratories around the world [438]. The proposed algorithm is called MELVIN [436] and is presented in Fig. 7.17.

To apply MELVIN, the user needs to specify a toolbox, that is, a set of available optical lab components. Moreover, the user defines the target properties and all pos-

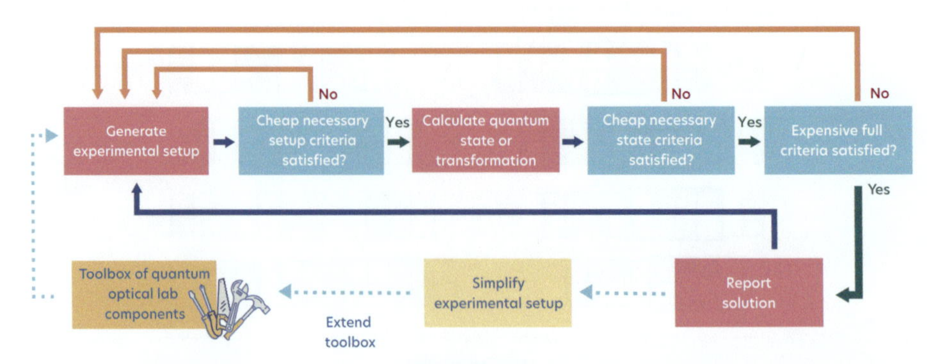

Figure 7.17 Example of an algorithm for computer-inspired quantum-optical experiments called MELVIN. Adapted from [438] with permission from Springer Nature.

sible conditions that characterize a final quantum state. The MELVIN algorithm first generates an experimental setup by randomly arranging the available optical components. Each optical component is a known symbolic modification of the input state. Then, the resulting quantum state and its properties are computed, as we know the initial quantum state and the symbolic transformations applied to it. If the quantum state meets all the criteria and exhibits a target property, then MELVIN reports the setup to the user.[16] More often, the generated quantum state does not match the target one, so MELVIN starts again by generating another setup. Therefore, MELVIN is heavily based on random search.

However, there are two characteristics of MELVIN that grant a significant speedup compared to a fully random search. First, the user can divide the required criteria into cheap and expensive ones, as presented in Fig. 7.17. The expensive criteria are then calculated only if the cheap ones are met first. Second, MELVIN is allowed to expand its initial toolbox by adding already tested configurations and use them as basic building elements in subsequent trials. This expansion of the available tools can be thought of as a learning component of MELVIN.

For example, MELVIN has been used to find experimental setups generating high-dimensional multipartite entangled states, as mentioned above. In [436], MELVIN identified setups that lead to states entangled in different ways. In particular, it found the first experimentally realizable scheme leading to a so-called high-dimensional Greenberger–Horne–Zeilinger state [436]. Moreover, as the authors of [436] admit, the resulting experiments contained interesting novel experimental techniques previously unknown to them.

Finally, studying MELVIN showed that each optical setup and initial state can be represented as weighted graphs. The successor of MELVIN, called THESEUS [437], takes advantage of a graph representation that allows replacing random search with a gradient-based search for optimal weights. Not only does THESEUS outperforms MELVIN in terms of discovery speed by a few orders of magnitude, but it also provides interpretable solutions as long as the graphs representing the discovered experimental setups are small enough.

7.3.5 Outlook and open problems

To conclude Section 7.3, we can use ML to speed up, optimize, validate, and design experiments, as well as analyze the collected data. Proposals to apply ML to speed up and optimize experimental work date back to 2009 [575], which may be why such applications pose one of the most widely accepted roles for ML in experimental physics. A fascinating direction is the so-called self-driving labs [576], which combine automated experimentation platforms with AI methods to enable autonomous experimentation. They promise an accelerated discovery rate and the liberation of experimentalists from tedious tasks.

When it comes to modern quantum technologies, the central challenges are the efficient characterization of quantum systems, the verification of quantum devices,

[16] Before reporting the solution, optionally, the setup is simplified using deterministic methods predefined by a user. For example, they may include iterative removal of a random optical component and check whether it changes the final quantum state.

and the validation of the underpinning physical models. ML is expected to improve the computational cost of these tasks. As a result, ML-based Hamiltonian learning is becoming a widely used technique to verify quantum experiments. Interesting examples are its application to nitrogen-vacancy center setups [577] and to nuclear magnetic resonance measurements [578].

Scaling of ML approaches to larger sizes of quantum devices remains an important challenge. Although the ML algorithms perform exceptionally well on large experimental datasets, adding more qubits (and, therefore, tuning parameters) generates learning difficulties. These problems are especially daunting in quantum dot systems. There are efforts toward tuning multiple parameters at once [579, 580] or toward reducing the amount of experimental data needed for tuning [581]. However, efficient tuning of large-scale quantum devices with hundreds of parameters requires new methods.

Moreover, AI promises breakthroughs when it comes to designing novel experiments. In particular, AI is argued to provide out-of-the-box solutions when unaware of existing human approaches [436, 438, 582]. So far, AI-guided design has been explored mainly in quantum optics with significant successes. However, the discussed approaches (MELVIN and THESEUS) are readily extendable only to experiments where we can calculate how each modification in the setup influences the generated quantum system and its desired properties. Applying MELVIN or THESEUS to experiments with very expensive (or nonexistent) theoretical descriptions requires novel ideas. Another example of AI-guided discovery of experimental setups is the use of graph-based search to automatically identify laser cooling schemes for molecules based on spectroscopic data [582], which promises breakthroughs in ultracold chemistry and physics by extending the range of available ultracold species. Automatic search is again possible due to the well-understood physics underlying laser cooling. Finally, it is inspiring to think about combining the proposal of self-driving labs with AI designing novel experiments, which would create an ultimate robot scientist who never tires and never stops looking for new solutions and discoveries.

Further reading

- King, R. D. *et al.* (2009). *The automation of science.* Science 324, 85–89. Report on the building of one of the first "robot scientists" named "Adam" aiming at automating hypothesis formation and recording of experiments [575].
- Wiebe, N. *et al.* (2014). *Hamiltonian learning and certification using quantum resources.* Phys. Rev. Lett. 112, 190501. The first proposal of Hamiltonian learning that combined quantum simulators and Bayesian inference [583].
- Raccuglia, P. *et al.* (2016). *Machine-learning-assisted materials discovery using failed experiments.* Nature 533, 73–76. Example of ML use for discovery of materials that outperforms traditional human approaches [584].
- Häse, F., Roch, L. M., & Aspuru-Guzik, A. (2019). *Next-generation experimentation with self-driving laboratories.* Trends Chem. 1, 282–291. Perspective on self-driving laboratories and their role in scientific discovery [576].

8 Physics for deep learning

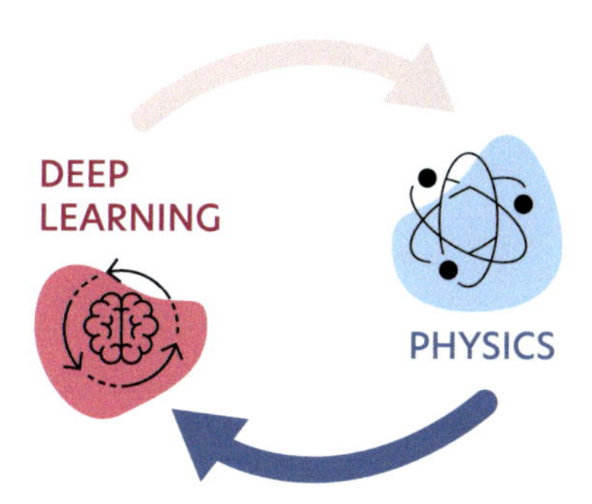

Figure 8.1 There exists a two-way influence between machine learning (ML) and physics. In this chapter, we focus on the less known approach, that is, physics for ML.

So far, we have discussed different applications of ML which aim at solving various problems in quantum science. In contrast, in this chapter, we focus on how physics (in particular statistical and quantum physics) influences ML research (as shown in Fig. 8.1). In Section 8.1, we explain the fundamental theoretical challenges of ML and show how tools of statistical physics can shed some light on these problems. In Section 8.2, we discuss quantum computing and promises of quantum machine learning (QML).

8.1 Statistical physics for machine learning

In this section, we present how to apply concepts from physics (in particular, tools of statistical physics such as the thermodynamic limit or order parameters describing phase transitions) to develop a theory of ML (see Fig. 8.2) [585]. This idea was born already in the 1980s, but the DL revolution in the 2010s has caused a renewed surge of interest in this approach.

Indeed, help from statistical physics is very needed, as we do not understand many conundrums in ML! For example, modern NNs can have billions of trainable parameters.[1] How can we even find well-generalizing minima within such enormous, nonconvex loss landscapes? Another riddle relates to the so-called bias-variance trade-off, which we have shown in Section 2.2, and which indicates that in the regime of a high model complexity, models should heavily overfit their datasets as presented in Fig. 8.3(a). But in practice, we see that these gigantic overparametrized DL models generalize very well, as seen in Fig. 8.3(b). So how do they escape this traditional bias-variance trade-off? A related question concerns the capacity of DL models and the development of its useful measures. We have a long way toward a full understanding of these puzzles. A way of tackling them is to study simple, solvable models, following a traditional approach of physicists to study new systems. The results from toy problems can give us clues on how more complex models work.

[1]One of the latest champions is Microsoft's GPT-3 with over 175 billion parameters.

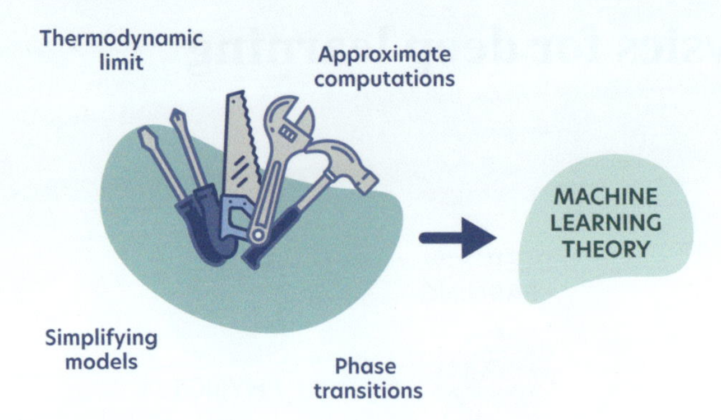

Figure 8.2 Statistical physics toolbox for understanding the ML theory.

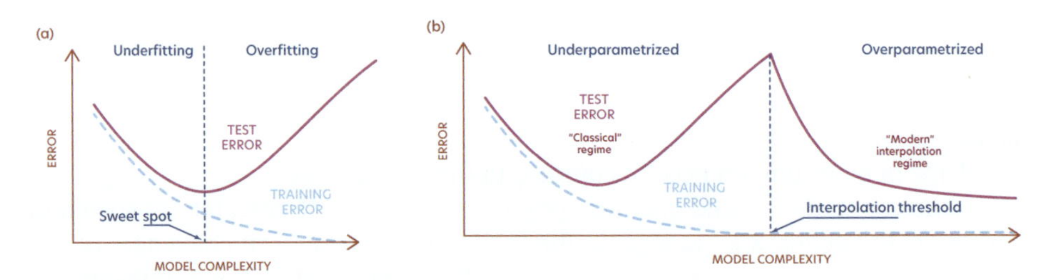

Figure 8.3 Classical and modern understanding of the generalization. (a) The classical U-shaped error curve arises from the bias-variance tradeoff. (b) The double descent error curve incorporates the classical U-shape in the classical regime and the low generalization error of modern overparametrized models. Adapted from [63] with publisher permission.

This section has four parts. First, in Section 8.1.1, we go through the seminal study on the capacity of the perceptron, which gives an idea of how statistical physics can be useful for learning problems. Then we discuss three directions of this interdisciplinary research, that is, the teacher–student paradigm for studying generalization in Section 8.1.2, how we can model the structure of data in Section 8.1.3, and study the dynamics of learning in Section 8.1.4.

8.1.1 Capacity of the perceptron

The simplest ML model we can think of is a single perceptron, f, already presented in Section 2.4.4 (see Fig. 2.6(b)). In this section, we focus on its capacity, that is, the question of how many data points it can fit. To answer it, let us make the additional assumption that the dataset is in *general position*.[2] The assumption is reasonable –

[2]The set of points in $\mathbb{R}^d$ is in general position if and only if every set of $(d+1)$ points are not in any possible hyperplane of dimension d. In other words, as long as there are no three data points on a single line or four points on a single plane, and so on, the set is in general position. Intuitively, any random dataset is in general position.

if we have many copies of the same training point, they should not contribute to the estimation of the model capacity.

A single perceptron is only capable of learning linearly separable patterns. Therefore, we can reformulate the question of its capacity to the question of whether randomly labeled datasets of size n with binary labels are linearly separable. The probability of such a linear separability, $p_R(\alpha)$, is a function of α, which is the ratio between the number of training points, n, and the number of data features (or data dimensionality), m. In the case of the perceptron, the number of features is equal to the number of perceptron weights, d,[3] therefore $\alpha = \frac{n}{d}$. In this problem, you can understand the parameter α as the difficulty of the classification task, which increases with the number of training points and decreases with the number of parameters.

To calculate p_R, we could resolve to geometric arguments. This approach was chosen by Thomas Cover in 1960s [586]. However, here we choose to rephrase this problem in the language of statistical physics as was done by Elizabeth Gardner in 1987 [587].

> Namely, we can take the space of all possible weights, so $\mathbb{R}^d$ and calculate the volume of those weights that fulfill all the constraints of the random labeling.

In other words, we calculate how many sets of weights could solve the problem of separating randomly labeled training data, $\mathcal{D} = \{\boldsymbol{x}^{(k)}, y^{(k)}\}_{k=1}^{n}$:

$$V_{n,d} = \int_{\mathbb{R}^d} d\theta \prod_{k=1}^{n} \delta(f(\boldsymbol{x}^{(k)}; \theta) - y^{(k)}). \tag{8.1}$$

The δ-function in Eq. (8.1) is 1 only when the ground-truth label is equal to the label predicted by the perceptron. With each new data point k, we are adding a new constraint, and the volume of possible weights shrinks. To have at least one set of such weights, the volume must be larger than zero, $V_{n,d} > 0$. Therefore, we define the critical task difficulty, α_c, as the value of α for which $V_{n,d}$ goes down to zero. If we can calculate this, we solve the problem of the perceptron capacity.

Let us make one modification to the equation that leads us closer to statistical physics. We introduce an effective Hamiltonian this counts the number of misclassified training data points,

$$H(\theta; \mathcal{D}) = \sum_{k=1}^{n} \Theta(-f(\theta; x^{(k)})y^{(k)}), \tag{8.2}$$

where the Heaviside function $\Theta(\cdot)$ is equal to 1 if its argument is positive and 0 otherwise. We can "relax" the Dirac δ-distribution above by the Boltzmann factor of $H(\theta; \mathcal{D})$. Up to a multiplicative constant, Eq. (8.1) becomes

$$V_{n,d} \propto \lim_{\beta \to +\infty} \beta \int_{\mathbb{R}^d} d\theta \, e^{-\beta \sum_{k=1}^{n} \Theta(-f(\theta; x^{(k)})y^{(k)})} = \lim_{\beta \to +\infty} \int_{\mathbb{R}^d} d\theta e^{-\beta H(\theta; \mathcal{D})}. \tag{8.3}$$

[3] In general, a perceptron is parametrized by weights $\boldsymbol{w}$ and a bias b. For the remainder of this section, we ignore biases; therefore, weights are all model parameters θ, of size d.

Suddenly, the volume $V_{n,d}$ in Eq. (8.3) resembles the canonical *partition function*[4] from statistical physics with β playing the role of an inverse temperature, defined as $\frac{1}{k_B T}$. Therefore, the limit $\beta \to \infty$ corresponds to the zero-temperature limit. The problem is that this integral is hard to calculate as it lives in a huge d-dimensional space of all real numbers.[5] Moreover, the "effective energies" in the exponent depend on the training set. As such, each training set requires a separate calculation of the volume $V_{n,d}$.

Fortunately, the physics of *disordered systems* comes to the rescue. It has been applied to learning theory since the 1980s [588–594]. Namely, if we recognize a disordered system in Eq. (8.3), we can use solutions from statistical physics to compute this high-dimensional integral. Let us give a brief introduction to disordered systems. A disordered system is described by two types of random variables. The first type concerns the states of the system $s \in \mathbb{R}^d$. For example, for a system of d spins $-\frac{1}{2}$, $s \in \{-1, 1\}^d$ because each spin can be up or down. The second type concerns interactions between degrees of freedom, which can be parametrized by couplings $J \in \mathbb{R}^n$. For example, J can describe whether the spins want to align or antialign. The distribution of states in disordered systems is then described by the Boltzmann distribution:

$$p(s \mid J) = \frac{1}{Z_J} e^{-\beta H(s;J)}, \tag{8.4}$$

where $H(s; J)$ is an energy function depending on both s and J, and $Z_J = \int_{\mathbb{R}^d} dse^{-\beta H(s;J)}$ is the partition function equal and plays the role of a normalization.

As an example of a disordered system, let us consider a *spin glass* [595, 596], where the energy function is $H(s; J) = -\sum_{<i,j>} J_{ij} s_i s_j$ (resembling an Ising-type interaction, see Eq. (3.1)), where couplings J_{ij} are i.i.d. according to the normal distribution $p(J_{ij}) \propto \exp\{-(J_{ij} - J_0)/2J^2\}$, where J_0 and J^2 are the mean and variance. If all the J_{ij} are positive, the system is ferromagnetic, and the ground state of the system is easy to find. With random couplings, complications arise along with the frustration of the system: At a given site, a spin can be encouraged by neighbors to point in conflicting directions. Finding the ground state of such systems is a numerical challenge of its own. While in one dimension the solution is trivial and can be solved by a deterministic algorithm whose cost scales as $\mathcal{O}(n)$, the complexity grows in two dimensions and reaches NP-completeness in three and more dimensions [597].[6]

Now, let us tackle the exponent in Eq. (8.3), which we treat as an energy function. If we do that, there is a property of the free energy,[7] which can help us in simplifying the calculations. Namely, free energy is self-averaging.

[4]A partition function for a many-body classical discrete system is equal to $Z = \sum_i e^{-\beta \varepsilon_i}$, where i iterates over all possible microstates and ε_i is the energy of the ith microstate. If we go to a continuous system with n identical particles described by properties θ, the partition function is $Z \propto \int \exp\left(-\beta \sum_{i=1}^{n} H(\theta_i)\right) d\theta_1 \cdots d\theta_n$, where H is a classical Hamiltonian.

[5]This is also a reason why computation of any interesting partition function is hard.

[6]There are proposals to tackle this challenge with reinforcement learning (RL) [598].

[7]In the thermodynamic limit, the free energy of the system is $F = U - TS = -\frac{1}{\beta} \ln Z$, where U is the energy of the system and S is its entropy.

If a random quantity is *self-averaging*, two conditions are met: Its mean value and the most probable value coincide in the thermodynamic limit, and fluctuations around this mean value are sufficiently small. In other words, the system *concentrates on typical states.*

This property often holds for the free energy of disordered systems. Consider the following argument: Imagine dividing the macroscopic system into many subsystems, and each subsystem is still large enough to be considered macroscopic. Their interaction can be viewed as a surface effect and is negligible compared to the bulk. Therefore, each subsystem has a well-defined free energy and the realization of disorder, even if the specific values vary between subsystems. In the limit of an infinite number of subsystems (whose interactions can be ignored to first order), the disorder average of the free energy is automatically the average free energy across the disordered subsystems [596, 599, 600]. That is, for d large enough, the physics of the system is independent of the disorder realization:

$$\frac{1}{d}\ln Z_J \approx \lim_{d\to\infty} \mathbb{E}_J\left[\frac{1}{d}\ln Z_J\right]. \tag{8.5}$$

With the free energy being extensive, note that the converging quantity in the thermodynamic limit is the free energy per spin. This result is highly nontrivial, and tools such as replica computations, variational mean-field methods, and high-temperature expansions are necessary to identify where self-averaging applies and to compute the disorder averages.[8] In the following paragraph, we provide the intuition behind only one of the concepts behind Eq. (8.5), namely, *the replica trick*. Readers interested in more detailed explanations should refer to the tutorial reviews [603, 604].

Replica trick. In statistical physics, calculating averages makes sense only for extensive observables. The replica method is a way to calculate these averages with respect to disorder variables. We are particularly interested in the averaged value of the system free energy $F_J = -\frac{1}{\beta}\ln Z_J$. To obtain the averaged free energy $\mathbb{E}_J\left[F_J\right]$, we have to obtain the averaged value of the logarithm of the partition function $\mathbb{E}_J\left[\ln Z_J\right]$. It turns out that averaging the logarithm is challenging, but the average of powers of the partition function, $\mathbb{E}_J[Z^n]$ for $n \in \mathbb{N}$, can be estimated. Then, by using the identity,

$$\ln x = \lim_{n\to 0}\frac{x^n - 1}{n}, \tag{8.6}$$

we can write

$$\mathbb{E}_J[\ln Z] = \lim_{n\to 0}\frac{\mathbb{E}_J[Z^n] - 1}{n}. \tag{8.7}$$

[8] It is interesting to note that these nonrigorous physical approaches for disordered systems developed in the 1970s [596, 599, 600] are now being put on a more rigorous footing by mathematicians [601, 602]!

As we can see, the limit $n \to 0$ requires $n \in \mathbb{R}$. However, what we can do is to calculate Z^n for $n \in \mathbb{N}$. The partition function Z is an integral of the form $\int e^{-\beta H(s,J)}$, thus we can write Z^n as

$$Z^n = \int ds^{(1)} \ldots ds^{(n)} \prod_{a=1}^{n} e^{-\beta H(s^a,J))} = \int ds^{(1)} \ldots ds^{(n)} e^{-\beta \sum_{a=1}^{n} H(s^a,J))}, \tag{8.8}$$

where the exponent contains a sum over n independent samples or replicas. The replica trick consists in defining a function $\phi(n)$ being an analytic continuation of the function in the exponent. As such, $n \in \mathbb{R}$ becomes a continuous variable, and we can take limit $n \to 0$ in Eq. (8.7). In summary, assuming that we can calculate the averaged value $\mathbb{E}_J[Z^n]$, we can calculate the averaged value of the free energy F_J.

Finally, having Eq. (8.5), we can return to the volume $V_{d,n}$ in Eq. (8.3).

We associate this volume $V_{d,n}$ now with the partition function of a spin system. Spins (s) are now model parameters (θ), and couplings (J) are training data ($\mathcal{D}$), which pose the constraints to learn.

Applying the same analysis as in the previous paragraph, we can state that the free energy for a given realization of the dataset is just the free energy averaged over the dataset distribution when we consider large datasets and large perceptron with fixed ratio $\alpha = n/d$:

$$V_{d,n} \simeq \lim_{n\to\infty} \mathbb{E}_{\mathcal{D}}[V_{d,n} \mid \mathcal{D}] = V(\alpha). \tag{8.9}$$

Therefore, if you fix the distribution of data (disorder realization), you can find the α_c for which $V_{d,n} = 0$ and, as a result, the perceptron capacity. To be more exact, we can calculate it only for the large ("thermodynamic") limit of n for an arbitrary fixed α as V is actually expressed in terms of α.

We remind you that α_c indicates the critical task difficulty for which the volume of perceptron weights satisfying the constraints of random labeling goes to zero. It means that for the lower task difficulty, $\alpha < \alpha_c$, the randomly labeled data are linearly separable, while for the higher task difficulty, $\alpha > \alpha_c$, the data are no longer linearly separable. The probability, p_{R}, is therefore a step function of α in the thermodynamic limit. We plot p_{R} for real-valued parameters coming from a Gaussian distribution in blue in Fig. 8.4. To show finite-size effects, we can also compute $p_{\mathrm{R}}(\alpha)$ below the thermodynamic limit following Cover's argument [586]. To vary α, we can change n or d. In the case of perceptron, it is easier to keep d fixed and calculate p_R as a function of α for increasing n.

In the equivalent of the thermodynamic limit, so $n \to \infty$, we see a phase transition for a critical $\alpha_c = 2$, which means that the most difficult task that the perceptron is able to solve is when the number of training points (in general position) is twice as large as the number of parameters.

Interestingly, the solution for α_c (for which $V_{d,n} = 0$) depends on the setup of the problem, namely, the random data distribution and the allowed values of parameters

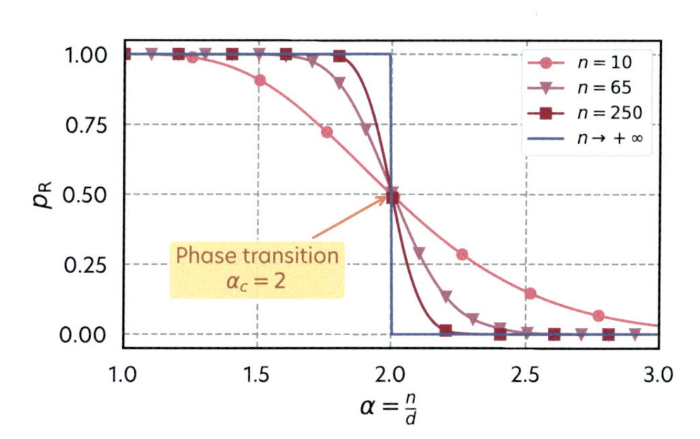

Figure 8.4 Probability of the randomly labeled data being linearly separable, p_{R}, as a function of the difficulty of the task, $\alpha = \frac{n}{d}$. Finite-size results were obtained analytically by Cover [586].

Table 8.1 The capacity of the perceptron depends on the distribution of the data and type of weights. The capacity is expressed as the minimal task difficulty, $\alpha_c = \frac{n}{d}$, for which the volume of possible solutions goes down to zero, $V(\alpha_c) = 0$.

	Distribution of data		Critical task difficulty, α_c
1	Gaussian inputs Real weights	$p(\boldsymbol{x}_i^{(k)}) = \mathcal{N}(\boldsymbol{x}_i; 0, 1)$ $\theta \in \mathbb{R}^d$	$\alpha_c = 2$
2	Binary inputs Binary weights	$p(\boldsymbol{x}_i^{(k)}) = \text{Bernoulli}(0.5)$ $\theta \in \{-1, 1\}^d$	$\alpha_c \approx 0.83$

(spin values). While the previous discussion has been conducted for Gaussian distribution of inputs and real perceptron parameters, $\boldsymbol{\theta}$, a different critical task difficulty is obtained for binary inputs and parameters, as presented in Table 8.1.

In this section, we have looked at the problem of perceptron capacity, which is well-known and decades old. As such, it serves the educational purpose well. In particular, we have seen that the statistical approach to learning focuses on simple solvable models (here, perceptrons). Moreover, we have seen that the statistical approach aims to express learning problems in terms of statistical problems, for example, disordered spin systems,[9] where physicists have already developed useful analytical tools.

In Sections 8.1.2–8.1.4, we briefly discuss selected modern results from the intersection of ML and statistical physics. For a more detailed review of this intersection, we refer to [5]. Moreover, an outstanding retrospective of these developments can be

[9] This also tells us that NNs with binary weights may be especially approachable for physicists. These are spin-1/2 problems!

found in the lecture titled "Statistical physics and ML: A 30-year perspective" of the late Naftali Tishby.

8.1.2 The teacher–student paradigm: A toy model to study the generalization

Our motivation for this section is to tackle the riddle of generalization, which is the ability of a model to make correct predictions on data unseen during training. However, our goal for this section is not to build new useful ML models or to distinguish between bad and good modern models in terms of generalization. Rather, we want to understand why useful modern ML models generalize so well. To do so, let us consider all elements of the learning task (such as model, optimization method, and data) in their simplest form. The toy model that helps us in this ambitious task falls under the teacher–student paradigm.

> The teacher–student paradigm consists of two main elements: a teacher which is a data-generating model and a student which is a model trying to learn the data generated by a teacher.

Teacher consists of an input distribution $p_X(\boldsymbol{x})$, for example, Gaussian or binary, and an input–output rule $p(y_{\mathrm{t}} \mid \boldsymbol{x}) = f_{\mathrm{t}}(\boldsymbol{x}, \theta^*)$. For now, let us assume that the teacher is a perceptron. In addition to the input–output rule, we may assume a ground-truth distribution on the weights $p_\Theta(\theta)$, from which the parameters θ^* of the teacher model were drawn. Once we decide on how a teacher looks like, it can generate training data: $\mathcal{D} = \{\boldsymbol{x}^{(k)}, y_{\mathrm{t}}^{(k)}\}_{k=1}^n = \{\boldsymbol{x}^{(k)}, f_{\mathrm{t}}(\boldsymbol{x}^{(k)}, \theta^*)\}_{k=1}^n$.

> The second element is the student, whose aim is to learn the distribution underlying the training data. In the teacher–student scheme, we know exactly what the data-generating distribution is. Therefore, we can easily distinguish between a student that simply fits the training data (limited generalization) and a student that recovers a teacher's input–output rule (perfect generalization). In other words, we can measure the generalization of the student.

To continue with the teacher–student strategy, we need to decide on a model for the student, $f_{\mathrm{s}}(\boldsymbol{x}, \theta)$, but also on a learning strategy. Let us start with the simplest scenario when a student is also a perceptron (like the teacher). To train, we could use the standard empirical loss minimization strategy, for example,

$$\theta^* = \operatorname{argmin}_\theta \left\{ \sum_{k=1}^{n} \mathcal{L}\left(y_{\mathrm{t}}^{(k)}, f_{\mathrm{s}}\left(\boldsymbol{x}^{(k)}, \theta\right)\right) \right\}, \tag{8.10}$$

where we aim to minimize a given distance between the teacher outputs $y_{\mathrm{t}}^{(k)}$ and student outputs $y_{\mathrm{s}}^{(k)} = f_{\mathrm{s}}(\boldsymbol{x}^{(k)}, \theta)$. Alternatively, we can consider the following Bayesian posterior distributions on the parameters and draw values of the parameters according to it:

$$p(\theta \mid \mathcal{D}) \propto \prod_{k=1}^{n} p\left(y_{\mathrm{t}}^{(k)} \mid \theta, \boldsymbol{x}^{(k)}\right) p(\theta). \tag{8.11}$$

Equation (8.11) denotes the posterior distribution, that is, the belief on the student model weights θ given the data set $\mathcal{D}$ and the prior assumption on the student weights $p(\theta)$.

Assuming, for example, an MSE loss, the student generalization error for given weights θ is defined as the expected error over the entire data distribution:

$$\mathcal{E}_g(\theta) = \mathbb{E}_{\boldsymbol{x},y}\left[(y - f_{\mathrm{s}}(\boldsymbol{x},\theta))^2\right]. \tag{8.12}$$

In the best possible scenario, the student model $f_s(\cdot,\cdot)$ is identical to the teacher model $f_t(\cdot,\cdot)$ underlying the generated data. When a student is identical to the teacher, we call the setting *Bayes optimal* and define the Bayes error (see Section 2.3) of the student as,

$$\mathcal{E}_g^{\mathrm{opt}}(\mathcal{D}) = \mathbb{E}_\theta\left[\mathbb{E}_{\boldsymbol{x},y}\left[(y - f_{\mathrm{s}}(\boldsymbol{x},\theta))^2\right] \mid \mathcal{D}\right], \tag{8.13}$$

which is a mean error for student parameters θ drawn from the posterior distribution in Eq. (8.11). This is a fundamental quantity from the point of view of information theory: It quantifies how much information on the weights θ the training dataset $\mathcal{D}$ provides, assuming that the student has perfect knowledge of the form of the problem. We can use the same tools as in Section 8.1.1 (disorder average, thermodynamic limit, and replica computation) to obtain:

$$\mathcal{E}_g^{\mathrm{opt}}(\mathcal{D}) \underset{n\to\infty}{\rightarrow} \mathbb{E}[\mathcal{E}_g^{\mathrm{opt}}(\mathcal{D}) \mid \mathcal{D}] = \mathcal{E}_g^{\mathrm{opt}}(\alpha). \tag{8.14}$$

Here again, the limiting generalization error takes the form of a function of the ratio $\alpha = \frac{n}{m}$ between the number of data points and the number of data features or weights. We no longer interpret this ratio as the difficulty of the classification task as in the capacity computation. Instead, in generalization problems, it is more useful to think of α as the sample complexity, that is, the amount of training data available to infer the input–output rule. In the following paragraphs, we examine generalization for a few different pairs of teachers and students.

Two perceptrons. The generalization error from Eq. (8.14) is shown in Fig. 8.5. We can compare the limiting Bayesian optimal generalization error (red line in panels (a) and (b)) with the training of a perceptron at finite m by minimizing a loss function, such as performing a logistic regression with gradient descent (blue squares). In panel (a), for binary weights, we have a first-order phase transition [592, 593]. In panel (b), for real-valued weights, there is a smooth decrease of the generalization error [605]. In both cases, there is a computational gap between the optimal generalization error and logistic regression with gradient descent.

Finally, the same generalization error of the student perceptron can be studied when learning occurs with algorithms called GAMP. For the introduction to these methods, see [604, 606, 607]. For our needs, it is enough to know that these algorithms

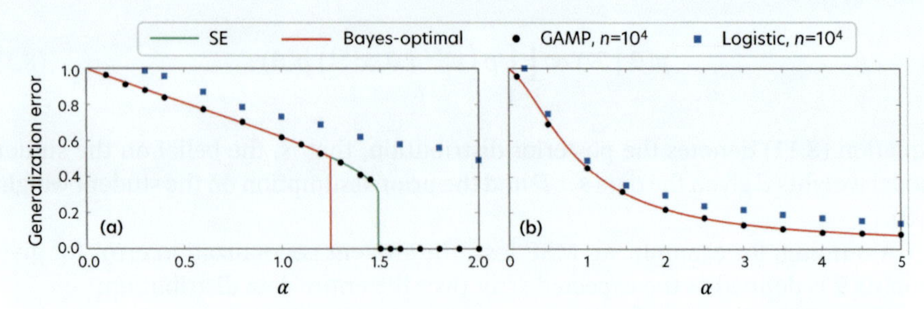

Figure 8.5 Generalization error as a function of the task difficulty α, which is the ratio between the number of training points and the number of (student) model parameters for perceptrons with (a) $\theta \in \mathbb{R}^d$ or (b) $\theta \in \{-1, 1\}$. The red line is the exact Bayes-optimal generalization error. The blue squares are for a fine-tuned perceptron with gradient-based minimization of the error. We see the computational gap between these results. Gap (a) gets smaller or (b) disappears for message-passing algorithms. Black circles are results for $n = 10^4$ obtained using generalized approximate message passing (GAMP), and the green line denotes the results of state evolution (SE), which approximates the limit $n \to \infty$. Adapted from [605] under the CC BY 4.0 DEED license.

provide an alternative to convex optimization and allow for efficient calculations of quantities based on graphs (like perceptrons or NNs), which are sampled from distributions like Eqs. (8.12) to (8.14). Moreover, they are remarkable in that their asymptotic ($n, d \to \infty$, $n/d = \alpha$) performance can be analyzed rigorously using the so-called state evolution (SE). Armed with this knowledge, we now see that the generalization error obtained using GAMP in Fig. 8.5 is much closer to the Bayes error compared to the optimization with gradient descent. In panel (b), the gap completely disappears. In panel (a), there is a remaining computational gap between GAMP and the exact Bayes error. This regime is called a *hard phase*. It comes from the fact that, in practice, our computational time is limited to the polynomial regime. Interestingly, there is no known efficient algorithm that would beat GAMPs in the hard phase of this perceptron learning [605].

Two-layer NNs. So far, both the teacher and the student have been modeled with perceptrons. We can switch to more complex models. For the remainder of this section, we use two special two-layer NNs with a rich history in statistical physics. We start with *committee machines* [608, 609] shown in Fig. 8.6(a). Their analytical treatment is possible in the limit of an infinite number of input features, m, and data size, n, while keeping a finite number of hidden units. In particular, we present here soft committee machines that allow for an even simpler analysis. In soft committee machines, we train only parameters belonging to the first layer of the machine, θ_1, of size $d_1 = d = mD$, where m is the number of features and D is the number of hidden units. The second layer is fixed and identical for both the teacher and the student. The second NN used in this section is a *random feature model* [610, 611] presented in Fig. 8.6(b). Interestingly, their analytical analysis is enabled by a fixed first layer whose parameters are set to random values. The number of these parameters is also

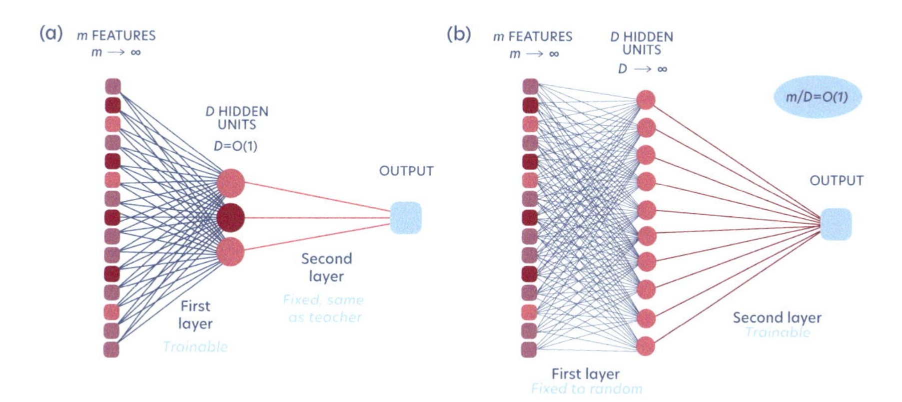

Figure 8.6 Schematic illustration of used two-layer NNs. (a) Soft committee machine. The parameters belonging to the first layer, $\theta_1 \in \mathbb{R}^{m\times D}$, are trainable, whereas the parameters of the second layer, $\theta_2 \in \mathbb{R}^{D\times 1}$, are chosen identical to the parameters of the teacher. Its analytical treatment is possible if $m \to \infty$ and $D = O(1)$. (b) Random feature model. Its first layer is fixed to random parameter values. The second layer is trainable. The number of hidden units can be varied to study overparametrization. In the analysis, the number of hidden units $D \to \infty$ scales linearly with the number of inputs $m \to \infty$, that is, $m/D = O(1)$.

$d_1 = d = mD$. Therefore, only the second layer parameters can be trained. The idea behind the random first layer is that projecting a lower-dimensional input onto a much higher-dimensional space leads to better separation of the data, which then can be successfully processed by a single-layer NN.[10] Also note that random feature models can have an arbitrary number of hidden units, in particular larger than the number of input features, which allows for a study of overparametrization.

Two committee machines. Now, we are ready to tackle generalization with more complex models. Here, we use soft committee machines. For now, a teacher and a student share the same architecture. The formulation of the problem stays the same. We calculate the generalization error from Eq. (8.14) of the student committee machine when learning data generated by the teacher committee machine [609]. We plot the generalization errors in Fig. 8.7(a) and (b) for committee machines with two hidden neurons. Similarly as before, we see in panel (a) that for real-valued weights, the generalization error (obtained with GAMP and SE) is equal to the Bayes one, while for binary weights in panel (b), there is a computational gap between both errors. This time, we also look at the overlap between hidden neurons of the student and of the teacher, which measures the similarity neuron-by-neuron between

[10]In other words, you can think of such a projection as mapping input data to a feature space as discussed in Section 4.2 on kernel methods. Interestingly, [610, 611] showed that random data projection onto a feature space is not much worse compared to projecting onto an optimized feature space. However, randomization is much cheaper than optimization.

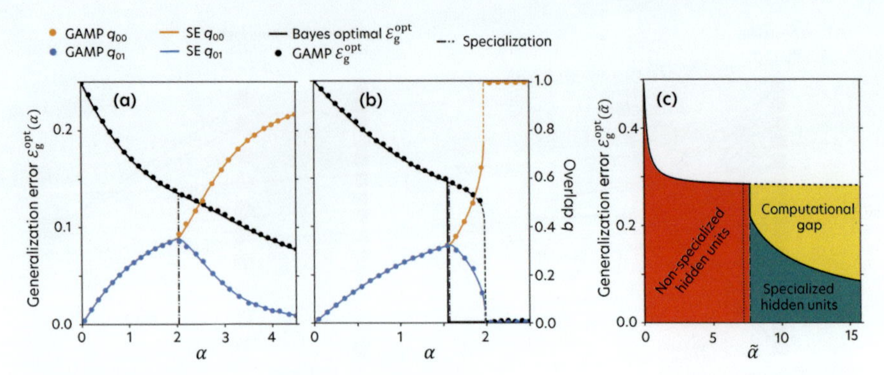

Figure 8.7 Generalization error and specialization in committee machines as functions of the task difficulty $\alpha = \frac{n}{m} \propto \frac{n}{d}$, which is the ratio of the number of training points and the number of input features. We consider the committee machines with (a) $\theta \in \mathbb{R}^{m \times D}$, $D = 2$, or with (b) $\theta \in \{-1, 1\}^{m \times D}$, $D = 2$. The black line is the exact Bayes-optimal generalization error, and the black dots are obtained by studying the committee machine with GAMP. The orange and blue lines and dots indicate the overlap of the two hidden neurons of the student committee machine with the two hidden neurons of the teacher committee machine, calculated with GAMP and SE, respectively. We see that specialization is responsible for the rapid decrease in generalization error. (c) Generalization of panel (a) to a large number of hidden neurons, D. Phase diagram calculated for the task difficulty, $\tilde{\alpha} = \frac{\alpha}{D}$. Adapted from [609] with permission of IOP Publishing, Ltd. Permission conveyed through Copyright Clearance Center, Inc.

the teacher and the student. To be more precise, we look at the matrix $\boldsymbol{Q} = [q_{jj'}] = \frac{1}{m}\sum_{i=1}^{m} \Theta^*_{1,ij}\theta_{1,ij'}$, where $\boldsymbol{\Theta}_1$ and θ_1 are the parameters of the first layers of the teacher and the student, respectively. It turns out that there is a so-called specialization phase transition [612, 613].

> In the regime of low task complexity, both hidden units of the student committee machine learn the same function. After crossing the critical α, when enough data is available, the hidden neurons of the student start to *specialize*. Each student neuron selects a different teacher neuron to converge to. The specialized phase is associated with lower generalization error than the nonspecialized one, see Fig. 8.7.

The specialization for teacher and student committee machines with two hidden neurons ($D = 2$) takes place for $\alpha_c \approx 2$ for real-value weights and for $\alpha_c \approx 1.5$ for binary weights, which means that specializing neurons require at least 2 and 1.5 times more training data than the number of data features, m, that is, approximately as much training data as the number of parameters in the first layer, $d_1 = 2m$. Similar observations hold if both the teacher and student committee machines have a large number of hidden neurons ($D \gg 2$). We can plot a phase diagram of the generalization error as a function of a rescaled task difficulty, $\tilde{\alpha} = \frac{\alpha}{D} = \frac{n}{Dm}$ for real-valued weights. It is presented in Fig. 8.7(c). In total, we find three distinct phases: two correspond to specialized and nonspecialized hidden neurons, and above the specialized phase, there is a computational gap where a model in principle has enough information to specialize but is unable to do so due to shortcomings of its optimization.

Overparametrization. As we have already mentioned in the introduction, one of the most puzzling phenomena in modern ML is the generalization capability of heavily overparametrized models. In real-world settings, it is natural to think of the level of the model overparametrization as the ratio between the number of model parameters, d, and the number of available training data points, n. Surprisingly, we see in practice that models with large d are able to extract meaningful relations from much fewer training data points. In turn, with the teacher–student scheme, we can make the definition of overparametrization more rigorous because we have direct access to the "ground-truth" number of parameters needed to describe the input–output rule, which is the number of teacher parameters. Therefore, the level of overparametrization can be understood as a ratio between the number of parameters of the student and the teacher. In particular, the student can have much more parameters than the teacher. To study overparametrization, it is then a necessity to have mismatched teacher–student architectures. Crucially, this mismatch of architectures means that the student cannot achieve a Bayes optimal error anymore.

For the remainder of this section, we study the generalization error of overparametrized student models. This time we employ as a student a random feature model, presented already in Fig. 8.6(b). The analysis requires the model's first-layer weights to be fixed to random values. The number of student parameters in the second layer can vary compared to the teacher.[11] As such, we have a full control over how overparametrized the student is. We come back to the study of overparametrization in committee machines in Section 8.1.4.

With a student random feature model, we are ready to study the generalization error as a function of overparametrization, $\frac{1}{\alpha} = \frac{d}{n}$, where $d = d_1$ is the number of parameters in the first fixed random layer of the student. As the student cannot achieve a Bayes optimal error anymore, we need to change the training objective, for example, to an MSE with ℓ_2 regularization. Using various analytical tools, we can still approximate the generalization error of the student and plot it as a function of overparametrization $\frac{d}{n}$. In [614], the generalization error in regression and classification tasks was analyzed for various strengths of regularization. Their results are shown in Fig. 8.8. The left (right) column shows the generalization error of the mismatched student for optimal and suboptimal regularization strengths in a regression (classification) problem.

[11] Ref. [614] interprets the same exact setting as a teacher generating labels with a perceptron, itself acting on a low-dimensional latent space, and input data generated with a one-layer generative NN from this latent space. A student perceptron is trained on the input data-label pairs.

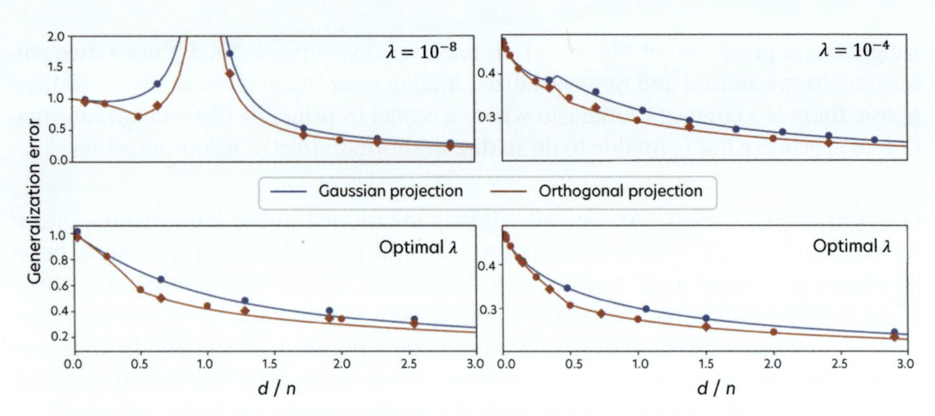

Figure 8.8 Generalization errors as functions of the ratio of the number of model parameters and number of training data points for mismatched teacher–student models where the student is a random feature model. The first (second) column shows the generalization error in the case of a regression (classification) problem. The upper row shows results for suboptimal regularization strengths, where the generalization error curves exhibit a double descent. The bottom row shows results for optimal regularization, where the double descent disappears. Adapted from [614] with permission of IOP Publishing, Ltd. Permission conveyed through Copyright Clearance Center, Inc.

Remarkably, in the case of mismatched student–teacher architectures, the generalization error curve exhibits a characteristic double descent. Moreover, the optimal choice of regularization cancels the first error descent, resulting in the generalization error steadily decreasing with the increasing number of model parameters.

Therefore, these results on toy models give us a hint on the origin of the double descent phenomenon. It occurs when the student and teacher have mismatched architectures, and the choice of regularization strength is suboptimal. Interestingly, [614] also showed that the magnitude of the initial generalization error ascent in the double descent phenomenon depends on whether the problem is a classification or regression task.

In summary, the study of toy models indicates that there are numerous reasons for the generalization error being larger than the Bayes optimal error. In general, the generalization capabilities depend on:

- whether a data-generating model (teacher) and learning model (student) have mismatched architectures,
- whether the model aims at solving a regression or classification problem,
- the choice of optimization method, target function, and available computation time, and
- the sample complexity (how much training data is available and, for teacher–student committee machines, the degree of specialization of the neurons).

8.1.3 Models of data structure

So far, while studying sources of generalization errors, we have mainly played with the architectures of teacher and student models, which specify the structure of the input–output rule underlying the data. In particular, we have only considered random input datasets where all input features are independent. Clearly, while such isotropic data simplify the analytical analysis, it is quite unrealistic. Ideally, we would like to study prototypical datasets, such as MNIST [31] or ImageNet [34], but these are difficult to treat analytically. Instead, let us move one step away from the datasets given by white noise and use the teacher–student paradigm to study the impact of data anisotropy on the generalization error. To this end, we employ *salient and weak feature models* [615]. Within these feature models, the data remain Gaussian (as in most of the previous sections), $\boldsymbol{x} \sim \mathcal{N}(0, \Sigma_{\boldsymbol{x}})$, but the covariance is not isotropic as if $\Sigma_{\boldsymbol{x}} = I_d$. Instead, it is anisotropic:

$$\Sigma_{\boldsymbol{x}} = \begin{bmatrix} \sigma_{\boldsymbol{x},1} I_{\phi_1 d} & 0 \\ 0 & \sigma_{\boldsymbol{x},2} I_{\phi_2 d} \end{bmatrix}, \tag{8.15}$$

where $\sigma_{\boldsymbol{x},1} \gg \sigma_{\boldsymbol{x},2}$, and $\phi_{1/2} d$ denotes the number of data features (equal to the number of perceptron parameters) that are affected by the variance $\sigma_{\boldsymbol{x},1/2}$ (with $\phi_1 + \phi_2 = 1$). Parameters affected by large variance, $\sigma_{\boldsymbol{x},1}$, form the salient subspace, whereas ones with a small variance, $\sigma_{\boldsymbol{x},2}$, form the weak subspace as presented in Fig. 8.9(a). We assume the weak subspace to be much larger than the salient one, $\phi_2 \gg \phi_1$.

If we add such a structure to our input data and run the teacher–student scheme with mismatched architectures (here, the teacher is a perceptron and the student is a random feature model), we can still compute the generalization error exactly in the high-dimensional limit [615]. Importantly, this generalization error now depends on the anisotropy of the input data. In particular, it depends on how the teacher perceptron is aligned with respect to the weak and salient data subspaces as presented in Fig. 8.9(a). The dashed hyperplanes mark the separation of the input space by the teacher perceptron and lie perpendicular to the perceptron parameter vector, $\boldsymbol{\theta}$. This vector can be aligned in various ways with the data anisotropy. If $\boldsymbol{\theta}$ is aligned with the salient subspace, the hyperplane cuts along the weak subspace, and the only subspace relevant to discriminate the data points is the salient subspace, in which the variance of the data is concentrated, and the weak subspace can be effectively ignored. In this case, due to the data structure, the problem has a small effective dimension corresponding to the salient space $\phi_1 d$, so it is easier to solve. In turn, if $\boldsymbol{\theta}$ is aligned with the weak subspace, the impact of the data structure is negligible since the student needs to discriminate along an axis where data has low variance compared to the typical variance of the data. Here, the problem closely resembles the (fully) isotropic case.

The impact of the data structure on the generalization curve as a function of the ratio of the number of parameters of the student model and the dimensionality of the data, $\frac{d}{m}$, is shown in Fig. 8.9(b).

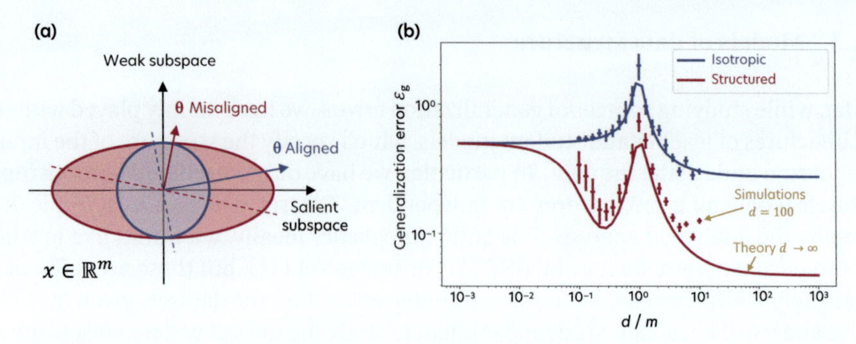

Figure 8.9 Data structure entering the teacher–student scheme. (a) The data space can be separated into weak and salient subspaces, where the data is characterized by a small and large variance, respectively. The teacher perceptron with parameters $\boldsymbol{\theta}$ can be aligned (blue vector) or misaligned (purple vector) with the salient subspace. The classification task specified by the teacher (represented as a line separating the data) is easier (compared to the isotropic case) if $\boldsymbol{\theta}$ is aligned with the salient subspace. (b) The model can detect the structure existing in the data. As a result, the generalization error is lower for the structured case than for the fully isotropic case. The results for the isotropic data are similar to the results for anisotropic data, where the teacher perceptron is misaligned. The generalization curve shows a double descent. Adapted from [615].

Interestingly, the structure in the data is detected during training before the generalization error peaks due to overfitting and improves the generalization error as compared to the isotropic case.

Why does the structure help? This is so because, in practice, the model can ignore the weak subspace and focus on the salient one, which lowers the dimensionality of the problem. The fact that the generalization error is lower in the anisotropic case compared to the isotropic case remains true even in the highly overparametrized regime ($\frac{d}{m} = 10^3$). Moreover, the double descent phenomenon is also exacerbated in the presence of data structure. Note that both these effects take place only when the teacher perceptron is aligned with the salient subspace. Otherwise, the setup closely resembles the isotropic case.

It turns out that many more questions can be addressed with the teacher–student paradigm using salient and weak feature models. In particular, the authors of [615] checked the interplay between the data structure and other elements of ML problems, like the choice of the loss function. Recalculating the quantities in Fig. 8.9(b) for the MSE and logistic loss, one observes that the overfitting peak is attenuated in the case of logistic loss. Therefore, it seems that logistic loss takes more advantage of the existing data structure. We can confirm this further by computing the generalization error for both loss functions as a function of the teacher–data alignment. As discussed earlier, this alignment determines how much data structure is *effectively* present in the problem. In agreement with the results described above, when the alignment is increased, the gap between the generalization error of MSE and logistic loss increases.

8.1.4 Dynamics of learning

Finally, we can investigate the dynamics of learning and its dependence on the model overparametrization using the teacher–student schemes described above. An example of a simplified model of learning is *online learning*, which has been analyzed since the 1990s. In online learning, the model is fed a stream of data, where the model sees each data point only once. We build a loss function based on this example and perform a parameter update according to the gradient of this loss function. In fact, we perform a parameter update after each data point encounter. Thus, the number of optimization steps is equal to the number of seen training data points. If we take the continuous time and high-dimensional limit and average over all random variables (which is doable with the replica method if we assume samples at distinct times are uncorrelated), we can again calculate the generalization error explicitly. In particular, we can calculate how it changes during training. In other words, we can track the quality of the model predictions over the course of the training.

Such an analysis has already been conducted in the 1990s for perceptrons and committee machines [617, 618]. It showed, for instance, that during online training, the generalization error decreases with different convergence rates given different learning rates. Recently, the same analysis was revisited for soft committee machines [616] considering the impact of overparametrization. In the simplified case of matching teacher–student models and training limited to only a single student layer, results show how the generalization error drops the moment the student neurons specialize and attain a large overlap with the teacher neurons. We can also investigate the effect of overparametrization on learning dynamics using committee machines as presented in Fig. 8.6(a) in the regime of the number of data features, $m \to \infty$, with the sigmoidal activation function, $g(x) = \mathrm{erf}(x/\sqrt{2})$. Here, we study the generalization error as a function of the ratio between the number of hidden units of the student D and the teacher T given by $\frac{D}{T}$. Figure 8.10 shows two cases of online learning of such overparametrized students. In the first case, shown in panels (a) and (b), only the first hidden layer of the student model can be trained, whereas the parameters of the second hidden layer are fixed and identical to the respective teacher layer. In the second case, presented in panels (c) and (d), both student layers are trained. Panel (a) shows that the generalization error actually increases with the size of the trainable student layer, proving that overparametrization can be detrimental in some scenarios. To understand why, we analyze the teacher–student overlaps at the end of the training in the form of $\boldsymbol{R} = [R_{it}]$ where each matrix element measures the similarity between the weights of the ith student node and the tth teacher node. We also study the overlap of the weights of different student nodes with each other ($\boldsymbol{Q} = [Q_{ij}]$). We plot both matrices in panel (b). We see that in the case of soft committee machines, only the number of student neurons that is equal to the number of teacher neurons specialize. The rest simply picks up the noise present in the available data, which impairs generalization. However, if we allow all layers to be trained, a very different behavior is observed. In panel (c), we see that the generalization error decreases as one overparametrizes the student model. This time, all neurons learn something related to the teacher neurons. Due to this, additional neurons are beneficial as each teacher neuron can be learned by an ensemble of student neurons that contributes to "denoising" the estimation of the teacher parameters.

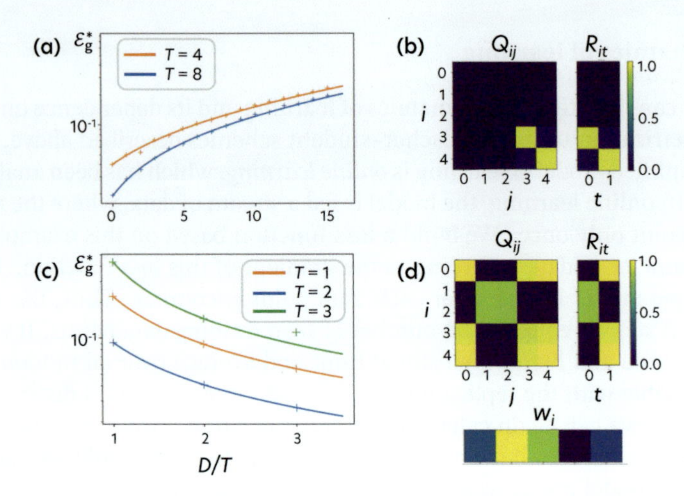

Figure 8.10 Dynamics of learning in overparametrized committee machines. In panels (a) and (b), we allow only a single trainable student layer, and another layer is fixed. In panels (c) and (d), we train the whole student model. (a), (c) Generalization error vs. student overparametrization which occurs when the number of student hidden units, D, is larger than the number of teacher hidden units, T. (b), (d) Self-overlaps of the student ($\boldsymbol{Q}$ matrix) and overlaps between the overparametrized student and the teacher ($\boldsymbol{R}$ matrix). Vector $\boldsymbol{w}$ contains the weights of the second layer of the student. Adapted from [616] under the CC BY 4.0 DEED license.

Outlook and open problems

In this section, we have seen how to use analytical tools from statistical physics to study problems in ML. In particular, we have discussed the seminal problem of perceptron capacity. Subsequently, we have focused on a powerful paradigm for studying the generalization error: the teacher–student scheme. This scheme can include various modifications that address all elements of the learning problem.

We can study different teacher and student architectures, and they can be mismatched. We have shown results for perceptrons, committee machines, and random feature models, but, in general, we can have, for example, a pretrained generative model (described in more detail in Section 7.2.2) playing a role of a teacher as it was done in [619]. One can also analyze more complex datasets than those provided by a salient and weak feature model. In particular, it is possible to confirm intuitions gained from the analytical analysis of simple models with simulations on standard benchmark datasets [615], such as MNIST [31] and CIFAR [33]. Finally, one can go beyond the online gradient descent and study the multi-pass SGD (which involves multiple encounters of the same data points) with the dynamical mean-field theory [620], bringing us closer and closer to modern optimization methods.

Moreover, one can investigate the capacities of large ML architectures (in contrast to simple perceptrons). Statistical tools also play an increasingly important role in the research on QML. For example, the Gardner approach was successfully applied to quantum perceptrons [621, 622] and quantum NNs [623]. In particular, it turned out that the quantum perceptron has some advantages over its classical counterparts when it comes to capacity [622]. Moreover, the teacher–student scheme was proposed to systematically compare different quantum NN architectures [624]. Finally, there are works studying phases in the learning dynamics of ML models [625, 626].

Further reading

- Gabrié, M. (2020). *Mean-field inference methods for neural networks.* J. Phys. A: Math. Theor. 53, 223002. Review on the mean-field methods mentioned within this section. In particular, it contains principles for derivations of high-temperature expansions, the replica method, and message-passing algorithms [604].
- Zdeborová, L. (2020). *Understanding deep learning is also a job for physicists.* Nat. Phys. 16, 602–604. A short and friendly introduction to how physics can help ML [585].
- Recordings of lectures of the Summer School on Statistical Physics of Machine Learning which took place on Jul 4–29, 2022 in Les Houches, France.
- Jupyter notebooks prepared as tutorials for the Summer School: Machine Learning in Quantum Physics and Chemistry [2].
- Castellani, T. & Cavagna, A. (2005). *Spin-glass theory for pedestrians.* J. Stat. Mech. P05012. Pedagogical review on mean-field methods for spin glasses [603].

8.2 Quantum machine learning

This section explores yet another direction: How quantum information and quantum hardware can be used to solve data-driven tasks. This recent field is called QML.[12] This field started with the development of quantum algorithms aiming for a potential fully quantum advantage. In recent years, there has been an increasing interest in another direction: studying hybrid quantum-classical algorithms (also often called quantum-enhanced algorithms), where part of the algorithm is performed on a quantum device. With the development of new experimental platforms for quantum computation, researchers are now looking for applications tailored to these hybrid algorithms and trying to determine if and how quantum advantage can arise in such systems. While the quantum advantage would represent a breakthrough, the study of the quantum-enhanced algorithms running on these hybrid devices is an interesting problem in itself and can potentially lead to the discovery of exciting physics.

In the following sections, we provide an overview of the recent advances in the field. We do not aim to provide a complete review but rather an introduction to selected topics. In the last section, we refer to recent reviews of the field for the interested reader.

[12]Often in literature, QML incorporates both quantum-enhanced ML and ML applied to quantum, for example, ML for quantum information processing. In this section, we use QML for quantum-enhanced ML. A detailed discussion about this convention can be found in Section 8.2.2.

8.2.1 Gate-based quantum computing

In the following sections, we focus on the description of *gate-based quantum computation*. These concepts are used throughout the whole section.

> The most common building blocks of gate-based quantum computation are *qubits* and *quantum gates*. A gate-based quantum algorithm, specified as a sequence of gate operations and measurements performed on qubits, can be conveniently represented as a *quantum circuit*.

Qubits are two-level quantum systems that can be realized by isolating two degrees of freedom in several experimental platforms, such as photonic platforms [627, 633], superconducting circuits [628], trapped ions [629], or Rydberg atoms in optical tweezers [630, 631]. When performing a calculation, a quantum computer modifies the state of the qubits or entangles them using quantum gates.

Quantum gates are unitary operations and can be represented by unitary matrices. The dimensions of these matrices depend on the number of qubits on which these gates act. The scaling of their dimension is exponential in the number of qubits.

Examples of single qubit gates are the Hadamard and Pauli-X gates, which read in the single qubit basis $\{|0\rangle, |1\rangle\}$

$$H = \frac{1}{\sqrt{2}}\begin{pmatrix} 1 & 1 \\ 1 & 1 \end{pmatrix}, X = \begin{pmatrix} 0 & 1 \\ 1 & 0 \end{pmatrix}, \tag{8.16}$$

or parametrized gates such as the single qubit rotation gate

$$R_X(\theta) = \frac{1}{\sqrt{2}}\begin{pmatrix} \cos\theta & -i\sin\theta \\ -i\sin\theta & \cos\theta \end{pmatrix}, \tag{8.17}$$

parametrized in terms of the angle θ. An example of a two-qubit gate is the controlled NOT gate (CNOT), which reads in the two qubit basis $\{|00\rangle, |01\rangle, |10\rangle, |11\rangle\}$

$$\text{CNOT} = \begin{pmatrix} 1 & 0 & 0 & 0 \\ 0 & 1 & 0 & 0 \\ 0 & 0 & 0 & 1 \\ 0 & 0 & 1 & 0 \end{pmatrix}. \tag{8.18}$$

In general, *quantum circuits* can be depicted with quantum diagrams, as exemplarily shown in Fig. 8.11. Each line corresponds to a qubit. This circuit has two gates acting on a single qubit (X and H) and an entangling gate (CNOT) acting on two qubits. The last part of the diagram is the measurement, which is an interaction with individual qubits that forces their collapse to one of the two levels. As the measurements are destructive, the careful choice of a set of measurements is necessary to properly extract the relevant information from the quantum circuit.

8.2.2 What is quantum machine learning?

To better understand QML, let us first have a look at Fig. 8.12(a). Generally, ML algorithms are run on classical data, for example, in image classification or natural language processing. We have thus a classical algorithm dealing with classical

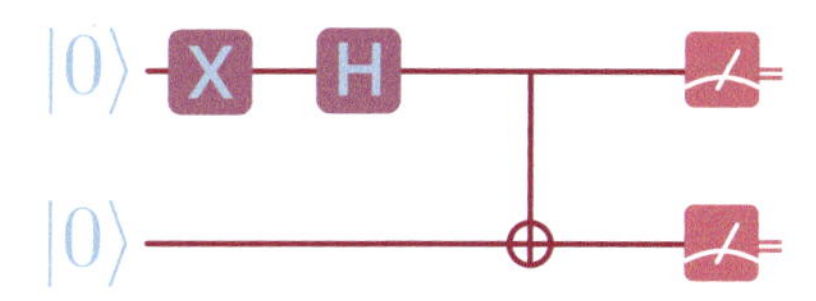

Figure 8.11 Illustration of a quantum circuit diagram with two initialized qubits, q_0 and q_1, and three different quantum gates: the Hadamard gate H and the σ_x gate X, both acting on q_0, and a two-qubit gate (CNOT). The final element is the measurement on q_0 and q_1.

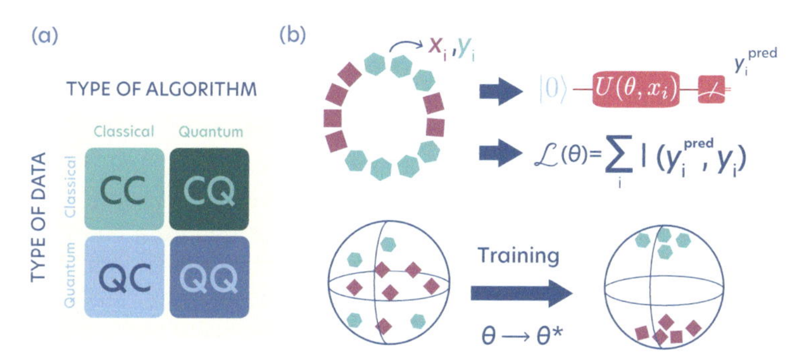

Figure 8.12 (a) Table of the different types of data and algorithms. (b) Sketch of an example of a classification task on one qubit.

data (CC). This book focuses mainly on the case of classical ML applied to quantum data (CQ), for example, quantum states. On the other hand, QML deals with the integration of quantum devices in ML algorithms. Therefore, the algorithms can be quantum, and the data can be either classical (QC) or quantum (QQ). In this section, we focus mainly on the QC side, as the QQ side is only at its early stage of development [633–636].

Let us discuss an elementary example to introduce the revised building blocks of ML in the context of the QC QML. We consider the classification problem of one-dimensional data on a ring. We intuitively sketch each step of this QML classification problem in Fig. 8.12(b). First, the classical data $(\boldsymbol{x}_i, y_i)$ is encoded in a quantum computer. Here, for example, we encode the data points on a single qubit through the action of a parametrized unitary $U_{\theta,\boldsymbol{x}_i}$, where $\boldsymbol{\theta}$ are the parameters of this unitary transformation (e.g., $R_X(\theta)$ introduced in Section 8.2.1). Then, a measurement is performed, and one can define the output of the measurement y_{pred} as a label (here 1 or 0). We then construct a loss function $\mathcal{L}(\boldsymbol{\theta})$ depending on the predicted and ground-truth labels. We can see here that the *quantum-enhanced* part corresponds to the evaluation of y_{pred} on a quantum computer. Once the loss function is defined, the minimization can be performed on a classical computer with the method of your choice, such as gradient descent or a gradient-free optimizer (e.g., Nelder–Mead). In this simple example, the training has a simple interpretation. Initially, the weights $\boldsymbol{\theta}$ of the unitary are randomly distributed. Consequently, the mapping of our classical data to the qubit is randomly distributed on the Hilbert space. The optimization procedure aims to push the two classes toward the opposite poles of the Bloch sphere. Therefore, for the weights after training θ^*, we expect that data on the Bloch sphere is much more ordered.

8.2.3 Ideal quantum computers

Computational complexity theory is a field of computer sciences that focuses on classifying computational problems in terms of the resources they need. In particular, classical computers are known to excel at solving problems belonging to two complexity classes: solvable in polynomial time (P) and bounded-error probabilistic polynomial time (BPP). Having an ideal quantum computer, a natural question arises: What types of problems, if any, can be solved in a polynomial time on a quantum computer while taking an exponential time on a classical computer? In this context, another complexity class was defined and, roughly, includes all problems which can be solved and verified with a quantum computer in polynomial time (BQP).

One of the first quantum algorithm with an exponential speedup has been proposed in the context of discrete Fourier transform. The quantum Fourier transform algorithm [637] performs the discrete Fourier transform on 2^n amplitudes using a quantum circuit consisting of only $\mathcal{O}(n \log(n))$ quantum gates. The classical algorithm needs $\mathcal{O}(n2^n)$ operations to perform the same task. Another example of an algorithm with such a speedup is the Shor algorithm for efficient number factorization [638]. It uses building blocks from the quantum phase estimation algorithm [639] and the quantum Fourier transform to gain an exponential speedup with respect to the best classical algorithm for this task. The Harrow–Hassidim–Lloyd (HHL) algorithm [640] is another very famous algorithm that was designed to solve a system of linear equations

$$\boldsymbol{A}\boldsymbol{x} = \boldsymbol{b}, \tag{8.19}$$

where $\boldsymbol{A}$ is an $n \times n$ sparse matrix with condition number k. The algorithm is able to find the vector $\boldsymbol{x}$ in $\mathcal{O}(\log(n)\, k^2)$ time instead of the typical $\mathcal{O}(n^2)$ for standard algorithms. This is an exponential speed-up in the size of the system; however, one crucial remark to keep in mind is that the classical algorithm returns the solution directly, while in the HHL algorithm, the solution is encoded in a quantum state that needs to be repeatedly measured to read it out.

Machine learning algorithms largely rely on linear algebra, which generally constitutes much of ML computational cost. For example, the classification problem with a support vector machine (SVM) generally requires quadratic programming (see Section 2.4.3 for the general idea of SVM) but a special form of SVM[13] boils down to solving a system of linear equations. In this context, quantum computers might speed up such costly operation. One application of the HHL algorithm has been proposed, for example, in the context of SVMs [641] (see [642] for a recent experimental realization on a four-qubit quantum computer) and data fitting [643]. It is worth noticing that the exact amount of quantum speedup provided by the HHL algorithm with respect to classical algorithms is under debate [644]. Another caveat consists in the

[13]For the special case of least-squares SVM, the problem can be written as a solving a linear system of equations.

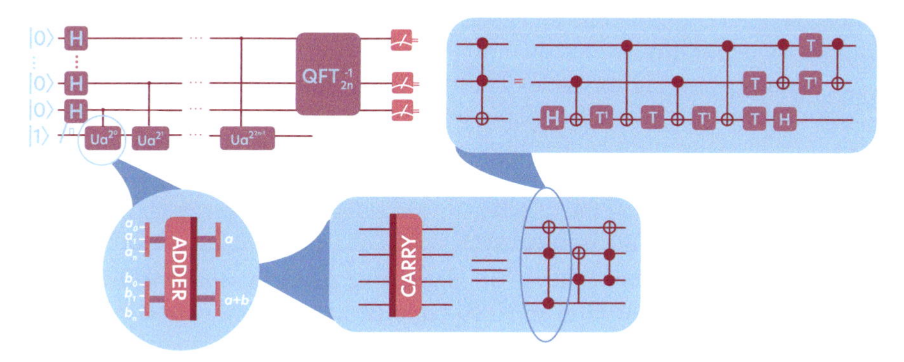

Figure 8.13 Realization of the famous Shor algorithm in a real quantum computer. Top left diagram presents a concise theoretical circuit of this algorithm. Due to the limitation to certain gates (CNOT and SWAP), generic gates have to be decomposed and the circuit requires more gates and higher depths.

fact that the classical data should be efficiently encoded in the quantum algorithm efficiently. This is another issue that must be solved by the community.

8.2.4 Noisy intermediate-scale quantum era

Until now, we have only considered ideal quantum computers to run quantum algorithms, such as the Shor, the quantum Fourier transform, and HHL algorithms. However, the realization of these algorithms for a number of qubits where such advantage matters is not yet feasible in near-term quantum computers. The main reasons are that (i) quantum computers currently contain *too few qubits* (nowadays in the order of hundreds) and (ii) they perform imperfect operations (noisy).

Furthermore, algorithms such as the Shor algorithm have to be compiled on real devices. This means that unitaries acting on several qubits have to be *decomposed in elementary gates* that can be physically realized in the experimental platform. Such a transformation might lead to complex quantum circuits with native gates [645]. For example, for the IBM-Q Washington platform, only the CNOT, ID, RZ, $\sqrt{X}$, and X gates are native gates. Any other gate must be decomposed into these gates. Since these gates form a set of universal quantum gates [646], this is, in theory, sufficient but, in practice, it can lead to very complex quantum circuits with a large number of gates. For example, we consider the decomposition of the Shor algorithm into these gates, as shown in Fig. 8.13. A simple-appearing circuit consists, in practice, of many operations on real quantum devices. The latter might be especially problematic due to *noise and decoherence* that are intrinsically present in the physical devices.

In modern quantum computers, we can identify three primary sources of errors: gate errors (generated by a non-precise application of the desired gate), decoherence errors (loss of coherence of the wave function as a function of time), and readout errors (erroneous readout of the qubits state during the measurement procedure).

Due to many different noise and error sources in real quantum computers, we are far from the fault-tolerant quantum computation. Instead, we are in the so-called noisy intermediate-scale quantum (NISQ) era [425]. The qubits of the current quantum processors are noisy and require quantum error correction. Nevertheless, the study of the physics of such systems is interesting in itself. In particular, there might be applications with a quantum speedup within this regime, as in the case of the recent quantum advantage experiment [628].

It is now clear that we cannot run algorithms requiring many gates or implement gates with low error rates in NISQ circuits. If the circuit contains too many gates, the coherence gets lost as well as the superposition and entanglement between different qubits. A natural question arises: Can we design algorithms that perform well on NISQ devices and do not require fully error-corrected quantum computers? This means one has to find clever ways to explore the exponentially big Hilbert space without exact algorithms. One approach is using quantum computers to generate variational states and to find a procedure to converge iteratively to the solution instead of taking a direct deterministic path (e.g., by performing the optimization on a classical computer). We go into more detail into these variational approaches in Section 8.2.6. Before, in Section 8.2.5, we present how NISQ devices can be used for SVM with kernels.

To sum up, the NISQ era still has many open problems in experimental quantum computing and in quantum error correction. As of 2024, state-of-the-art devices include 100+ qubits with error rates of less than 0.2%.[14] Nonetheless, recent years showed many examples of useful variational quantum simulations that can be performed with the near-term devices, for example, see [647]. Moreover, many error mitigation routines have been developed to ease the noise effects in quantum computers, allowing for extraction of useful information from noisy devices in the near term [648–651]. NISQ devices are also an excellent trial field to study physics without building a fault-tolerant quantum computer. Finally, useful applications of NISQ devices can still be found, and they can be considered as a step toward fault-tolerant quantum computing.

8.2.5 Support vector machines with quantum kernels

We have seen in Section 8.2.3 that ideal quantum computers could allow one to accelerate the numerically costly parts of the SVM algorithm by implementing the HHL algorithm. There, the key element has been to use the quantum computer to solve the linear system of equations. In 2018, three independent works [652–654] followed an interesting alternative direction: using kernels evaluated directly on quantum devices, while performing the rest of the SVM algorithm classically.

The idea is sketched in Fig. 8.14(a). Let us consider a dataset that is not linearly separable. We therefore want to nonlinearly embed it in a higher-dimensional space

[14] For example, in the random circuit sampling experiment from 2024 [734], the authors reported single-qubit gate errors of 0.04% and two-qubit gates errors of 0.14% for superconducting qubits. These errors are of the same order of magnitudes as the ones recently reported for a neutral atom quantum processing unit [735].

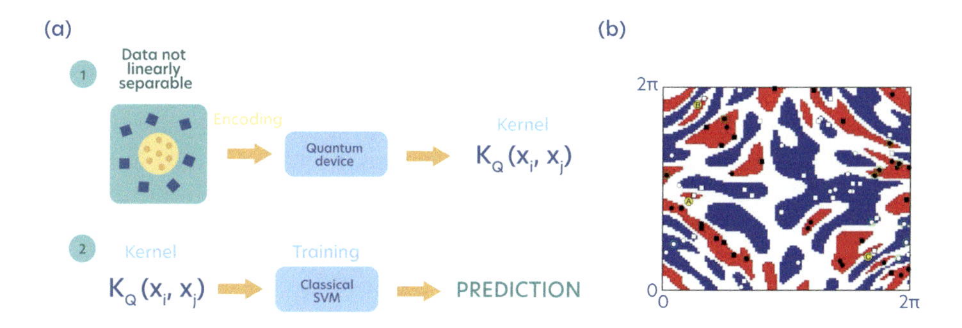

Figure 8.14 Quantum SVM enhanced by a quantum device. (a) Sketch of the steps of the SVM enhanced by a quantum kernel. The data are encoded in a quantum device, such as a quantum circuit, which computes the kernel. These kernels are then used in classical SVM. (b) Example of a dataset used in [652] to show the capacity of quantum kernels. Blue (red) regions correspond to label 1 (0). Taken from [652] with permission from Springer Nature.

such that the data is linearly separable in this space (see Section 4.2 for more detail). We here use a quantum device to encode classical data $\boldsymbol{x}$ into a high-dimensional Hilbert space $|\psi(\boldsymbol{x})\rangle$, or even infinite in the case of squeezed states considered in [653]. In this case, the choice of the encoding of the classical state into the quantum state is crucial as it determines the quality of the feature map. More importantly, quantum devices and in particular quantum circuits can allow for the efficient computation of the scalar product between two quantum states, which allows one to define a quantum kernel

$$K_Q(\boldsymbol{x}_i, \boldsymbol{x}_j) = |\langle\psi(\boldsymbol{x}_i)|\psi(\boldsymbol{x}_j)\rangle|^2 = \sum_n \lambda_n \phi_n(\boldsymbol{x}_i)\phi_n(\boldsymbol{x}_j), \tag{8.20}$$

which has all the properties of a classical kernel with a feature map ϕ and defines an RKHS (see Section 4.1.3).[15] As such, quantum kernels can be directly used in classical algorithms such as kPCA or kSVM or Gaussian processes (GPs) [655] rendering them quantum algorithms.

To be more concrete, we explain the main ingredients of the quantum kernel introduced in [652]. Given a dataset of points $\{\boldsymbol{x}_i\}$ with labels $\{y_i\}$, the feature map is defined in terms of the unitary transformation $U(\boldsymbol{x_i})$ that can be realized in a quantum circuit of qubits

$$\boldsymbol{x}_i \mapsto |\psi(\boldsymbol{x}_i)\rangle = U(\boldsymbol{x_i})|0\rangle, \tag{8.21}$$

where $|0\rangle$ stands for the product state $|0\rangle^{\otimes n}$. Typically, the classical data encoding into the quantum circuit can be done through parametric local rotations of single qubits. The unitary is then built through repeated application of these data-dependent gates

[15]The careful reader may notice that Eq. (8.20) is the norm squared of the inner product instead of the typical inner product expected for kernels. This becomes clearer when writing the kernel in terms of density matrices $K_Q(\boldsymbol{x}_i, \boldsymbol{x}_j) = \mathrm{Tr}(\rho_i\rho_j)$. The kernel is then corresponding to the Frobenius inner product of density matrices ρ_i and ρ_j.

and other nonparametric gates, such as entangling gates and Hadamard gates. We do not enter into the details of the construction of the circuit, but the interested reader can have a look at the following Qiskit tutorial for more details [656].

The quantum kernel can then be computed on a quantum circuit with the compute–uncompute trick: One basically implements the following quantum circuit $U^\dagger(\boldsymbol{x}_j)U(\boldsymbol{x}_i)|0\rangle$ and measures it in the z basis. The frequency of the all-zero outcome, therefore, gives an estimate of the kernel $K_Q(\boldsymbol{x}_i, \boldsymbol{x}_j)$.

Given these kernels, the optimization of the parameters of the SVM can be performed on a classical computer (see Section 4.2.2) using, for example, Bayesian optimization [655] presented in Section 4.3. [652] generated a complex classification problem, shown in Fig. 8.14(b), where the blue (red) region corresponds to label 0 (1). They then generated a training set by selecting random points in these regions and performed the SVM enhanced by the quantum kernel. The algorithm yields very good results with around 95% of accuracy on the test set for this synthetic dataset.

The previous example shows that quantum kernels can represent complex datasets. However, a quantum advantage has yet to be observed for a general dataset [657, 658]. A recent important step in this direction has been achieved in [633], where the authors constructed datasets that cannot be classified efficiently on a classical computer and in [659] where the authors have studied supervised learning of handwritten images on quantum computers with an improved scaling using randomized measurements. Along the same lines, finding optimal ways to construct quantum kernels [660], that is, how one should perform the encoding of the inputs $\boldsymbol{x} \mapsto |\psi(\boldsymbol{x})\rangle$, is still an active line of research [652]. In [661], for example, the authors construct quantum kernels for SVM algorithms based on the Bayesian information criterion (BIC) (see Section 4.4.1) as a selection metric. Using the quantum kernels constructed in this fashion, the SVMs achieve significantly higher performance in selected classification problems compared to optimized classical models with conventional kernels.

8.2.6 Variational approaches

This section deals with the optimization of quantum circuits that can be realized in NISQ devices. In particular, we focus here on the so-called variational quantum algorithms. This idea generalizes the toy example we introduced in Section 8.2.2. We define a parametrized quantum circuit (PQC), a circuit that depends on a set of parameters $\{\boldsymbol{\theta}\}$ ($\boldsymbol{\theta}$ can be, for example, the angles of single qubit rotations). Then, one defines an objective function $C(\boldsymbol{\theta})$ that we aim to minimize. Such an objective function can always be written as a function depending on a set of observables and on the PQC. Our goal is then to find the optimal set of parameters $\boldsymbol{\theta}^*$ minimizing the objective function. Such variational approach has applications in many fields such as ML (classification, generative models), many-body physics and quantum chemistry (ground state finding), and combinatorial optimization.

Variational quantum eigensolver (VQE). We illustrate the building blocks of the variational approach with the example of the VQE [662]. Given a Hamiltonian $\hat{H}$, the goal of the VQE is to minimize the energy $E = \langle\psi|\hat{H}|\psi\rangle$. The general principle of the VQE is shown in Fig. 8.15. The process starts with an initial state which is easy

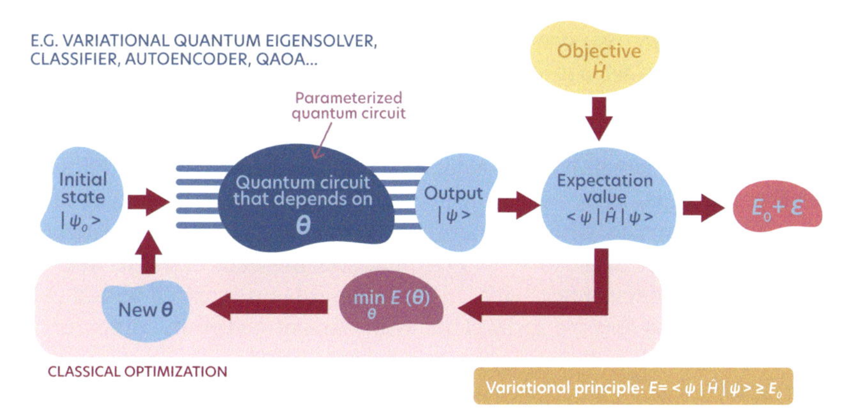

Figure 8.15 Variational optimization of quantum circuits, including the initial state, parametrized quantum circuit (PQC), output, objective, measurement, and classical optimization.

to prepare on the quantum computer, for example, the product state $|0\rangle^{\otimes n}$, which for simplicity we denote by $|0\rangle$. This is followed by the PQC including, for example, all the parameters θ of the quantum gates and which can be seen, at this point, as a black box that prepares a quantum state. In the first iteration, this state is just a random state and is used as the initial state for the expectation function, which is in general the Hamiltonian of the physical system we want to study, up to a global phase. The Hamiltonian can describe the interaction within a molecule or a spin system, but can, in general, be any kind of cost function in operational form that can be written in the computational basis of the quantum hardware we are using. The next step is the minimization of the cost function with a classical subroutine to converge toward the lowest energy state of the physical system in the space of the quantum states that can be reached by our PQC in a self-consistent manner. We also know that, if the system is gapped and the ground state unique, the minimal value of the expectation value of the Hamiltonian is the ground-state energy and the corresponding eigenvalue is the ground-state wave function.[16]

Going a bit more into detail, for the PQCs we are computing the energy of the ansatz

$$E_0 = \min_{\theta} \langle \psi(\theta) | \hat{H} | \psi(\theta) \rangle = \min_{\theta} \langle 0 | U^{\dagger}(\theta) \hat{H} U(\theta) | 0 \rangle ,$$

where θ are the parameters of the gates that are optimized to minimize the expectation value. The variational state is the unitary state, that is, our PQC applied to the initial state $|0\rangle$. However, to run and work with the PQC, we have to make several assumptions: First, we are assuming the existence of a set of parameters that approximates the ground state and that our PQC can represent that specific solution. Second, that it is possible to converge to the solution without being stuck in local minima and, finally, that the circuit can be run on a NISQ computer. Taking all these assumptions

[16]The energy measured at each iteration is an upper bound of the ground-state energy, according to the variational principle. The ability to reach the global minimum, of course, depends on the capacity of the circuit. If the circuit does not contain the solution, such an ansatz will never reach the minimal energy [663].

into account, there are two ways to design a PQC. The first option is the problem-inspired design which we can use when we exploit some physical properties of the system we want to represent, for example, by using the Hamiltonian representation to design the unitary operation as happens in the VQE algorithm. However, these kinds of ansatz require, in general, many gates or a particular qubit connectivity, making it unfeasible for bigger systems in current quantum computers. Another way is the hardware-efficient ansatz, which is a heuristic method that requires way less quantum gates and that consists of preparing a PQC that uses the native gate set and respects the quantum computer connectivity. In general, problem-inspired ansatz use to be more precise but less feasible to implement in current quantum computers, while this happens otherwise with the hardware-efficient ansatz.

The next step after defining the PQC is the choice of the *objective function* which can be everything that encodes our problem in a quantum operator, for example, a Hamiltonian as shown in Fig. 8.15. The objective function is then decomposed into Pauli strings because their individual expectation values can be measured with the quantum computer. This measurement procedure the next important step in the process. In this step, we extract the value of the objective function from our quantum computer or our quantum devices. This in itself is a challenging task as the objective function often cannot be measured directly. Instead, the expectation value of a Pauli string is computed on the quantum hardware by making the wave function collapse in the corresponding Pauli basis. From this measurement, we can extract bit strings (i.e., lists of 0 and 1, for example, $[(0,0,1,1,0,0),(1,0,1,0,0,0),(0,0,1,0,1,0)]$). Combining many bitstrings from individual Pauli measurements, we can reconstruct the expectation value of any objective function that can be written as the tensor product of Pauli matrices. Finding a suitable measurement strategy that requires the least amount of bit strings to accurately yield the value of the objective function is an ongoing research endeavor [664].

The last step is the classical optimization in which we have to navigate through the PQC parameter space by using, for example, a gradient-based approach. The gradients can be written in terms of expectation values of the quantum circuit derivatives with respect to a parameter, and we do not have direct access to the gradients of the quantum state. In NISQ devices, the gradient of the PQC can be computed with the parameter shift rule [665]: For each parameter θ, one can compute exactly its partial derivative by evaluating two PQCs.[17] The problem in this step is the number of required measurements. To run the classical minimization algorithm, measurements of all gradients are required. Therefore, this method can be very expensive and includes a large number of variables and multiple iterations to converge to the ground state, and that is why other gradient-free methodologies are exploited, such as genetic algorithms or RL strategies. After all, the combination of the variational optimization of the quantum circuit with classical optimization algorithms is an efficient way to use NISQ devices for real-world problems.

[17] In comparison, computation of the gradients of the loss of classical NNs requires only a single evaluation of the model, thanks to the existence of reverse mode differentiation, also known as backpropagation; see Section 2.5.

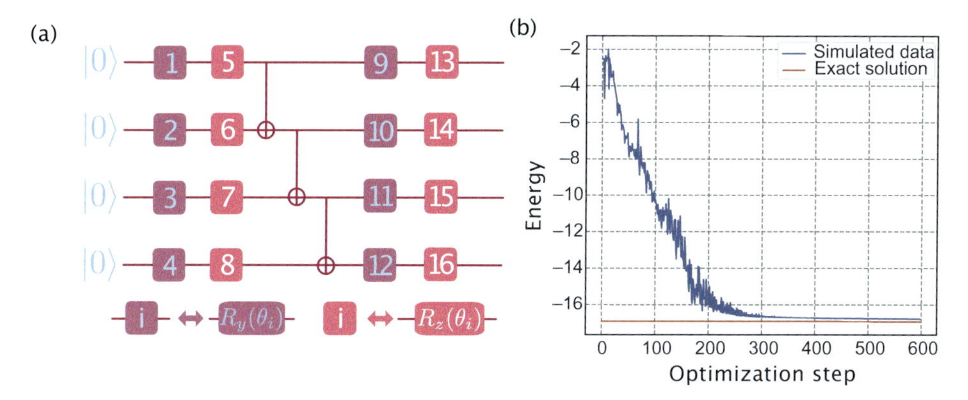

Figure 8.16 An example of a variational quantum simulation. (a) The parametric quantum circuit used for the simulation. (b) The energy of the system during the optimization algorithm.

Let us discuss a concrete example: the Heisenberg Hamiltonian acting on four spins, which reads

$$\hat{H} = \sum_{i=1}^{3}(J_1\sigma_i^x\sigma_{i+1}^x + J_2\sigma_i^y\sigma_{i+1}^y + J_3\sigma_i^z\sigma_{i+1}^z) + \sum_{i=1}^{4}(h_1\sigma_i^x + h_2\sigma_i^y + h_3\sigma_i^z). \tag{8.22}$$

We fix the parameters of the Hamiltonian to $\boldsymbol{J} = [1, 1, -1]$ and $\boldsymbol{h} = [1, 1.5, 3]$. These parameters are chosen to be far away from any phase transition point, not to make the problem too difficult. We can use this Hamiltonian as a benchmark for our algorithm. Let us also consider a very easy quantum ansatz for the four-qubit case. The ansatz consists in two rotation gates, applied to every qubits in Y and Z direction, three CNOT gates that connect every qubits, and two more rotations in Y and Z direction. The circuit is sketched in Fig. 8.16(a).

In VQEs, we want to use the variational circuit to minimize the energy of the system. If we did this operation on a classical computer, the computational cost of the evaluation of the energy would scale exponentially with the number of qubits. The expectation value of the Hamiltonian of the system can be computed efficiently on a perfect quantum computer. The computational cost is linear in the number of qubits. We can feed this cost function to an optimizer; in this case, we use the Nelder–Mead optimization routine. In Fig. 8.16(b), we plot the value of the energy of the system as a function of the optimization step. We can find a relatively good approximation of the ground-state energy with few variational parameters and polynomial computational cost in the number of qubits. Many things can be improved in these kinds of simulations, both on the design of the variational ansatz, and on the optimization routine. In the design of variational quantum circuits, we can, for example, impose symmetries of the system we are studying. In the optimization routine, we could use gradient-based methods, such as SGD, or gradient-free methods, such as Bayesian optimization (BO) from Section 4.3. For instance, in [230] the authors derive a novel kernel, inspired by the parameter shift rule to update the VQE parameters accurately with less measurements on the quantum computer. We refer to Section 4.5.4 for more details.

8.2.7 Parametrized quantum circuits for quantum machine learning

Classification tasks. Variational quantum circuits can be used to perform the classification of classical data. The first nontrivial task in the construction of such a quantum algorithm is the loading of the classical data on the quantum hardware. Moreover, the algorithm must be able to process efficiently these data and have a way to perform the classification. In [666], the authors have shown that PQC with data reuploading can lead to a good classifier. In [667], the authors have introduced the concept of quantum convolutional neural networks. As for classical CNN, these variational quantum circuits have more capacity. In particular, the authors have shown how these circuits can be used to perform classification on symmetry-protected topological phase in the Haldane chain directly from the quantum state, that can be obtained with a VQE. This idea has been realized experimentally in a recent work [634]. Moreover, it is worth to notice that the abovementioned classifiers are closely related to quantum kernels [668, 669]. Finally, the authors of [670] have developed a recurrent quantum NN that has been used for classification and generation of handwritten digits.

Quantum reinforcement learning (RL). Parametrized quantum circuits can also be used to realize action-value functions, or RL policies themselves (see Section 6.2 for Q-learning and deep Q-learning) policies (see Section 6.3 for an introduction to policy gradient). Two examples of algorithms that take advantage of PQCs in the context of RL can be found in [671, 672]. In [671], the quantum circuit is trained using a policy gradient algorithm and is used to solve classical environments, that is, to find the optimal policy for the task at hand. The choice of the action, the probability of which occurring we want to fix, is going to be encoded in the measured observable – if we have a certain set of actions, we define a certain set of observables. The RL architecture states that the agent observes a quantum state that the circuit will produce and the expectation value of such observable is going to encode $\pi(a|s)$ – the probability of the action a in a given RL state s, that is, it corresponds to the policy π. In [672], the authors have used the OpenAI Gym [673] examples as benchmark environments for the variational quantum algorithm for deep Q-learning, as for example, the *Cartpole* game (a cart that can move left and right and the agent is trying to balance the pole attached to the cart). Compared to the classical models (for which the exemplary environments were created), the quantum models can reach a similar accuracy using much fewer parameters (which alone does not imply better models but highlights their difference). An interesting and open question would be to understand whether these results can be generalized to other environments.

Quantum autoencoders. In the same spirit as their classical counterpart (see Section 2.4.5), quantum autoencoders [674] are used to compress quantum data on a quantum computer. Quantum autoencoders act directly on data encoded in qubits and can thus be used to compress quantum data without needing to have a classical representation that would have an exponential cost. Since there are patterns that classical computation cannot generate, for example, entanglement, the quantum version of an autoencoder might be able to recognize patterns beyond classical capabilities. The encoding done with quantum autoencoder transforming quantum data into a latent space with fewer qubits. Having a set of quantum data that encoded into n qubits, we aim to find the representation of the n-qubit state in a latent space formed by $k = n - p$ qubits. The p qubits are so-called *trash qubits*. This encoding is done via a variational map represented by a quantum circuit with a polynomial number of parameters $U(\theta)$. Since the encoding procedure is a unitary (quantum circuit), the decoding operator is just represented by the Hermitian conjugate of $U^{\dagger}(\theta)$. It is still under debate whether this quantum algorithm can have a computational advantage with respect to their classical counterpart. Variational quantum autoencoders have also been proposed for various applications, such as quantum data denoising [675], phase classification [124], clustering of the Hilbert space [125], or quantum error correction [676].

Generative models. Generative models are algorithms learning the distribution of a dataset. In quantum mechanics, the inherent quantum nature of the devices can be of great help for learning probability distributions and in particular for quantum wave functions. Recently, diverse QML architectures have been proposed: these include quantum Hamiltonian-based models [677], quantum GANs [678, 679], and quantum Born machines [680–682]. In particular, quantum circuit Born machines are generative model that can represent classical distribution of data, represented as pure quantum states. In this context, variational quantum circuits can provide a useful tool to represent these probabilities distribution and, moreover, an efficient way to sample from these distributions. The algorithm has comparable performance to its classical counterpart. In [682], the authors have proposed yet another quantum circuit Born machine that can learn the probability distribution of coherent thermal states, where the probability distribution is given by the Boltzmann weights. Such algorithms can be run on NISQ devices and are good candidates for quantum advantage in near term.

8.2.8 Current experimental and theoretical limitations

In this last section, we discuss some experimental and theoretical open problems which have to be overcome for successful applications and use of NISQ devices. One important topic is the *quantum error mitigation*, that is, reducing or compensating errors. This includes classical postprocessing techniques as well as active operations on the hardware itself. The former approach includes techniques as *stabilizer-based approaches* which rely on information associated with conserved quantities as spin or particle number [683] and mitigation scheme based on classical postprocessing of data [648–651]. These methods are, however, only postprocessing tools after we ran the circuit. Another way of error mitigation are *active mitigation techniques* or quantum optimal control strategies. In contrast to the postprocessing techniques, these methods are directly related to experiments and the quantum hardware [684–689].

We are not only facing experimental but also theoretical open problems that have to be solved or overcome. One of these problems is the *barren plateau problem* [690] which appears for global cost functions of quantum circuits parametrized with local unitaries. Without prior knowledge about the solution, the parameters θ of the PQC are initialized randomly. As a consequence, we obtain a barren plateau: the expected value of the gradient as well as the expected value of the variance are exponentially vanishing with the number of qubits and/or the circuit depth. In other words, this means that the loss landscape is mainly flat, with a narrow gorge hosting the global minimum [691]. Possible solutions to the barren plateau problem consist in using parameters close to the solution, using a local cost function instead of global ones, or introducing correlations between parameters [690, 691]. The downside of the latter solutions are that these methods do not work well for strongly correlated systems. A general solution to the barren plateau remains still an open theoretical problem. Moreover, it is totally unclear that there are any natural problems where a PQC will outperform a classical learning engine [669]. The loss landscape is, furthermore, characterized by the appearance of many local minima that can be far away from the global minimum [692]. A similar situation is known for the training of classical NNs [43] – it is an open question whether this observation poses an actual challenge in practical applications of PQCs.

Another theoretical obstacle includes the capacity of the PQC. When setting a PQC ansatz, we have to be careful not to narrow the Hilbert space accessible by the PQC too much. If we do so, we might end up in a wrong area of the Hilbert space and we cannot reach a good approximation of the solution [663, 693]. There are some measures (e.g., Haar distributions), but the capacity remains an open problem for the PQC ansatz. However, even if a suitable ansatz class for a given task is known in advance, already the optimization of its hyperparameters (such as the circuit depth) constitutes a hard task, that is, there exist problem instances where we cannot expect to even find approximately good solutions after optimization [694].

Circuit compilation is another important challenge, involving both theory and experimental parts: the theoretical circuit, the decomposition into native gates, the simplification, and finally the mapping to the hardware and real qubit system. The circuit compilation relies on the Solovay–Kitaev theorem [646, 695] which states that with a universal gate set it is possible to approximate any SU(N) with a circuit of polynomial depth up to a certain accuracy. However, when it comes to the specific hardware implementation, some gates are easier to control than others. In PQC, one always tries to use as many native gates as possible. This solution can make the quantum circuit shorter and simpler.

As a concluding remark, much effort has been devoted toward applications with a quantum advantage, that is, where a quantum computer is required using less resources than the classical counterpart. The study of such algorithms is crucial and will probably require the integration of quantum devices in high-performance computing facilities.

Outlook and open problems

We are currently in the NISQ era. Despite the complex theoretical and experimental challenges toward fault-tolerant quantum computation, the general objective in the NISQ era is to understand the possible algorithms that can be implemented in current experimental platforms. While the reduction of error rates affecting qubits and gates efficiency developments is a hard task demanding fundamental scientific and technological advances, there is a need to develop software tools to control quantum computers, develop error mitigation techniques and define quantum optimal control strategies in the meantime, as well as tools to characterize variational quantum algorithms such as the study of the loss landscape [491, 696] and the entanglement properties [697]. Moreover, the development of algorithms taking advantage of quantum computers without having a direct classical counterpart is a very interesting but challenging direction beyond the CQ paradigm.

Further reading

- Biamonte, J. *et al.* (2017). *Quantum machine learning.* Nature 549, 195. Very famous review on the subject of QML.
- Bharti, K. *et al.* (2022). *Noisy intermediate-scale quantum (NISQ) algorithms.* Rev. Mod. Phys. 94, 015004. Review paper focused on the variational quantum circuits simulations in the NISQ era.
- Li, W. & Deng, D.-L. (2022). *Recent advances for quantum classifiers.* Sci. China: Phys. Mech. Astron. 65, 220301. Recent review paper on classification algorithms with quantum computers.
- Cerezo, M. *et al.* (2021). *Variational quantum algorithms.* Nat. Rev. Phys. 3, 625. Very nice review paper on the recent advances in variational quantum algorithms and their applications.
- Qiskit tutorial on quantum machine learning.
- Pennylane tutorials and demos on quantum machine learning.
- TensorFlow quantum tutorial on quantum reinforcement learning.

9 Conclusion and outlook

In the last decade, ML (and DL in particular) has been intensively studied and has revolutionized many topics, including computer vision and natural language processing.

The new toolbox and set of ideas coming from this field have also found successful applications in the sciences. In particular, ML and DL have been used to tackle problems in the physical and chemical sciences, both in the classical and quantum regimes. Their applications range from particle physics, fluid dynamics, cosmology, many-body quantum systems [4–6], to quantum computing and quantum information theory [698]. On the other hand, physicists have started applying tools from statistical physics to try to understand the dynamics related to training of DL [585] and are also exploring potential hardware based on quantum physics [699].

This book aims to introduce physicists and chemists to selected topics in ML and some of their applications in physics and chemistry. As this field is relatively new and rapidly growing, we have decided to focus on explaining key concepts in ML for scientists with a physics or chemistry background and briefly reviewed some of the possible applications. We have also discussed how physics can help in gaining a deeper understanding of the intrinsic mechanisms governing DL and how quantum technologies can be used for data-driven tasks. The list of topics and applications covered in this book is, of course, not exhaustive. We hope, nevertheless, that we have conveyed our enthusiasm for ML applied to quantum sciences and that we have properly introduced the necessary building blocks needed for the keen reader to dive into this field.

Finally, we summarize the directions explored in this book and share our view on potential exciting developments. We note that our views date to 2022, so we are very curious whether they will pass the test of time in the reader's hands.

Characterization and classification of trajectories and phases. Researchers have intensively studied different ML and DL methods to tackle the characterization and classification of trajectories and phases. While many of these techniques have been very successful [6, 700], many challenges remain. Majority of the works focused on reproducing known phase diagrams with supervised learning schemes. The ability to process unlabeled data and apply unsupervised learning or self-supervised learning constitutes a big step forward in assisting physicists in the discovery of new exotic phases of matter. Furthermore, since the most powerful models are black boxes, interpretability techniques are essential to help physicists to discover relevant physical concepts learned by these models. For example, [149, 156] were able to discover physical concepts or recover conservation laws from trajectories. Another very interesting direction is the classification of phases directly from experimental data [106, 140, 141, 143, 701, 702], especially if it was accompanied by information about the corresponding order parameters. It would also be interesting to understand the effect of experimental noise on the classification with respect to the simulated data.

Gaussian processes and kernel methods. Gaussian processes and kernel-based regression methods are ML algorithms that are not considered suitable for large-dimensional systems due to their cubic scaling with the size of the training data set [241]. Nonetheless, kernel-based methods have proven to be robust regression tools with accuracy comparable to DL methods without the caveat of hyperparameter optimization. They have also played a significant role in the field of optimization thanks to the success of Bayesian optimization [182]. The advancement of kernel-based methods has been focused on two main challenges: (i) numerical routines for matrix inversion and (ii) more robust kernel functions. The rise of GPUs has enabled the development of efficient algorithms for kernel-based methods (e.g., `GPyTorch` [236]). The accuracy of kernel-based methods is based on the learning capacity of the underlying kernel function. Although algorithms similar to the Bayesian information criterion have proven to be very useful for the construction of kernels well suited for data, the rise of the automatic differentiation (AD) and the ability to parametrize more complex kernels, along the lines of [234], offer exciting alternative directions. Finally, kernel-based methods have also been expanded to molecular systems where a string-based comparison is carried as the kernel function [703] and used to calculate the similarities between the Fock states [704].

Neural network quantum states. Neural network representation of the many-body wave function appeared to be very successful in predicting the properties of the ground state of the system (such as energy), even outperforming state-of-the-art techniques (PEPS) for the $J_1 - J_2$ model [705]. There is also great interest in using NQS for time evolution of the many-body wave function [283, 287, 706], especially in dimensions greater than one. Current challenges are the generalization of NQS to mixed states, for example, for open quantum systems, and the implementation of symmetries in NQS. There is a particular interest in finding strategies to extend NQS to fermionic systems. Furthermore, the applications of NQS to ab initio studies of interacting electrons in continuous space is a promising direction for quantum chemistry and physics [293, 294]. An interesting other direction is to gain a better understanding of the internal structure and the capacity of NQS [707]. Finally, the application of NQS for quantum state reconstruction is a very active field.

Reinforcement learning. Reinforcement learning (RL) provides a powerful framework with a broad range of applications in the development of quantum technologies. There is an ongoing effort to combine RL techniques with experimental setups, which opens a variety of research avenues. An interesting direction is the real-time control of quantum simulators [441, 447, 708], which may allow us to prepare and study complex phases of matter beyond our current capabilities. Similarly, we can enhance NISQ devices with RL-based control to progress toward fault-tolerant quantum computation [422, 431, 442, 709]. Another possible direction is the design of experimental platforms with RL, with which we may discover new approaches for quantum experiments. In [375], the authors discover new optical setups to prepare highly entangled quantum states. In a similar fashion, we could explore new quantum computing architectures or design new technical devices, for example. On a more theoretical level, RL is a powerful optimization tool that can help us solve challenging problems either on its own or in combination with other established techniques [440]. An interesting general direction is to bridge theoretical and experimental advances, for which we could use RL, for instance, to design Hamiltonians with desired properties of interest [710].

Differentiable programming. The application of differentiable programming (∂P) might be very beneficial in physics. It can be applied to different techniques such as variational Monte Carlo (for neural quantum state (NQS), see Chapter 5), tensor network [462], or mean field [467]. These works show that such algorithms can remove the tedious part of calculating derivatives while retaining state-of-the-art results. The integration of AD into other tasks such as solving differential equations is also very promising.

Machine learning for scientific discovery. Is AI capable of scientific discovery and understanding? The hopes and prospects coming from the use of ML in science are gigantic, but so far achievements that can be called "scientific discoveries" have been rare. In particular, the DeepMind AlphaFold [711, 712] algorithm may truly revolutionize biology and medicine thanks to its ability to predict the three-dimensional structure of a protein based solely on its genetic sequence (known as the protein folding problem). Automated and self-driving labs can change how we do experiments [576]. Artificial intelligence may also address the dire need for scalable and efficient verification of quantum devices, as well as validation of the underpinning physical Hamiltonians. Another example is an ML-guided selection of preformulated hypotheses presented in [143]. The authors first trained a model on numerical data from two different theories that were hypothesized to underlie the physical system in question and then asked the model which theory described the experimental snapshots of the system better. An exciting approach is using ML to guide scientists to interesting regimes of the problem as was done in phase classification [126], mathematics [713], and quantum information [714, 715]. Finally, note that having an omniscient oracle that can predict the outcome of any process does not a priori provide us with or prove scientific understanding [40]. Again, this points to the key challenge of ML interpretability.

Statistical physics for machine learning. Statistical mechanics and the physicist's view can help shed light on the inner workings of ML. Using computation methods from the physics of disordered systems and the teacher–student modeling of learning problems have already proven to be a powerful paradigm for studying a central puzzle of modern ML: generalization of overparametrized models [614, 615]. In parallel, the dynamics of learning can be studied with these same methods [616, 620], as well as with the Langevin equation [625]. The statistical mechanics tools can also be of great help to improve training of ML architectures. Recent works in this direction improved the training of restricted Boltzmann machines (RBMs) using a physical approach [626, 716]. Both the Gardner program and the teacher–student paradigm have been successfully used to study the capacity of quantum architectures [621–623] and the generalization of quantum NNs [624].

Potential hardware accelerators based on physical processes. Today, NNs are run on classical devices. While GPUs have become a game changer in the last decade in this field, the memory, computation time, and energy used in the current NN architectures are constantly growing and will eventually become a bottleneck. As such, there is a great effort to find new devices implementing NNs in physical devices [717].

The main goal of this research direction is to construct a physical realization of NNs performing fully parallel and fast operations. Examples of such platforms are optical implementations of NNs [718–725] or exciton–polaritons [726–730]. Another direction has been discussed in Section 8.2: the use of hybrid classical-quantum devices to perform data-driven tasks. A crucial point in this direction is the integration of quantum devices in high-performance computing facilities. Finally, recent works have explored yet another direction coming back to ideas from the early days of classical ML: designing quantum generalizations of Hopfield networks [731, 732].

A Mathematical details on principal component analysis

We can motivate PCA from two different perspectives: The first one is sketched in the main text and is based on retaining the largest possible data variance when reducing the dimensionality of the data. As such, it corresponds to a constrained maximization problem. PCA can also be motivated as the algorithm which finds a low-rank approximation $\boldsymbol{Z}$ to the design matrix $\boldsymbol{X}$ such that the distance between the two matrices is minimized up to a projection $\boldsymbol{V}$. We will prove that these two approaches are equivalent. That is, we show that the projection matrix $\boldsymbol{V}$ is the solution to

$$\boldsymbol{V} = \underbrace{\arg\min_{\boldsymbol{V}'} \min_{\boldsymbol{Z}} \frac{1}{n}\left\|\boldsymbol{X} - \boldsymbol{V}'\boldsymbol{Z}\right\|^2}_{\text{min. error in high-dim. space}} = \underbrace{\arg\max_{\boldsymbol{V}'} \frac{1}{n}\left\|\boldsymbol{V}'^{\mathsf{T}}\boldsymbol{X}\right\|^2}_{\text{max. variance in low-dim. space}}. \tag{A.1}$$

Because we deal with matrices, the norm refers to the Frobenius norm $\|\boldsymbol{A}\|^2 := \operatorname{tr}[\boldsymbol{A}^{\mathsf{T}}\boldsymbol{A}]$.

Let us recap the approach from the main text, that is, the variance maximization on the right side of Eq. (A.1). We define the design matrix $\boldsymbol{X}$ from the n m-dimensional data points by stacking them together. However, for the mathematical proofs, we define it in its transposed version as an $(m \times n)$ matrix. We construct the empirical covariance matrix $\boldsymbol{\Sigma}$ (assuming zero mean in the data) as $\boldsymbol{\Sigma} = (\boldsymbol{X}\boldsymbol{X}^{\mathsf{T}})/n = \boldsymbol{\Sigma}^{\mathsf{T}}$. It contains all covariances between any two input features. We wish to find a linear transformation $\boldsymbol{V}$ that preserves the maximal variance of the data. As we assume data with zero mean, the projected data $\boldsymbol{V}^{\mathsf{T}}\boldsymbol{X}$ also has zero mean. The empirical variance of the input data is given by $\sum_{i=1}^{n} \boldsymbol{x}_i^2/n$. We want to find the columns of the projection matrix iteratively. The first column $\boldsymbol{v}_1$ of $\boldsymbol{V}$ is obtained by maximizing the variance $\sum_{i=1}^{n}(\boldsymbol{v}_1^{\mathsf{T}}\boldsymbol{x}_i)^2/n$. This can be summarized by the following optimization problem:

$$\max_{\boldsymbol{v}_1} \frac{1}{n}\left\|\boldsymbol{v}_1^{\mathsf{T}}\boldsymbol{X}\right\|^2 = \max_{\boldsymbol{v}_1} \frac{1}{n}\boldsymbol{v}_1^{\mathsf{T}}\boldsymbol{X}\boldsymbol{X}^{\mathsf{T}}\boldsymbol{v}_1 = \max_{\boldsymbol{v}_1} \boldsymbol{v}_1^{\mathsf{T}}\boldsymbol{\Sigma}\boldsymbol{v}_1 \quad \text{s.t. } \boldsymbol{v}_1^{\mathsf{T}}\boldsymbol{v}_1 = 1. \tag{A.2}$$

Here, the constraint enforces a finite value for the maximum and we have inserted the definition of the norm (of vectors here) and of the covariance matrix. We solve the constrained optimization problem using the method of Lagrange multipliers. To this end, we define the Lagrangian $\mathcal{L}(\boldsymbol{v}_1) = \boldsymbol{v}_1^{\mathsf{T}}\boldsymbol{\Sigma}\boldsymbol{v}_1 - \mu_1(\boldsymbol{v}_1^{\mathsf{T}}\boldsymbol{v}_1 - 1)$ and calculate its differential as $d\mathcal{L} = 2(\boldsymbol{v}_1^{\mathsf{T}}\boldsymbol{\Sigma} - \mu_1\boldsymbol{v}_1^{\mathsf{T}})d\boldsymbol{v}_1$. The optimal solution requires $d\mathcal{L} = 0$. This is fulfilled, if $\boldsymbol{\Sigma}\boldsymbol{v}_1 = \mu_1\boldsymbol{v}_1$. We identify the eigenvalue problem, where $\boldsymbol{v}_1$ must be an eigenvector of $\boldsymbol{\Sigma}$ with eigenvalue μ_1. Choosing μ_1 to be the largest eigenvalue λ_1 of $\boldsymbol{\Sigma}$ then maximizes our objective.

For the next column $\boldsymbol{v}_2$ of $\boldsymbol{V}$, we start from Eq. (A.2) and enforce orthogonality between $\boldsymbol{v}_1$ and $\boldsymbol{v}_2$ by adding the constraint $\boldsymbol{v}_1^{\mathsf{T}}\boldsymbol{v}_2 = 0$. We can write the modified Lagrangian and calculate its differential with respect to $d\boldsymbol{v}_2$. This leaves us with the condition that $2\boldsymbol{\Sigma}\boldsymbol{v}_2 - 2\mu_2\boldsymbol{v}_1^{\mathsf{T}}\boldsymbol{v}_2 - \kappa\boldsymbol{v}_1 = 0$ with κ being the Lagrange multiplier for the orthogonality condition. Multiplying both sides with $\boldsymbol{v}_1^{\mathsf{T}}$ from the left and applying the orthogonality condition, we find that $\kappa = 0$. Plugging this into the previous condition, we again arrive at $\boldsymbol{\Sigma}\boldsymbol{v}_2 = \mu_2\boldsymbol{v}_2$. With the same reasoning as before, we see that μ_2 has to be the second-largest eigenvalue λ_2 of $\boldsymbol{\Sigma}$ with its corresponding eigenvector $\boldsymbol{v}_2$. Iteratively, we can identify the other entries of $\boldsymbol{V}$ as the remaining eigenvectors of $\boldsymbol{\Sigma}$ ordered by their eigenvalues.

We now understand the reason behind the procedure discussed in the main text and why we can drop the eigenvectors that carry the least variance to achieve a dimensionality reduction. The dimensionality-reduced data now has a variance spread along each axis according to the respective PCs. This spread can also be transformed to unit variance along each axis by modifying the projected design matrix as $\boldsymbol{X}_{\text{white}} = \boldsymbol{\Lambda}^{-1/2}\boldsymbol{V}^{\mathsf{T}}\boldsymbol{X}$ which is called *whitening* of the data. Here, $\boldsymbol{\Lambda} = \operatorname{diag}(\lambda_1, \lambda_2, \dots)$ is the diagonal matrix with the k largest eigenvalues of $\boldsymbol{\Sigma}$ in descending order. To see this, consider the eigenvalue decomposition of $\boldsymbol{X}\boldsymbol{X}^{\mathsf{T}}/n = \boldsymbol{V}\boldsymbol{\Lambda}\boldsymbol{V}^{\mathsf{T}}$ with $\boldsymbol{V}^{\mathsf{T}}\boldsymbol{V} = \mathbb{1} = \boldsymbol{V}\boldsymbol{V}^{\mathsf{T}}$. Rearranging terms then leads to $\boldsymbol{\Lambda}^{-1/2}\boldsymbol{V}^{\mathsf{T}}\boldsymbol{X}\boldsymbol{X}^{\mathsf{T}}\boldsymbol{V}\boldsymbol{\Lambda}^{-1/2} = n\mathbb{1}$ and, thus, to the identity as the corresponding covariance matrix of $\boldsymbol{X}_{\text{white}}$.

As suggested by Eq. (A.1), there is another, equivalent approach to find $\boldsymbol{V}$ by minimizing the approximation error between the design matrix $\boldsymbol{X}$ and its low-rank reconstruction $\boldsymbol{Z}$. The mean-squared error (MSE) between the two matrices up to projection $\boldsymbol{V}$ is given by the following constrained optimization problem:

$$\min_{\boldsymbol{V},\boldsymbol{Z}} \|\boldsymbol{X} - \boldsymbol{V}\boldsymbol{Z}\|^2 = \min_{\boldsymbol{V},\boldsymbol{Z}} \operatorname{tr}\left[(\boldsymbol{X} - \boldsymbol{V}\boldsymbol{Z})^{\mathsf{T}}(\boldsymbol{X} - \boldsymbol{V}\boldsymbol{Z})\right], \quad \text{s.t. } \boldsymbol{V}^{\mathsf{T}}\boldsymbol{V} = \mathbb{1} \tag{A.3}$$

where we inserted the definition of the Frobenius norm. The constraint can be placed without loss of generality: assume that $\boldsymbol{V}^{\mathsf{T}}\boldsymbol{V} = \boldsymbol{A} \neq \mathbb{1}$. Consider the eigenvalue decomposition of $\boldsymbol{A} = \boldsymbol{W}\boldsymbol{\Lambda}\boldsymbol{W}^{\mathsf{T}}$ with $\boldsymbol{W}^{\mathsf{T}}\boldsymbol{W} = \mathbb{1}$. Thus, $\boldsymbol{V}^{\mathsf{T}}\boldsymbol{V} = \boldsymbol{A} = \boldsymbol{W}\boldsymbol{\Lambda}\boldsymbol{W}^{\mathsf{T}} \Rightarrow \boldsymbol{\Lambda} = (\boldsymbol{V}\boldsymbol{W})^{\mathsf{T}}(\boldsymbol{V}\boldsymbol{W})$ and we recover our constraint by setting $\tilde{\boldsymbol{V}} = \boldsymbol{\Lambda}^{-1/2}\boldsymbol{V}\boldsymbol{W}$ and minimize over $\tilde{\boldsymbol{V}}$ instead.

We solve this again with Lagrange multipliers. However, we now have a matrix constraint and therefore introduce the matrix-valued Lagrange multiplier $\boldsymbol{M}^{\mathsf{T}}$. Since distances between matrices are given by the Frobenius norm, the Lagrangian reads as

$$\mathcal{L}(\boldsymbol{V},\boldsymbol{Z}) = \operatorname{tr}\left[(\boldsymbol{X} - \boldsymbol{V}\boldsymbol{Z})^{\mathsf{T}}(\boldsymbol{X} - \boldsymbol{V}\boldsymbol{Z})\right] - \operatorname{tr}\left[\boldsymbol{M}(\boldsymbol{V}^{\mathsf{T}}\boldsymbol{V} - \mathbb{1})\right].$$

Using matrix calculus, the differential is $d\mathcal{L} = -2\operatorname{tr}\left[(\boldsymbol{X} - \boldsymbol{V}\boldsymbol{Z})^{\mathsf{T}}\boldsymbol{V}d\boldsymbol{Z}\right]$. Setting it to zero, we require that $\boldsymbol{X}^{\mathsf{T}}\boldsymbol{V} = \boldsymbol{Z}^{\mathsf{T}}\boldsymbol{V}^{\mathsf{T}}\boldsymbol{V} = \boldsymbol{Z}^{\mathsf{T}}$ and thus $\boldsymbol{Z} = \boldsymbol{V}^{\mathsf{T}}\boldsymbol{X}$ which we plug into the objective as $\operatorname{tr}\left[(\boldsymbol{X} - \boldsymbol{V}\boldsymbol{Z})^{\mathsf{T}}(\boldsymbol{X} - \boldsymbol{V}\boldsymbol{Z})\right] = \operatorname{tr}\left[\boldsymbol{X}^{\mathsf{T}}\boldsymbol{X} - \boldsymbol{X}^{\mathsf{T}}\boldsymbol{V}\boldsymbol{V}^{\mathsf{T}}\boldsymbol{X}\right]$. The first term can be dropped as it does not depend on the minimization parameter $\boldsymbol{V}$. We can rewrite the second term as $-\left\|\boldsymbol{V}^{\mathsf{T}}\boldsymbol{X}\right\|$ and absorb the minus sign by turning the minimization into a maximization of $\left\|\boldsymbol{V}^{\mathsf{T}}\boldsymbol{X}\right\|$. This is exactly the objective of the variance maximization principle in Eq. (A.2), and we see their equivalence. In our derivation, we did not discuss how the Lagrange multiplier $\boldsymbol{M}$ disappears. The reasoning, however, is similar to before, where the eigenvalue decomposition of $\boldsymbol{M}$ has to be considered. This effectively only adds a rotation of $\boldsymbol{V}$, which can again be absorbed by redefining $\boldsymbol{V}$. Finally, let us remark on the consequence of this second derivation: we can obtain a suitable low-rank approximation of the design matrix $\boldsymbol{X}$ by selecting only the $k \ll n$ eigenvectors $(\boldsymbol{v}_i)_i$ of the largest corresponding eigenvalue to compose $\boldsymbol{V} = [\boldsymbol{v}_1, \dots, \boldsymbol{v}_k]$. This way, we ensure that the approximation error vanishes when $k = n$. This yields $\boldsymbol{X} \approx \boldsymbol{V}\boldsymbol{V}^{\mathsf{T}}\boldsymbol{X}$ as an approximation, justifying the procedure in Algorithm 2 in Section 3.2.1.

B Derivation of the kernel trick

Here, we present a derivation of the kernel trick. The training data $\mathcal{D} = \{(\boldsymbol{X}, \boldsymbol{y})\}$ is defined as

$$\boldsymbol{X} = \begin{bmatrix} \boldsymbol{x}_1^\mathsf{T} \\ \boldsymbol{x}_2^\mathsf{T} \\ \vdots \\ \boldsymbol{x}_n^\mathsf{T} \end{bmatrix} = \begin{bmatrix} x_{1,1} & x_{1,2} & \cdots & x_{1,m} \\ x_{2,1} & x_{2,2} & \cdots & x_{2,m} \\ \vdots & \vdots & \ddots & \vdots \\ x_{n,1} & x_{n,2} & \cdots & x_{n,m} \end{bmatrix} \text{ and } \boldsymbol{y} = \begin{bmatrix} y_1 \\ y_2 \\ \vdots \\ y_n \end{bmatrix}, \tag{B.1}$$

where each row of $\boldsymbol{X}$ (i.e., $\boldsymbol{x}_i$) is one data point associated with an observable y_i, n is the number of data points, and m is the number of features. Let us consider a linear model $f(\boldsymbol{x}, \theta) = \boldsymbol{x}^\mathsf{T}\theta$ as in Section 2.4.1. In ridge regression, the loss function is then given as

$$\begin{aligned} \mathcal{L}(\theta, \boldsymbol{X}, \boldsymbol{y}) &= \|f(\boldsymbol{X}, \theta) - \boldsymbol{y}\|_2^2 + \lambda \|\theta\|_2^2 \\ &= \|\boldsymbol{X}\theta - \boldsymbol{y}\|_2^2 + \lambda \|\theta\|_2^2, \end{aligned} \tag{B.2}$$

where $\|\cdot\|_2$ denotes the ℓ_2-norm. The optimal set of parameters θ^* is found by minimizing $\mathcal{L}(\theta, \boldsymbol{X}, \boldsymbol{y})$ with respect to θ,

$$\begin{aligned} \theta^* &= \arg\min_\theta \mathcal{L}(\theta, \boldsymbol{X}, \boldsymbol{y}) \\ &= \arg\min_\theta \|\boldsymbol{X}\theta - \boldsymbol{y}\|_2^2 + \lambda \|\theta\|_2^2. \end{aligned} \tag{B.3}$$

To validate that θ^* is the optimal solution of $\mathcal{L}$, we can verify that $\nabla_\theta \mathcal{L}\big|_{\theta^*} = \boldsymbol{0}$. Given the linear dependence of θ in f, $\nabla_\theta \mathcal{L}$ has a closed-form solution. Before we proceed with the derivation, let us first expand Eq. (B.2):

$$\begin{aligned} \mathcal{L}(\theta, \boldsymbol{X}, \boldsymbol{y}) &= (\boldsymbol{X}\theta - \boldsymbol{y})^\mathsf{T} (\boldsymbol{X}\theta - \boldsymbol{y}) + \lambda\theta^\mathsf{T}\theta \\ &= \theta^\mathsf{T}\boldsymbol{X}^\mathsf{T}\boldsymbol{X}\theta - \theta^\mathsf{T}\boldsymbol{X}^\mathsf{T}\boldsymbol{y} - \boldsymbol{y}^\mathsf{T}\boldsymbol{X}\theta + \boldsymbol{y}^\mathsf{T}\boldsymbol{y} + \lambda\theta^\mathsf{T}\theta. \end{aligned} \tag{B.4}$$

Solving for θ^* by setting the gradient of $\mathcal{L}$ w.r.t. θ to zero, we get

$$\nabla_\theta \mathcal{L} = \boldsymbol{0} = 2\boldsymbol{X}^\mathsf{T}\boldsymbol{X}\theta - 2\boldsymbol{X}^\mathsf{T}\boldsymbol{y} + 2\lambda\theta. \tag{B.5}$$

Please consult [733] for the derivative identities needed to derive Eq. (B.5). Solving for θ, we obtain

$$\left(\boldsymbol{X}^\mathsf{T}\boldsymbol{X} + \lambda\mathbb{1}\right)\theta = \boldsymbol{X}^\mathsf{T}\boldsymbol{y}, \tag{B.6}$$

where

$$\theta^* = \left(\boldsymbol{X}^\mathsf{T}\boldsymbol{X} + \lambda\mathbb{1}\right)^{-1}\boldsymbol{X}^\mathsf{T}\boldsymbol{y}. \tag{B.7}$$

Before we proceed further, let us examine the $\boldsymbol{X}^\mathsf{T}\boldsymbol{X}$ term

$$\boldsymbol{X}^\mathsf{T}\boldsymbol{X} = \begin{bmatrix} \boldsymbol{x}_1 & \boldsymbol{x}_2 & \cdots & \boldsymbol{x}_n \end{bmatrix} \begin{bmatrix} \boldsymbol{x}_1^\mathsf{T} \\ \boldsymbol{x}_2^\mathsf{T} \\ \vdots \\ \boldsymbol{x}_n^\mathsf{T} \end{bmatrix} = \begin{bmatrix} x_{1,1} & \cdots & x_{n,1} \\ x_{1,2} & \cdots & x_{n,2} \\ \vdots & \ddots & \vdots \\ x_{1,m} & \cdots & x_{n,m} \end{bmatrix} \begin{bmatrix} x_{1,1} & \cdots & x_{1,m} \\ x_{2,1} & \cdots & x_{2,m} \\ \vdots & \ddots & \vdots \\ x_{n,1} & \cdots & x_{n,m} \end{bmatrix}. \tag{B.8}$$

Here, $\boldsymbol{X}^\intercal\boldsymbol{X}$ is a $(m \times m)$ matrix, where the matrix elements represent the dot-product in the "number-of-data-points" space.

The optimal solution θ^* can furthermore be rewritten as

$$\theta^* = \boldsymbol{X}^\intercal \left(\boldsymbol{X}\boldsymbol{X}^\intercal + \lambda\mathbb{1}\right)^{-1} \boldsymbol{y}, \tag{B.9}$$

where we used the following matrix identity [733]

$$(\boldsymbol{A}\boldsymbol{B} + \mathbb{1})^{-1}\boldsymbol{A} = \boldsymbol{A}(\boldsymbol{B}\boldsymbol{A} + \mathbb{1})^{-1}. \tag{B.10}$$

In the same manner, let us examine the term $\boldsymbol{X}\boldsymbol{X}^\intercal$ given by

$$\boldsymbol{X}\boldsymbol{X}^\intercal = \begin{bmatrix} x_{1,1} & \cdots & x_{1,m} \\ x_{2,1} & \cdots & x_{2,m} \\ \vdots & \ddots & \vdots \\ x_{n,1} & \cdots & x_{n,m} \end{bmatrix} \begin{bmatrix} x_{1,1} & \cdots & x_{n,1} \\ x_{1,2} & \cdots & x_{n,2} \\ \vdots & \ddots & \vdots \\ x_{1,m} & \cdots & x_{n,m} \end{bmatrix}. \tag{B.11}$$

As we can observe, the matrix elements now represent the standard dot-product between two points of training data, $\boldsymbol{x}_i^\intercal\boldsymbol{x}_j$. A disadvantage of using Eq. (B.9) is that inverting $\boldsymbol{X}\boldsymbol{X}^\intercal$, a $n{\times}n$ matrix, becomes computationally more expensive when $n \gg m$.

By using Eq. (B.10) to rewrite θ^* (Eq. (B.7)) into Eq. (B.9), the prediction of a new point $\boldsymbol{x}_{\text{new}}$ becomes

$$\begin{aligned} f(\boldsymbol{x}_{\text{new}}, \theta^*) &= \left(\theta^*\right)^\intercal \boldsymbol{x}_{\text{new}} = \boldsymbol{x}_{\text{new}}^\intercal\theta^* \\ &= \underbrace{\boldsymbol{x}_{\text{new}}^\intercal\boldsymbol{X}^\intercal}_{\text{kernel}} \underbrace{\left(\boldsymbol{X}\boldsymbol{X}^\intercal + \lambda\mathbb{1}\right)^{-1} \boldsymbol{y}}_{\text{parameters}}, \end{aligned} \tag{B.12}$$

where the term $\left(\boldsymbol{X}\boldsymbol{X}^\intercal + \lambda\mathbb{1}\right)^{-1} \boldsymbol{y}$ represents the optimal parameters of the model, and $\boldsymbol{x}_{\text{new}}^\intercal\boldsymbol{X}^\intercal$ is the representation of $\boldsymbol{x}_{\text{new}}$ in feature space of the training data. For a linear model, $\boldsymbol{x}_{\text{new}}^\intercal\boldsymbol{X}^\intercal$ is computed by the dot product between the point where the function is evaluated and the training data,

$$\boldsymbol{x}_{\text{new}}^\intercal\boldsymbol{X}^\intercal = \begin{bmatrix} x_1^{\text{new}} & x_2^{\text{new}} & \cdots & x_m^{\text{new}} \end{bmatrix} \begin{bmatrix} x_{1,1} & \cdots & x_{n,1} \\ x_{1,2} & \cdots & x_{n,2} \\ \vdots & \ddots & \vdots \\ x_{1,m} & \cdots & x_{n,m} \end{bmatrix} = \begin{bmatrix} \boldsymbol{x}_{\text{new}}^\intercal\,\boldsymbol{x}_1 \\ \boldsymbol{x}_{\text{new}}^\intercal\,\boldsymbol{x}_2 \\ \vdots \\ \boldsymbol{x}_{\text{new}}^\intercal\,\boldsymbol{x}_n \end{bmatrix}^\intercal. \tag{B.13}$$

From Eq. (B.12) and Eq. (B.13), we can observe that a second linear model over the $\{\boldsymbol{x}_{\text{new}}^\intercal\boldsymbol{x}_i\}_{i=1}^n$ feature space could be defined,

$$f(\boldsymbol{x}_{\text{new}}, \tilde{\theta}^*) = \left(\tilde{\theta}^*\right)^\intercal (\boldsymbol{X}\boldsymbol{x}_{\text{new}}), \tag{B.14}$$

where $\tilde{\theta}^* = \left(\boldsymbol{X}\boldsymbol{X}^\intercal + \lambda\mathbb{1}\right)^{-1} \boldsymbol{y}$ corresponds to an n-dimensional vector.

The initial model considered is a linear model on $\boldsymbol{x}$, $f(\boldsymbol{x}, \theta) = \boldsymbol{x}^\intercal\theta$. However, we could consider a linear model over a basis-set $\Phi = [\phi_0, \phi_1, \ldots, \phi_\ell]$ spanning an alternative *feature space*. If we replace in our derivation $\boldsymbol{x}$ for $\Phi(\boldsymbol{x}) =$

$[\phi_0(\boldsymbol{x}), \phi_1(\boldsymbol{x}), \dots, \phi_\ell(\boldsymbol{x})]$, that is, we transform our data into the corresponding feature space, we elevate Eq. (B.12) to

$$\begin{aligned} f(\Phi(\boldsymbol{x}_{\text{new}}), \theta^*) &= (\theta^*)^\intercal \Phi(\boldsymbol{x}_{\text{new}}) = \Phi(\boldsymbol{x}_{\text{new}})^\intercal \theta^* \\ &= \Phi(\boldsymbol{x}_{\text{new}})^\intercal \Phi(\boldsymbol{X})^\intercal \left(\Phi(\boldsymbol{X})\Phi(\boldsymbol{X})^\intercal + \lambda \mathbb{1}\right)^{-1} \boldsymbol{y}. \end{aligned}$$

Here, $\Phi(\boldsymbol{x}_{\text{new}})^\intercal \Phi(\boldsymbol{X})^\intercal$ corresponds to the dot product in the basis-set expansion between $\boldsymbol{x}_{\text{new}}$ and the training data $\boldsymbol{X}$. Moreover, $\Phi(\boldsymbol{X})\Phi(\boldsymbol{X})^\intercal$ corresponds to the dot product in the basis-set expansion between all training data points, that is, $[\Phi(\boldsymbol{X})\Phi(\boldsymbol{X})^\intercal]_{ij} = \Phi(\boldsymbol{x}_i)^\intercal \Phi(\boldsymbol{x}_j)$.

In the context of kernel methods, $\Phi(\boldsymbol{X})\Phi(\boldsymbol{X})^\intercal$ is known as the design matrix $\boldsymbol{K}$. It should be stressed that, the computation of $f(\Phi(\boldsymbol{x}_{\text{new}}), \theta^*)$ does only depend on the basis-set expansion via the dot product $\Phi(\boldsymbol{x}_i)^\intercal \Phi(\boldsymbol{x}_j)$. This enables the kernel trick presented in Section 4.1. Finally, our derivation was done for KRR, however, the logarithm of the likelihood of a GP (Eq. (4.42)) has the same algebraic form. Therefore, our derivation also illustrates how GP models operate via kernels.

C Choosing the kernel matrix as the covariance matrix for a Gaussian process

The covariance function is a crucial quantity in the context of Gaussian process regression (GPR) as it encodes some preexisting assumptions on the target function we aim to learn or on the noise affecting the targets. In Chapter 4, we have discussed how the covariance function of a GP can be expressed in terms of the kernel function. Within this context, the notion of *similarity* between data points is of great relevance, and kernels are the most suitable tool to incorporate this notion of *proximity* of the data in the covariance function of the regressor. As such, one can choose the covariance of the GP prior to be the kernel matrix $\boldsymbol{K}$ given by the kernel function K as

$$\mathrm{Cov}(\boldsymbol{x}_i, \boldsymbol{x}_j) = \boldsymbol{K}_{ij} = K(\boldsymbol{x}_i, \boldsymbol{x}_j), \tag{C.1}$$

which shall overwrite our initial choice of the Gaussian prior see Eq. (4.41)) over the parameters

$$p_{\mathrm{prior}}(\theta) \sim \mathcal{N}(0, \boldsymbol{\Sigma}_{m+1}). \tag{C.2}$$

To see this, let us start with the linear model of Eq. (2.23) and map the input $\boldsymbol{x}$ to the feature space using the feature map ϕ such that

$$f(\boldsymbol{x}) = \phi(\boldsymbol{x})^{\mathsf{T}}\theta. \tag{C.3}$$

The expectation value over the parameters θ can be computed as

$$\mathbb{E}[f(\boldsymbol{x})] = \phi(\boldsymbol{x})^{\mathsf{T}}\,\mathbb{E}[\theta] = 0. \tag{C.4}$$

Using this result, the covariance can be expressed as

$$\begin{aligned}
\mathrm{Cov}(\boldsymbol{x}_i, \boldsymbol{x}_j) &= \mathbb{E}\left[f(\boldsymbol{x}_i)f(\boldsymbol{x}_j)\right] \\
&= \phi(\boldsymbol{x}_i)^{\mathsf{T}}\,\mathbb{E}\left[\theta\theta^{\mathsf{T}}\right]\phi(\boldsymbol{x}_j) \\
&= \phi(\boldsymbol{x}_i)^{\mathsf{T}}\boldsymbol{\Sigma}_{m+1}\phi(\boldsymbol{x}_j) \\
&= \phi'(\boldsymbol{x}_i)^{\mathsf{T}}\phi'(\boldsymbol{x}_j).
\end{aligned} \tag{C.5}$$

The last step uses the fact that $\boldsymbol{\Sigma}_{m+1}$ allows for an eigenvalue decomposition, that is, $\boldsymbol{\Sigma}_{m+1} = \boldsymbol{U}\Lambda\boldsymbol{U}^{\mathsf{T}}$ which, in turn, allows to define $\phi'(\boldsymbol{x}) := \Lambda^{1/2}\boldsymbol{U}^{\mathsf{T}}\phi(\boldsymbol{x})$. Hence, we can apply the kernel trick to the covariance matrix and promote it to the kernel matrix $\boldsymbol{K}$ as done in Eq. (C.1). Taking Eq. (C.4) together with Eq. (C.5), we come back to the same form as in Eq. (C.2).

Our goal is to evaluate the marginal likelihood $p(\boldsymbol{y} \mid \boldsymbol{X})$:

$$p(\boldsymbol{y} \mid \boldsymbol{X}) = \int_{\mathbb{R}^{m+1}} p(\boldsymbol{y} \mid \theta, \boldsymbol{X})p(\theta \mid \boldsymbol{X})\,\mathrm{d}\theta. \tag{C.6}$$

By *marginal* likelihood, we refer to the marginalization over the model's parameters θ, that is, performing the integral in Eq. (C.6). As we see shortly, the marginal likelihood can be expressed in terms of the kernel matrix. Under the GP model, the prior is Gaussian with zero mean and variance now given by the kernel matrix $\boldsymbol{K}$, that is, $\theta \mid \boldsymbol{X} \sim \mathcal{N}(\mathbf{0}, \boldsymbol{K})$. We can write the logarithm of the prior as

$$\log p(\theta \mid \boldsymbol{X}) = \log \frac{\exp\left\{-\frac{1}{2}\,\theta^{\intercal}\boldsymbol{K}^{-1}\theta\right\}}{\sqrt{(2\pi)^n\,|\boldsymbol{K}|}} = -\frac{1}{2}\,\theta^{\intercal}\boldsymbol{K}^{-1}\theta - \frac{1}{2}\log|\boldsymbol{K}| - \frac{n}{2}\log 2\pi\,, \tag{C.7}$$

where $|\boldsymbol{K}|$ denotes the determinant of $\boldsymbol{K}$.

Since the likelihood, the first term in the integrand in Eq. (C.6), is a factorized Gaussian with $\boldsymbol{y} \mid \theta \sim \mathcal{N}(\theta, \sigma^2\mathbb{1})$, the integrand reduces to a product of two Gaussians, that is, becomes itself Gaussian and we can readily apply well-known relations for a product of two Gaussians. In particular, given two Gaussian distributions $\mathcal{N}(\boldsymbol{a}, \boldsymbol{A})$ and $\mathcal{N}(\boldsymbol{b}, \boldsymbol{B})$ the following result holds:

$$\mathcal{N}(\boldsymbol{a}, \boldsymbol{A})\,\mathcal{N}(\boldsymbol{b}, \boldsymbol{B}) = Z^{-1}\mathcal{N}(\boldsymbol{c}, \boldsymbol{C}), \tag{C.8}$$

where $\boldsymbol{c} = \boldsymbol{C}\left(\boldsymbol{A}^{-1}\boldsymbol{a} + \boldsymbol{B}^{-1}\boldsymbol{b}\right)$ and $\boldsymbol{C} = (\boldsymbol{A}^{-1} + \boldsymbol{B}^{-1})^{-1}$. In the equation above, the matrices $\{\boldsymbol{A}, \boldsymbol{B}\}$ represent the variances of the two Gaussian while $\{\boldsymbol{a}, \boldsymbol{b}\}$ are the corresponding means. The resulting Gaussian thus has a variance equal to the inverse of the sum of the inverse variances and a mean equal to the convex sum of the means weighted by their precision matrices (inverse of the variance). The normalizing constant Z^{-1} looks itself like a Gaussian as

$$Z^{-1} = (2\pi)^{-\frac{n}{2}}\,|\boldsymbol{A} + \boldsymbol{B}|^{-\frac{1}{2}}\exp\left\{-\frac{1}{2}(\boldsymbol{a} - \boldsymbol{b})^{\intercal}(\boldsymbol{A} + \boldsymbol{B})^{-1}(\boldsymbol{a} - \boldsymbol{b})\right\}. \tag{C.9}$$

Leveraging this result, we perform the integration in Eq. (C.6) and refer to [241] for further details and proofs. For applying the results from Eqs. (C.8) and (C.9), the set of means and variances from Eq. (C.8), that is, $\{\boldsymbol{a}, \boldsymbol{b}\}$ and $\{\boldsymbol{A}, \boldsymbol{B}\}$, for our specific problem become $\{\theta, \mathbf{0}\}$ and $\{\sigma^2\mathbb{1}, \boldsymbol{K}\}$, respectively. Following the derivation, we obtain a closed-form solution for the marginal likelihood as

$$\log p(\boldsymbol{y} \mid \boldsymbol{X}) = -\frac{1}{2}\boldsymbol{y}^{\intercal}\left(\boldsymbol{K} + \sigma^2\mathbb{1}\right)^{-1}\boldsymbol{y} - \frac{1}{2}\log\left|\boldsymbol{K} + \sigma^2\mathbb{1}\right| - \frac{n}{2}\log 2\pi\,. \tag{C.10}$$

For completeness, we should mention that this same result could have been obtained by noticing that $\boldsymbol{y} \mid \boldsymbol{X} \sim \mathcal{N}(\mathbf{0}, \boldsymbol{K} + \sigma^2\mathbb{1})$. In conclusion, we find that training a GP reduces to finding the parameters of the kernel function K, which maximize the logarithm of the marginal log-likelihood from Eq. (4.60).

References

[1] B. Blaiszik, *blaiszik/ml_publication_charts: AI/ML Publication Statistics for 2022*, doi:10.5281/zenodo.7713954 (2023).

[2] Summer School: Machine Learning in Science and Technology, GitHub repository with selected tutorials from the school (2021), doi:10.5281/zenodo.13959917.

[3] A. Dawid, J. Arnold, B. Requena, *et al.*, GitHub repository with figures prepared for these Lecture Notes (2022), doi:10.5281/zenodo.13959927.

[4] V. Dunjko and H. J. Briegel, *Machine learning & artificial intelligence in the quantum domain: A review of recent progress*, Rep. Prog. Phys. **81**(7), 074001 (2018), doi:10.1088/1361-6633/aab406.

[5] G. Carleo, I. Cirac, K. Cranmer, *et al.*, *Machine learning and the physical sciences*, Rev. Mod. Phys. **91**, 045002 (2019), doi:10.1103/RevModPhys.91.045002.

[6] J. Carrasquilla, *Machine learning for quantum matter*, Adv. Phys. X **5**(1), 1797528 (2020), doi:10.1080/23746149.2020.1797528.

[7] C. Williams, *A brief introduction to artificial intelligence*, In *Proc. OCEANS '83*, pp. 94–99, doi:10.1109/OCEANS.1983.1152096 (1983).

[8] R. Dearden and C. Boutilier, *Abstraction and approximate decision-theoretic planning*, Artif. Intell. **89**(1), 219 (1997), doi:10.1016/S0004-3702(96)00023-9.

[9] J.-D. Zucker, *A grounded theory of abstraction in artificial intelligence*, Phil. Trans. R. Soc. Lond. B **358**(1435), 1293 (2003), doi:10.1098/rstb.2003.1308.

[10] L. Saitta and J.-D. Zucker, *Abstraction in Artificial Intelligence and Complex Systems*, Springer, New York, NY, doi:10.1007/978-1-4614-7052-6 (2013).

[11] M. Mitchell, *Abstraction and analogy-making in artificial intelligence*, Ann. N.Y. Acad. Sci. **1505**(1), 79 (2021), doi:10.1111/nyas.14619.

[12] H. Moravec, *Mind Children: The Future of Robot and Human Intelligence*, Harvard University Press, Cambridge, MA, doi:10.2307/1575314, P. 15 (1988).

[13] I. Goodfellow, Y. Bengio and A. Courville, *Deep Learning*, The MIT Press (2016).

[14] G. Marcus, *Deep learning is hitting a wall*, Nautilus, Accessed: 03-11-2022 (2022).

[15] T. J. Sejnowski, *The Deep Learning Revolution: Machine Intelligence Meets Human Intelligence*, The MIT Press, doi:10.7551/mitpress/11474.001.0001 (2018).

[16] Y. LeCun, Y. Bengio and G. Hinton, *Deep learning*, Nature **521**(7553), 436 (2015), doi:10.1038/nature14539.

[17] J. Schmidhuber, *Deep learning in neural networks: An overview*, Neural Netw. **61**, 85 (2015), doi:10.1016/j.neunet.2014.09.003.

[18] G. E. Hinton, S. Osindero and Y.-W. Teh, *A fast learning algorithm for deep belief nets*, Neural Comput. **18**(7), 1527 (2006), doi:10.1162/neco.2006.18.7.1527.

[19] V. Volkov and J. W. Demmel, *Benchmarking GPUs to tune dense linear algebra*, In *SC '08: Proc. 2008 ACM/IEEE Conf. Supercomput.*, pp. 1–11, doi:10.1109/SC.2008.5214359 (2008).

[20] R. Raina, A. Madhavan and A. Y. Ng, *Large-scale deep unsupervised learning using graphics processors*, In *Proc. 26th Annu. Int. Conf. Mach. Learn.*, ICML '09, pp. 873–880. Association for Computing Machinery, New York, NY, USA, doi:10.1145/1553374.1553486 (2009).

[21] B. Marr, *How much data do we create every day? The mind-blowing stats everyone should read*, Forbes, Accessed: 05-21-2018 (2018).

[22] SeedScientific, *Volume of data/information created, captured, copied, and consumed worldwide from 2010 to 2025*, SeedScientific, Accessed: 01-28-2022 (2021).
[23] Statista Research Department, *Volume of data/information created, captured, copied, and consumed worldwide from 2010 to 2025*, Statista, Accessed: 03-18-2022 (2022).
[24] A. Dawid and Y. LeCun, *Introduction to latent variable energy-based models: A path towards autonomous machine intelligence* J. Stat. Mech. **2024**, 104011 (2024), doi:10.1088/1742-5468/ad292b.
[25] M. F. Dixon, I. Halperin and P. Bilokon, *Machine Learning in Finance: From Theory to Practice* (Vol. 1170), Springer International Publishing, New York, NY, doi:10.1007/978-3-030-41068-1 (2020).
[26] J. Eisenstein, *Introduction to Natural Language Processing*, The MIT Press, Cambridge, MA (2019).
[27] S. Polu, J. M. Han, K. Zheng, M. Baksys, I. Babuschkin and I. Sutskever, *Formal mathematics statement curriculum learning* In ICLR 2023 – Int. Conf. Learn. Represent. (2023), arXiv:2202.01344.
[28] V. Mnih, K. Kavukcuoglu, D. Silver, *et al.*, *Human-level control through deep reinforcement learning*, Nature **518**(7540), 529 (2015), doi: 10.1038/nature14236.
[29] O. Vinyals, I. Babuschkin, W. M. Czarnecki, *et al.*, *Grandmaster level in StarCraft II using multi-agent reinforcement learning*, Nature **575**(7782), i350 (2019), doi:10.1038/s41586-019-1724-z.
[30] D. Silver, A. Huang, C. J. Maddison, *et al.*, *Mastering the game of Go with deep neural networks and tree search*, Nature **529**(7587), 484 (2016), doi:10.1038/nature16961.
[31] Y. Lecun, L. Bottou, Y. Bengio and P. Haffner, *Gradient-based learning applied to document recognition*, Proc. IEEE **86**(11), 2278 (1998), doi:10.1109/5.726791.
[32] R. A. Fisher, *The use of multiple measurements in taxonomic problems*, Ann. Eug. **7**, 179 (1936), doi:10.1111/j.1469-1809.1936.tb02137.x.
[33] A. Krizhevsky, *Learning multiple layers of features from tiny images*, Tech. rep., MIT & NYU, CiteSeer^X 10.1.1.222.9220 (2009).
[34] O. Russakovsky, J. Deng, H. Su, *et al.*, *ImageNet large scale visual recognition challenge*, Int. J. Comput. Vis. **115**(3), 211 (2015), doi:10.1007/s11263-015-0816-y.
[35] D. Gissin, *Active learning review*, GitHub.io, Accessed: 04-08-2022 (2020).
[36] P. Ren, Y. Xiao, X. Chang, *et al.*, *A survey of deep active learning*, ACM Comput. Surv. **54**(9) (2021), doi:10.1145/3472291.
[37] J. E. van Engelen and H. H. Hoos, *A survey on semi-supervised learning*, Mach. Learn. **109**(2), 373 (2020), doi:10.1007/s10994-019-05855-6.
[38] M. Krenn, J. Landgraf, T. Foesel and F. Marquardt, *Artificial intelligence and machine learning for quantum technologies*, Phys. Rev. A **107**, 010101 (2023), doi:10.1103/PhysRevA.107.010101.
[39] F. Chollet, *On the measure of intelligence* (2019), arXiv:1911.01547.
[40] M. Krenn, R. Pollice, S. Y. Guo, *et al.*, *On scientific understanding with artificial intelligence*, Nat. Rev. Phys. **4**, 761 (2022), doi:10.1038/s42254-022-00518-3.
[41] R. Bagheri, *Weight initialization in deep neural networks*, Towards Data Science, Accessed: 02-16-2022 (2020).

[42] T. Akiba, S. Sano, T. Yanase, T. Ohta and M. Koyama, *Optuna: A next-generation hyperparameter optimization framework*, In *Proc. 25th ACM SIGKDD Int. Conf. Knowl. Discov. Data Min.*, KDD '19, pp. 2623–2631. Association for Computing Machinery, New York, NY, USA, doi:10.1145/3292500.3330701 (2019).
[43] A. L. Blum and R. L. Rivest, *Training a 3-node neural network is NP-complete*, Neural Netw. **5**(1), 117 (1992), doi:10.1016/S0893-6080(05)80010-3.
[44] H. Li, Z. Xu, G. Taylor, C. Studer and T. Goldstein, *Visualizing the loss landscape of neural nets*, In *NeurIPS 2018 – Adv. Neural Inf. Process. Syst.* (2018), arXiv:1712.09913.
[45] L. Bottou, *Large-Scale Machine Learning with Stochastic Gradient Descent*, In Y. Lechevallier and G. Saporta, eds., *Proc. COMPSTAT'2010*, pp. 177–186. Physica-Verlag HD, Heidelberg, doi:10.1007/978-3-7908-2604-3_16 (2010).
[46] Y. Feng and Y. Tu, *The inverse variance–flatness relation in stochastic gradient descent is critical for finding flat minima*, Proc. Natl. Acad. Sci. U.S.A. **118**(9) (2021), doi:10.1073/pnas.2015617118.
[47] J. D. Lee, M. Simchowitz, M. I. Jordan and B. Recht, *Gradient descent only converges to minimizers*, In V. Feldman, A. Rakhlin and O. Shamir, eds., *29th Annu. Conf. Learn. Theory*, vol. 49 of *Proc. Mach. Learn. Res.*, pp. 1246–1257. PMLR, Columbia University, New York, New York, USA (2016), arXiv:1602.04915.
[48] A. Choromanska, M. Henaff, M. Mathieu, G. Ben Arous and Y. LeCun, *The loss surfaces of multilayer networks*, In *AISTATS 2015 – Int. Conf. Artif. Intell. Stat.*, vol. 38, pp. 192–204. PMLR (2015), arXiv:1412.0233.
[49] Y. N. Dauphin, R. Pascanu, C. Gulcehre, K. Cho, S. Ganguli and Y. Bengio, *Identifying and attacking the saddle point problem in high-dimensional non-convex optimization*, In *NIPS 2014 – Adv. Neural Inf. Process. Syst.* (2014), arXiv:1406.2572.
[50] L. Sagun, L. Bottou and Y. LeCun, *Eigenvalues of the Hessian in deep learning: Singularity and beyond* (2016), arXiv:1611.07476.
[51] G. Alain, N. Le Roux and P. A. Manzagol, *Negative eigenvalues of the Hessian in deep neural networks*, In *ICLR 2018 – Int. Conf. Learn. Represent.* (2018), arXiv:1902.02366.
[52] I. Sutskever, J. Martens, G. Dahl and G. Hinton, *On the importance of initialization and momentum in deep learning*, In *ICML 2013 – 30th Int. Conf. Mach. Learn.*, vol. 28, pp. 1139–1147 (2013).
[53] Y. Liu, Y. Gao and W. Yin, *An improved analysis of stochastic gradient descent with momentum*, In NeurIPS 2020 – Adv. Neural Inf. Process. Syst. (2020), arXiv:2007.07989.
[54] J. Duchi, E. Hazan and Y. Singer, *Adaptive subgradient methods for online learning and stochastic optimization*, J. Mach. Learn. Res. **12**, 2121 (2011), doi:10.5555/1953048.2021068.
[55] D. P. Kingma and J. Ba, *Adam: A method for stochastic optimization* In ICLR 2015 – Int. Conf. Learn. Represent. (2015), (2014), arXiv:1412.6980.
[56] Z. Zhang, *Improved Adam optimizer for deep neural networks*, In *2018 IEEE/ACM 26th Int. Symp. Qual. Serv. IWQoS 2018*, pp. 1–2, doi:10.1109/IWQoS.2018.8624183 (2018).
[57] C. Zhu, R. H. Byrd, P. Lu and J. Nocedal, *Algorithm 778: L-BFGS-B*, ACM Trans. Math. Softw. **23**(4), 550 (1997), doi:10.1145/279232.279236.
[58] L. M. Rios and N. V. Sahinidis, *Derivative-free optimization: A review of algorithms and comparison of software implementations*, J. Glob. Optim. **56**(3), 1247 (2012), doi:10.1007/s10898-012-9951-y.

[59] W. Liu, X. Wang, J. Owens and Y. Li, *Energy-based out-of-distribution detection*, In *NeurIPS 2020 – Adv. Neural Inf. Process. Syst.* (2020), arXiv:2010.03759.
[60] C. Zhang, S. Bengio, M. Hardt, B. Recht and O. Vinyals, *Understanding deep learning requires rethinking generalization*, In *ICLR 2017 – Int. Conf. Learn. Represent.* (2017), arXiv:1611.03530.
[61] D. H. Wolpert, *What Is Important about the No Free Lunch Theorems?*, pp. 373–388, Springer International Publishing, Cham, doi:10.1007/978-3-030-66515-9_13 (2021).
[62] P. Domingos, *A unified bias-variance decomposition for zero-one and squared loss*, In *NCAI 2000 – 17th Nat. Conf. Artif. Intell.*, pp. 564–569 (2000).
[63] M. Belkin, D. Hsu, S. Ma and S. Mandal, *Reconciling modern machine-learning practice and the classical bias–variance trade-off*, Proc. Natl. Acad. Sci. U.S.A. **116**(32), 15849 (2019), doi:10.1073/pnas.1903070116.
[64] K. Kawaguchi, L. P. Kaelbling and Y. Bengio, Generalization in deep learning, In *Mathematical Aspects of Deep Learning*. Cambridge University Press, doi:10.1017/9781009025096.003 (2022)
[65] J. Frankle and M. Carbin, *The lottery ticket hypothesis: Finding sparse, trainable neural networks*, In *ICLR 2019 – Int. Conf. Learn. Represent.* (2019), arXiv:1803.03635.
[66] L. Devroye, L. Györfi and G. Lugosi, *The Bayes Error*, pp. 9–20, Springer, New York, NY, doi:10.1007/978-1-4612-0711-5_2 (1996).
[67] C. R. Rao, *Generalized Inverse of a Matrix and Its Applications*, pp. 601–620, University of California Press, Berkeley, doi:10.1525/9780520325883-032 (1972).
[68] R. Tibshirani, *Regression shrinkage and selection via the lasso*, J. R. Stat. Soc. Ser. B Methodol. **58**(1), 267 (1996), doi:10.1111/j.2517-6161.1996.tb02080.x.
[69] Z. Zhou, X. Li and R. N. Zare, *Optimizing chemical reactions with deep reinforcement learning*, ACS Cent. Sci **3**(12), 1337 (2017), doi:10.1021/acscentsci.7b00492.
[70] A. Chervonenkis, *Early History of Support Vector Machines*, pp. 13–20, Springer, Berlin, Heidelberg, doi:10.1007/978-3-642-41136-6_3 (2013).
[71] B. E. Boser, I. M. Guyon and V. N. Vapnik, *A Training Algorithm for Optimal Margin Classifiers*, In *Proc. Fifth Ann. Workshop Compu. Learn. Theo.*, COLT '92, pp. 144–152. Association for Computing Machinery, doi:10.1145/130385.130401 (1992).
[72] J. Platt, *Sequential minimal optimization: A fast algorithm for training support vector machines*, Tech. rep. MSR-TR-98-14, Microsoft (1998).
[73] M. Minsky and S. Papert, *Perceptrons: An Introduction to Computational Geometry*, MIT Press, doi:10.7551/mitpress/11301.001.0001 (1969).
[74] F. Rosenblatt, *The perceptron: A probabilistic model for information storage and organization in the brain*, Psychol. Rev. **65**(6), 386 (1958), doi:10.1037/h0042519.
[75] A. N. Kolmogorov, *On the representation of continuous functions of many variables by superposition of continuous functions of one variable and addition, Dokl. Akad. Nauk* **114**, 953–956 (1957).
[76] G. Cybenko, *Approximation by superpositions of a sigmoidal function*, Math. Control Signals Syst. **2**(4), 303 (1989), doi:10.1007/BF02551274.
[77] K. Hornik, *Approximation capabilities of multilayer feedforward networks*, Neural Netw. **4**(2), 251 (1991), doi:10.1016/0893-6080(91)90009-T.
[78] D. E. Rumelhart, G. E. Hinton and R. J. Williams, *Learning representations by back-propagating errors*, Nature **323**(6088), 533 (1986), doi:10.1038/323533a0.

[79] D. P. Kingma and M. Welling, *Auto-encoding variational Bayes*, ICLR 2014 – Int. Conf. Learn. Represent. (2014), arXiv:1312.6114.

[80] D. J. Rezende, S. Mohamed and D. Wierstra, *Stochastic backpropagation and approximate inference in deep generative models*, In *ICML 2014 – Int. Conf. Mach. Learn.*, vol. 32, pp. 1278–1286 (2014), arXiv:1401.4082.

[81] A. Ng, *Sparse autoencoder*, CS294A Lecture notes, Stanford University (2011).

[82] A. Makhzani and B. Frey, *K-sparse autoencoders*, ICLR 2014 – Int. Conf. Learn. Represent. (2014), arXiv:1312.5663.

[83] P. Vincent, H. Larochelle, Y. Bengio and P.-A. Manzagol, *Extracting and composing robust features with denoising autoencoders*, In *ICML 2008 – 25th Int. Conf. Mach. Learn.*, pp. 1096–1103, doi:10.1145/1390156.1390294 (2008).

[84] Y. Burda, R. Grosse and R. Salakhutdinov, *Importance weighted autoencoders*, ICLR 2016 – Int. Conf. Learn. Represent. (2016), arXiv:1509.00519.

[85] B. Uria, M.-A. Côté, K. Gregor, I. Murray and H. Larochelle, *Neural autoregressive distribution estimation*, J. Mach. Learn. Res. **17**(1), 7184 (2016), doi:10.5555/2946645.3053487.

[86] S. Hochreiter and J. Schmidhuber, *Long short-term memory*, Neural Comput. **9**(8), 1735 (1997), doi:10.1162/neco.1997.9.8.1735.

[87] K. Cho, B. van Merriënboer, D. Bahdanau and Y. Bengio, *On the properties of neural machine translation: Encoder–decoder approaches*, In *SSST-8 – 8th Workshop on Syntax, Semantics and Structure in Statistical Translation*, pp. 103–111, doi:10.3115/v1/W14-4012 (2014).

[88] D. Wu, L. Wang and P. Zhang, *Solving statistical mechanics using variational autoregressive networks*, Phys. Rev. Lett. **122**(8), 080602 (2019), doi:10.1103/PhysRevLett.122.080602.

[89] K. A. Nicoli, S. Nakajima, N. Strodthoff, W. Samek, K.-R. Müller and P. Kessel, *Asymptotically unbiased estimation of physical observables with neural samplers*, Phys. Rev. E **101**(2), 023304 (2020), doi:10.1103/PhysRevE.101.023304.

[90] J.-G. Liu, L. Mao, P. Zhang and L. Wang, *Solving quantum statistical mechanics with variational autoregressive networks and quantum circuits*, Mach. Learn.: Sci. Technol. **2**(2), 025011 (2021), doi:10.1088/2632-2153/aba19d.

[91] J. Carrasquilla, G. Torlai, R. G. Melko and L. Aolita, *Reconstructing quantum states with generative models*, Nat. Mach. Intell. **1**(3), 155–161 (2019), doi:10.1038/s42256-019-0028-1.

[92] O. Sharir, Y. Levine, N. Wies, G. Carleo and A. Shashua, *Deep autoregressive models for the efficient variational simulation of many-body quantum systems*, Phys. Rev. Lett. **124**(2), 020503 (2020), doi:10.1103/PhysRevLett.124.020503.

[93] C. M. Bishop, *Pattern Recognition and Machine Learning*, Springer, Berlin, Heidelberg (2006).

[94] P. Mehta, M. Bukov, C.-H. Wang, *et al.*, *A high-bias, low-variance introduction to machine learning for physicists*, Phys. Rep. **810**, 1 (2019), doi:10.1016/j.physrep.2019.03.001. https://d2l.ai/

[95] A. Zhang, Z. C. Lipton, M. Li and A. J. Smola, *Dive into deep learning* (2021), arXiv:2106.11342.

[96] T. Neupert, M. H. Fischer, E. Greplova, K. Choo and M. Denner, *Introduction to machine learning for the sciences* (2021), arXiv:2102.04883.

[97] J. Carrasquilla and G. Torlai, *How to use neural networks to investigate quantum many-body physics*, PRX Quantum **2**, 040201 (2021), doi:10.1103/PRXQuantum.2.040201.

[98] S. Sachdev, *Quantum Phase Transitions*, Cambridge University Press, doi:10.1017/cbo9780511973765 (2011).

[99] N. Goldenfeld, *Lectures on Phase Transitions and the Renormalization Group*, CRC Press, doi:10.1201/9780429493492 (2018).

[100] L. Onsager, *Crystal statistics. I. A two-dimensional model with an order-disorder transition*, Phys. Rev. **65**, 117 (1944), doi:10.1103/PhysRev.65.117.

[101] F. J. Wegner, *Duality in generalized Ising models and phase transitions without local order parameters*, J. Math. Phys. **12**(10), 2259 (1971), doi:10.1063/1.1665530.

[102] L. D. Landau, *On the theory of phase transitions. I.*, Phys. Z. Sowjet. **11**, 26 (1937), Reprinted in Collected Papers of L. D. Landau.

[103] L. D. Landau, *On the theory of phase transitions. II.*, Phys. Z. Sowjet. **11**, 545 (1937), Reprinted in Collected Papers of L. D. Landau.

[104] X.-G. Wen, *Topological orders in rigid states*, Int. J. Mod. Phys. B **4**(02), 239 (1990), doi:10.1142/S0217979290000139.

[105] B. Bernevig and T. Hughes, *Topological Insulators and Topological Superconductors*, Princeton University Press, Princeton, doi:10.1515/9781400846733 (2013).

[106] N. Käming, A. Dawid, K. Kottmann, *et al.*, *Unsupervised machine learning of topological phase transitions from experimental data*, Mach. Learn.: Sci. Technol. **2**, 035037 (2021), doi:10.1088/2632-2153/abffe7.

[107] N. Sun, J. Yi, P. Zhang, H. Shen and H. Zhai, *Deep learning topological invariants of band insulators*, Phys. Rev. B **98**, 085402 (2018), doi:10.1103/PhysRevB.98.085402.

[108] P. Zhang, H. Shen and H. Zhai, *Machine learning topological invariants with neural networks*, Phys. Rev. Lett. **120**, 066401 (2018), doi:10.1103/PhysRevLett.120.066401.

[109] M. D. Caio, M. Caccin, P. Baireuther, T. Hyart and M. Fruchart, *Machine learning assisted measurement of local topological invariants* (2019), arXiv:1901.03346.

[110] N. L. Holanda and M. A. R. Griffith, *Machine learning topological phases in real space*, Phys. Rev. B **102**, 054107 (2020), doi:10.1103/PhysRevB.102.054107.

[111] P. Baireuther, M. Płodzień, T. Ojanen, K. Tworzydło and T. Hyart, *Identifying Chern numbers of superconductors from local measurements* SciPost Phys. Core **6**, 087 (2023), doi:10.21468/SciPostPhysCore.6.4.087.

[112] P. Huembeli, A. Dauphin and P. Wittek, *Identifying quantum phase transitions with adversarial neural networks*, Phys. Rev. B **97**, 134109 (2018), doi:10.1103/PhysRevB.97.134109.

[113] C. Fefferman, S. Mitter and H. Narayanan, *Testing the manifold hypothesis*, J. Am. Math. Soc. **29**(4), 983 (2016), doi:10.1090/jams/852.

[114] L. Wang, *Discovering phase transitions with unsupervised learning*, Phys. Rev. B **94**, 195105 (2016), doi:10.1103/PhysRevB.94.195105.

[115] S. J. Wetzel, *Unsupervised learning of phase transitions: From principal component analysis to variational autoencoders*, Phys. Rev. E **96**, 022140 (2017), doi:10.1103/PhysRevE.96.022140.

[116] W. Hu, R. R. Singh and R. T. Scalettar, *Discovering phases, phase transitions, and crossovers through unsupervised machine learning: A critical examination*, Phys. Rev. E **95**(6), 062122 (2017), doi:10.1103/PhysRevE.95.062122.

[117] B. Schölkopf, A. Smola and K.-R. Müller, *Nonlinear component analysis as a kernel eigenvalue problem*, Neural Comput. **10**(5), 1299 (1998), doi:10.1162/089976698300017467.

[118] L. Van der Maaten and G. Hinton, *Visualizing data using t-SNE*, J. Mach. Learn. Res. **9**(11) (2008).

[119] L. McInnes, J. Healy and J. Melville, *UMAP: Uniform manifold approximation and projection for dimension reduction* (2018), arXiv:1802.03426.

[120] G. E. Hinton and S. T. Roweis, *Stochastic neighbor embedding*, In *NIPS 2002 – Adv. Neural Inf. Process. Syst.* (2002).

[121] E. Greplova, A. Valenti, G. Boschung, F. Schäfer, N. Lörch and S. D. Huber, *Unsupervised identification of topological phase transitions using predictive models*, New J. Phys. **22**(4), 045003 (2020), doi:10.1088/1367-2630/ab7771.

[122] J. Arnold, F. Schäfer, M. Žonda and A. U. J. Lode, *Interpretable and unsupervised phase classification*, Phys. Rev. Res. **3**, 033052 (2021), doi:10.1103/PhysRevResearch.3.033052.

[123] J. Carrasquilla and R. G. Melko, *Machine learning phases of matter*, Nat. Phys. **13**(5), 431 (2017), doi:10.1038/nphys4035.

[124] K. Kottmann, F. Metz, J. Fraxanet and N. Baldelli, *Variational quantum anomaly detection: Unsupervised mapping of phase diagrams on a physical quantum computer*, Phys. Rev. Res. **3**, 043184 (2021), doi:10.1103/PhysRevResearch.3.043184.

[125] T. Szołdra, P. Sierant, M. Lewenstein and J. Zakrzewski, *Unsupervised detection of decoupled subspaces: Many-body scars and beyond*, Phys. Rev. B **105**, 224205 (2022), doi:10.1103/PhysRevB.105.224205.

[126] K. Kottmann, P. Huembeli, M. Lewenstein and A. Acín, *Unsupervised phase discovery with deep anomaly detection*, Phys. Rev. Lett. **125**, 170603 (2020), doi:10.1103/PhysRevLett.125.170603.

[127] T. Szołdra, P. Sierant, K. Kottmann, M. Lewenstein and J. Zakrzewski, *Detecting ergodic bubbles at the crossover to many-body localization using neural networks*, Phys. Rev. B **104**, L140202 (2021), doi:10.1103/PhysRevB.104.L140202.

[128] E. P. Van Nieuwenburg, Y.-H. Liu and S. D. Huber, *Learning phase transitions by confusion*, Nat. Phys. **13**(5), 435 (2017), doi:10.1038/nphys4037.

[129] Y.-H. Liu and E. P. L. van Nieuwenburg, *Discriminative cooperative networks for detecting phase transitions*, Phys. Rev. Lett. **120**, 176401 (2018), doi:10.1103/PhysRevLett.120.176401.

[130] S. S. Lee and B. J. Kim, *Confusion scheme in machine learning detects double phase transitions and quasi-long-range order*, Phys. Rev. E **99**, 043308 (2019), doi:10.1103/PhysRevE.99.043308.

[131] M. Richter-Laskowska, M. Kurpas and M. M. Maśka, *Learning by confusion approach to identification of discontinuous phase transitions*, Phys. Rev. E **108**, 024113 (2023), doi:10.1103/PhysRevE.108.024113.

[132] F. Schäfer and N. Lörch, *Vector field divergence of predictive model output as indication of phase transitions*, Phys. Rev. E **99**, 062107 (2019), doi:10.1103/PhysRevE.99.062107.

[133] P. Ronhovde, S. Chakrabarty, D. Hu, *et al.*, *Detecting hidden spatial and spatio-temporal structures in glasses and complex physical systems by multiresolution network clustering*, Eur. Phys. J. E **34**(9), 1 (2011), doi:10.1140/epje/i2011-11105-9.

[134] P. Ronhovde, S. Chakrabarty, D. Hu, *et al.*, *Detection of hidden structures for arbitrary scales in complex physical systems*, Sci. Rep. **2**(1), 1 (2012), doi:10.1038/srep00329.

[135] R. A. Vargas-Hernández, J. Sous, M. Berciu and R. V. Krems, *Extrapolating quantum observables with machine learning: Inferring multiple phase transitions from properties of a single phase*, Phys. Rev. Lett. **121**, 255702 (2018), doi:10.1103/PhysRevLett.121.255702.

[136] A. A. Shirinyan, V. K. Kozin, J. Hellsvik, M. Pereiro, O. Eriksson and D. Yudin, *Self-organizing maps as a method for detecting phase transitions and phase identification*, Phys. Rev. B **99**, 041108 (2019), doi:10.1103/PhysRevB.99.041108.

[137] T. Mazaheri, B. Sun, J. Scher-Zagier, *et al.*, *Stochastic replica voting machine prediction of stable cubic and double perovskite materials and binary alloys*, Phys. Rev. Mater. **3**, 063802 (2019), doi:10.1103/PhysRevMaterials.3.063802.

[138] O. Balabanov and M. Granath, *Unsupervised learning using topological data augmentation*, Phys. Rev. Res. **2**, 013354 (2020), doi:10.1103/PhysRevResearch.2.013354.

[139] S.-J. Gu, *Fidelity approach to quantum phase transitions*, Int. J. Mod. Phys. B **24**(23), 4371 (2010), doi:10.1142/S0217979210056335.

[140] B. S. Rem, N. Käming, M. Tarnowski, *et al.*, *Identifying quantum phase transitions using artificial neural networks on experimental data*, Nat. Phys. **15**, 917 (2019), doi:10.1038/s41567-019-0554-0.

[141] A. Bohrdt, S. Kim, A. Lukin, *et al.*, *Analyzing nonequilibrium quantum states through snapshots with artificial neural networks*, Phys. Rev. Lett. **127**, 150504 (2021), doi:10.1103/PhysRevLett.127.150504.

[142] Z. C. Lipton, *The mythos of model interpretability*, Commun. ACM **61**(10), 35 (2018), doi:10.1145/3233231.

[143] A. Bohrdt, C. S. Chiu, G. Ji, *et al.*, *Classifying snapshots of the doped Hubbard model with machine learning*, Nat. Phys. **15**(9), 921 (2019), doi:10.1038/s41567-019-0565-x.

[144] Y. Zhang, P. Ginsparg and E.-A. Kim, *Interpreting machine learning of topological quantum phase transitions*, Phys. Rev. Res. **2**, 023283 (2020), doi:10.1103/PhysRevResearch.2.023283.

[145] M. Cranmer, A. Sanchez-Gonzalez, P. Battaglia, *et al.*, *Discovering symbolic models from deep learning with inductive biases*, In *NeurIPS 2020 – Adv. Neural Inf. Process. Syst.* (2020), arXiv:2006.11287.

[146] P. Ponte and R. G. Melko, *Kernel methods for interpretable machine learning of order parameters*, Phys. Rev. B **96**, 205146 (2017), doi:10.1103/PhysRevB.96.205146.

[147] J. Greitemann, K. Liu and L. Pollet, *Probing hidden spin order with interpretable machine learning*, Phys. Rev. B **99**, 060404 (2019), doi:10.1103/PhysRevB.99.060404.

[148] K. Liu, J. Greitemann and L. Pollet, *Learning multiple order parameters with interpretable machines*, Phys. Rev. B **99**, 104410 (2019), doi:10.1103/PhysRevB.99.104410.

[149] R. Iten, T. Metger, H. Wilming, L. Del Rio and R. Renner, *Discovering physical concepts with neural networks*, Phys. Rev. Lett. **124**(1), 010508 (2020), doi:10.1103/PhysRevLett.124.010508.

[150] S. J. Wetzel and M. Scherzer, *Machine learning of explicit order parameters: From the Ising model to SU(2) lattice gauge theory*, Phys. Rev. B **96**(18), 184410 (2017), doi:10.1103/PhysRevB.96.184410.

[151] S. J. Wetzel, R. G. Melko, J. Scott, M. Panju and V. Ganesh, *Discovering symmetry invariants and conserved quantities by interpreting Siamese neural networks*, Phys. Rev. Res. **2**, 033499 (2020), doi:10.1103/PhysRevResearch.2.033499.

[152] C. Miles, A. Bohrdt, R. Wu, *et al.*, *Correlator convolutional neural networks: An interpretable architecture for image-like quantum matter data*, Nat. Commun. **12**(1), 1 (2021), doi:10.1038/s41467-021-23952-w.

[153] S. K. Radha and C. Jao, *Generalized quantum similarity learning* (2022), arXiv:2201.02310.

[154] Z. Patel, E. Merali and S. J. Wetzel, *Unsupervised learning of Rydberg atom array phase diagram with Siamese neural networks*, New J. Phys. **24**(11), 113021 (2022), doi:10.1088/1367-2630/ac9c7a.

[155] X.-Q. Han, S.-S. Xu, Z. Feng, R.-Q. He and Z.-Y. Lu, *A simple framework for contrastive learning phases of matter* Chin. Phys. Lett. **40**, 027501 (2023), doi:10.1088/0256-307X/40/2/027501.

[156] Z. Liu and M. Tegmark, *Machine learning conservation laws from trajectories*, Phys. Rev. Lett. **126**, 180604 (2021), doi:10.1103/PhysRevLett.126.180604.

[157] Z. Liu, V. Madhavan and M. Tegmark, *Machine learning conservation laws from differential equations*, Phys. Rev. E **106**, 045307 (2022), doi:10.1103/PhysRevE.106.045307.

[158] S. Ha and H. Jeong, *Discovering invariants via machine learning*, Phys. Rev. Res. **3**, L042035 (2021), doi:10.1103/PhysRevResearch.3.L042035.

[159] N. S. Keskar, J. Nocedal, P. T. P. Tang, D. Mudigere and M. Smelyanskiy, *On large-batch training for deep learning: Generalization gap and sharp minima*, In *ICLR 2017 – Int. Conf. Learn. Represent.* (2017), arXiv:1609.04836.

[160] L. Wu, Z. Zhu and E. Weinan, *Towards understanding generalization of deep learning: Perspective of loss landscapes* (2017), arXiv:1706.10239.

[161] P. Izmailov, D. Podoprikhin, T. Garipov, D. Vetrov and A. G. Wilson, *Averaging weights leads to wider optima and better generalization*, In *UAI 2018 – 34th Conf. Uncertain. Artif. Intell.*, vol. 2, pp. 876–885 (2018), arXiv:1803.05407.

[162] H. He, G. Huang and Y. Yuan, *Asymmetric valleys: Beyond sharp and flat local minima*, In *NeurIPS 2019 – Adv. Neural Inf. Process. Syst.* (2019), arXiv:1902.00744.

[163] L. Dinh, R. Pascanu, S. Bengio and Y. Bengio, *Sharp minima can generalize for deep nets*, In *ICML 2017 – 34th Int. Conf. Mach. Learn.*, vol. 3, pp. 1705–1714 (2017), arXiv:1703.04933.

[164] A. Dawid, P. Huembeli, M. Tomza, M. Lewenstein and A. Dauphin, *Hessian-based toolbox for reliable and interpretable machine learning in physics*, Mach. Learn.: Sci. Technol. **3**, 015002 (2022), doi:10.1088/2632-2153/ac338d.

[165] P. W. Koh and P. Liang, *Understanding black-box predictions via influence functions*, In *ICML 2017 – 34th Int. Conf. Mach. Learn.*, vol. 70, pp. 1885–1894. PMLR (2017), arXiv:1703.04730.

[166] P. Schulam and S. Saria, *Can you trust this prediction? Auditing pointwise reliability after learning*, In *AISTATS 2019 – Int. Conf. Artif. Intell. Stat.*, vol. 89, pp. 1022–1031. PLMR (2020), arXiv:1901.00403.

[167] D. Madras, J. Atwood and A. D'Amour, *Detecting extrapolation with local ensembles*, In *ICLR 2020 – Int. Conf. Learn. Represent.* (2020), arXiv:1910.09573.

[168] A. Dawid, P. Huembeli, M. Tomza, M. Lewenstein and A. Dauphin, *Phase detection with neural networks: Interpreting the black box*, New J. Phys. **22**(11), 115001 (2020), doi:10.1088/1367-2630/abc463.

[169] J. Arnold and F. Schäfer, *Replacing neural networks by optimal analytical predictors for the detection of phase transitions*, Phys. Rev. X **12**, 031044 (2022), doi:10.1103/PhysRevX.12.031044.

[170] J. Arnold, F. Schäfer, A. Edelman and C. Bruder, *Mapping out phase diagrams with generative classifiers* Phys. Rev. Lett. **132**, 207301 (2024), doi:10.1103/PhysRevLett.132.207301.

[171] C. Molnar, *Interpretable Machine Learning: A Guide for Making Black Box Models Explainable*, GitHub.io (2019).

[172] K.-R. Müller, S. Mika, K. Tsuda and K. Schölkopf, *An introduction to kernel-based learning algorithms*, In *Handbook of Neural Network Signal Processing*, pp. 94–133. CRC Press, Boca Raton, doi:10.1201/9781315220413 (2018).

[173] B. Schölkopf and A. J. Smola, *Learning with Kernels*, The MIT Press, Cambridge, MA, doi:10.7551/mitpress/4175.001.0001 (2018).

[174] T. Hofmann, B. Schölkopf and A. J. Smola, *Kernel methods in machine learning*, Ann. Statist. **36**(3), 1171 (2008), doi:10.1214/009053607000000677.

[175] G. Bachman and L. Narici, *Functional analysis*, Dover Publications, Mineola, NY (2000).

[176] J. Mercer, *XVI. Functions of positive and negative type, and their connection the theory of integral equations*, Philos. Trans. Royal Soc. A **209**(441–458), 415 (1909), doi:10.1098/rsta.1909.0016.

[177] N. Aronszajn, *Theory of reproducing kernels*, Trans. Am. Math. Soc. **68**(3), 337 (1950), doi:10.2307/1990404.

[178] B. Schölkopf, R. Herbrich and A. J. Smola, *A Generalized Representer Theorem*, In *Computational Learning Theory*, pp. 416–426. Springer, doi:10.1007/3-540-44581-1_27 (2001).

[179] B. Schölkopf, A. Smola and K.-R. Müller, *Kernel principal component analysis*, In *ICANN 1997 – Int. Conf. Neural Netw.*, pp. 583–588. Springer, doi:10.1007/BFb0020217 (1997).

[180] C. Saunders, A. Gammerman and V. Vovk, *Ridge Regression Learning Algorithm in Dual Variables*, In *Proc. 15th Int. Conf. Mach. Learn.*, ICML '98, pp. 515–521. Morgan Kaufmann Publishers Inc., San Francisco, CA, USA, doi:10.5555/645527.657464 (1998).

[181] A. J. Smola and B. Schölkopf, *On a kernel-based method for pattern recognition, regression, approximation, and operator inversion*, Algorithmica **22**(1), 211 (1998), doi:10.1007/PL00013831.

[182] R. Garnett, *Bayesian Optimization*, Cambridge University Press, in preparation (2022).

[183] R. M. Neal, *Bayesian Learning for Neural Networks*, vol. 118 of *Lecture Notes in Statistics*, Springer, doi:10.1007/978-1-4612-0745-0 (2012).

[184] N. Cressie, *The origins of kriging*, Math. Geol. **22**(3), 239 (1990), doi:10.1007/BF00889887.

[185] P. I. Frazier, *A tutorial on Bayesian optimization* (2018), arXiv:1807.02811.

[186] G. Schwarz, *Estimating the dimension of a model*, Ann. Stat. **6**(2), 461–464 (1978), doi:10.1214/aos/1176344136.

[187] P. Stoica and Y. Selen, *Model-order selection: A review of information criterion rules*, IEEE Signal Process. Mag. **21**(4), 36 (2004), doi:10.1109/MSP.2004.1311138.

[188] H. Akaike, *A new look at the statistical model identification*, IEEE Trans. Automat. Contr. **19**(6), 716 (1974), doi:10.1109/TAC.1974.1100705.

[189] D. Duvenaud, J. Lloyd, R. Grosse, J. Tenenbaum and G. Zoubin, *Structure discovery in nonparametric regression through compositional kernel search*, In *ICML 2013 – Int. Conf. Mach. Learn.*, vol. 28, pp. 1166–1174. PMLR (2013), arXiv:1302.4922.

[190] D. Duvenaud, H. Nickisch and C. E. Rasmussen, *Additive Gaussian processes*, In *NIPS 2011 – Adv. Neural Inf. Process. Syst.* (2011), arXiv:1112.4394.

[191] J. Dai and R. V. Krems, *Interpolation and extrapolation of global potential energy surfaces for polyatomic systems by Gaussian processes with composite kernels*, J. Chem. Theory Comput. **16**(3), 1386 (2020), doi:10.1021/acs.jctc.9b00700.

[192] R. A. Vargas-Hernández and J. R. Gardner, *Gaussian processes with spectral delta kernel for higher accurate potential energy surfaces for large molecules* (2021), arXiv:2109.14074.

[193] N. Q. Su, J. Chen, Z. Sun, D. H. Zhang and X. Xu, *$H + H_2$ quantum dynamics using potential energy surfaces based on the XYG3 type of doubly hybrid density functionals: Validation of the density functionals*, J. Chem. Phys. **142**, 084107 (2015), doi:10.1063/1.4913196.

[194] R. A. Vargas-Hernández, Y. Guan, D. H. Zhang and R. V. Krems, *Bayesian optimization for the inverse scattering problem in quantum reaction dynamics*, New J. Phys. **21**, 22001 (2019), doi:10.1088/1367-2630/ab0099.

[195] Z. Deng, I. Tutunnikov, I. S. Averbukh, M. Thachuk and R. V. Krems, *Bayesian optimization for inverse problems in time-dependent quantum dynamics*, J. Chem. Phys. **153**(16), 164111 (2020), doi:10.1063/5.0015896.

[196] J. T. Cantin, G. Alexandrowicz and R. V. Krems, *Transfer-matrix theory of surface spin-echo experiments with molecules*, Phys. Rev. A **101**(6), 062703 (2020), doi:10.1103/PhysRevA.101.062703.

[197] N. Sugisawa, H. Sugisawa, Y. Otake, R. V. Krems, H. Nakamura and S. Fuse, *Rapid and mild one-flow synthetic approach to unsymmetrical sulfamides guided by Bayesian optimization*, Chem. Methods **1**(11), 484 (2021), doi:10.1002/cmtd.202100053.

[198] A. Jasinski, J. Montaner, R. C. Forrey, *et al.*, *Machine learning corrected quantum dynamics calculations*, Phys. Rev. Res. **2**(3), 32051 (2020), doi:10.1103/PhysRevResearch.2.032051.

[199] R. A. Vargas Hernandez, *Bayesian optimization for calibrating and selecting hybrid-density functional models*, J. Phys. Chem. A **124**(20), 4053 (2020), doi:10.1021/acs.jpca.0c01375.

[200] J. Proppe, S. Gugler and M. Reiher, *Gaussian process-based refinement of dispersion corrections*, J. Chem. Theory Comput. **15**(11), 6046 (2019), doi:10.1021/acs.jctc.9b00627.

[201] R. Tamura and K. Hukushima, *Bayesian optimization for computationally extensive probability distributions*, PLoS One **13**(3), 1 (2018), doi:10.1371/journal.pone.0193785.

[202] S. Carr, R. Garnett and C. Lo, *BASC: Applying Bayesian optimization to the search for global minima on potential energy surfaces*, In *ICML 2016 – Int. Conf. Mach. Learn.*, vol. 48, pp. 898–907. PMLR (2016).

[203] L. Chan, G. R. Hutchison and G. M. Morris, *Bayesian optimization for conformer generation*, J. Cheminformatics **11**(1), 32 (2019), doi:10.1186/s13321-019-0354-7.

[204] R. A. Vargas-Hernández, C. Chuang and P. Brumer, *Multi-objective optimization for retinal photoisomerization models with respect to experimental observables*, J. Chem. Phys. **155**(23), 234109 (2021), doi:10.1063/5.0060259.

[205] J. Duris, D. Kennedy, A. Hanuka, *et al.*, *Bayesian optimization of a free-electron laser*, Phys. Rev. Lett. **124**, 124801 (2020), doi:10.1103/PhysRevLett.124.124801.
[206] S. Jalas, M. Kirchen, P. Messner, *et al.*, *Bayesian optimization of a laser-plasma accelerator*, Phys. Rev. Lett. **126**, 104801 (2021), doi:10.1103/PhysRevLett.126.104801.
[207] R. J. Shalloo, S. J. D. Dann, J.-N. Gruse, *et al.*, *Automation and control of laser wakefield accelerators using Bayesian optimization*, Nat. Commun. **11**(1), 6355 (2020), doi:10.1038/s41467-020-20245-6.
[208] T. Ueno, T. D. Rhone, Z. Hou, T. Mizoguchi and K. Tsuda, *COMBO: An efficient Bayesian optimization library for materials science*, Mater. Discov. **4**, 18 (2016), doi:10.1016/j.md.2016.04.001.
[209] R. Jalem, K. Kanamori, I. Takeuchi, M. Nakayama, H. Yamasaki and T. Saito, *Bayesian-driven first-principles calculations for accelerating exploration of fast ion conductors for rechargeable battery application*, Sci. Rep. **8**(1), 5845 (2018), doi:10.1038/s41598-018-23852-y.
[210] S. Ju, T. Shiga, L. Feng, Z. Hou, K. Tsuda and J. Shiomi, *Designing nanostructures for phonon transport via Bayesian optimization*, Phys. Rev. X **7**, 021024 (2017), doi:10.1103/PhysRevX.7.021024.
[211] J. Kuhn, J. Spitz, P. Sonnweber-Ribic, M. Schneider and T. Böhlke, *Identifying material parameters in crystal plasticity by Bayesian optimization*, Optim. Eng. (2021), doi:10.1007/s11081-021-09663-7.
[212] R.-R. Griffiths and J. M. Hernández-Lobato, *Constrained Bayesian optimization for automatic chemical design using variational autoencoders*, Chem. Sci. **11**, 577 (2020), doi:10.1039/C9SC04026A.
[213] A. Deshwal, C. M. Simon and J. R. Doppa, *Bayesian optimization of nanoporous materials*, Mol. Syst. Des. Eng. **6**, 1066 (2021), doi:10.1039/D1ME00093D.
[214] F. Häse, L. M. Roch, C. Kreisbeck and A. Aspuru-Guzik, *Phoenics: A Bayesian optimizer for chemistry*, ACS Cent. Sci. **4**(9), 1134 (2018), doi:10.1021/acscentsci.8b00307.
[215] F. Häse, M. Aldeghi, R. J. Hickman, L. M. Roch and A. Aspuru-Guzik, *Gryffin: An algorithm for Bayesian optimization of categorical variables informed by expert knowledge*, Appl. Phys. Rev. **8**(3), 031406 (2021), doi:10.1063/5.0048164.
[216] A. Biswas, A. N. Morozovska, M. Ziatdinov, E. A. Eliseev and S. V. Kalinin, *Multi-objective Bayesian optimization of ferroelectric materials with interfacial control for memory and energy storage applications*, J. Appl. Phys. **130**(20), 204102 (2021), doi:10.1063/5.0068903.
[217] Y. Wang, T.-Y. Chen and D. G. Vlachos, *NEXTorch: A design and Bayesian optimization toolkit for chemical sciences and engineering*, J. Chem. Inf. Model. **61**(11), 5312 (2021), doi:10.1021/acs.jcim.1c00637.
[218] M. Aldeghi, F. Häse, R. J. Hickman, I. Tamblyn and A. Aspuru-Guzik, *Golem: An algorithm for robust experiment and process optimization*, Chem. Sci. **12**, 14792 (2021), doi:10.1039/D1SC01545A.
[219] Z. Vendeiro, J. Ramette, A. Rudelis, *et al.*, *Machine-learning-accelerated Bose-Einstein condensation*, Phys. Rev. Res. **4**, 043216 (2022), doi:10.1103/PhysRevResearch.4.043216.
[220] H. Sugisawa, T. Ida and R. V. Krems, *Gaussian process model of 51-dimensional potential energy surface for protonated imidazole dimer*, J. Chem. Phys. **153**(11), 114101 (2020), doi:10.1063/5.0023492.

[221] C. Puzzarini, J. Bloino, N. Tasinato and V. Barone, *Accuracy and interpretability: The devil and the holy grail. New routes across old boundaries in computational spectroscopy*, Chem. Rev. **119**(13), 8131 (2019), doi:10.1021/acs.chemrev.9b00007.
[222] F. Herrera, K. W. Madison, R. V. Krems and M. Berciu, *Investigating polaron transitions with polar molecules*, Phys. Rev. Lett. **110**(22), 223002 (2013), doi:10.1103/PhysRevLett.110.223002.
[223] P. Deglmann, A. Schäfer and C. Lennartz, *Application of quantum calculations in the chemical industry: An overview*, Int. J. Quantum Chem. **115**(3), 107 (2015).
[224] Y. Cao, J. Romero, J. P. Olson *et al.*, *Quantum chemistry in the age of quantum computing*, Chem. Rev. **119**(19), 10856 (2019).
[225] A. J. McCaskey, Z. P. Parks, J. Jakowski *et al.*, *Quantum chemistry as a benchmark for near-term quantum computers*, npj Quantum Inf. **5**(1), 99 (2019).
[226] A. N. Ciavarella and I. A. Chernyshev, *Preparation of the SU (3) lattice Yang-Mills vacuum with variational quantum methods*, Phys. Rev. D **105**(7), 074504 (2022).
[227] M. C. Banuls, R. Blatt, J. Catani *et al.*, *Simulating lattice gauge theories within quantum technologies*, Eur. Phys. J. D **74**, 1 (2020).
[228] G. Iannelli and K. Jansen, *Noisy Bayesian optimization for variational quantum eigensolvers*, (2021), arXiv:2112.00426.
[229] J. Mueller, W. Lavrijsen, C. Iancu and W. A. de Jong, *Accelerating noisy VQE optimization with Gaussian processes, 2022 IEEE Int. Conf. Quantum Comput. Eng. (QCE)*, pp. 215–225 (2022).
[230] K. A. Nicoli, C. J. Anders, L. Funcke, *et al.*, *Physics-informed Bayesian optimization of variational quantum circuits*, In *NeurIPS 2023 – Adv. Neural Inf. Process. Syst.* (2023).
[231] K. M. Nakanishi, K. Fujii and S. Todo, *Sequential minimal optimization for quantum-classical hybrid algorithms*, Phys. Rev. Res. **2**, 043158 (2020), doi:10.1103/PhysRevResearch.2.043158.
[232] J. Platt, *Sequential minimal optimization: A fast algorithm for training support vector machines*, Microsoft Research Technical Report (1998).
[233] K. Asnaashari and R. V. Krems, *Gradient domain machine learning with composite kernels: Improving the accuracy of PES and force fields for large molecules*, Mach. Learn.: Sci. Technol. **3**(1), 015005 (2021), doi:10.1088/2632-2153/ac3845.
[234] A. G. Wilson, Z. Hu, R. Salakhutdinov and E. P. Xing, *Deep kernel learning*, In *AISTATS 2016 – Int. Conf. Artif. Intell. Stat.* (2016), arXiv:1511.02222.
[235] S. Sun, G. Zhang, C. Wang, W. Zeng, J. Li and R. Grosse, *Differentiable compositional kernel learning for Gaussian processes*, In *ICML 2018 – Int. Conf. Mach. Learn.* (2018), arXiv:1806.04326.
[236] J. Gardner, G. Pleiss, K. Q. Weinberger, D. Bindel and A. G. Wilson, *GPyTorch: Blackbox matrix-matrix Gaussian process inference with GPU acceleration*, In *NeurIPS 2018 – Adv. Neural Inf. Process. Syst.* (2018), arXiv:1809.11165.
[237] B. Charlier, J. Feydy, J. A. Glaunès, F.-D. Collin and G. Durif, *Kernel operations on the GPU, with Autodiff, without memory overflows*, J. Mach. Learn. Res. **22**(74), 1 (2021), arXiv:2004.11127.
[238] A. G. d. G. Matthews, M. van der Wilk, T. Nickson, *et al.*, *GPflow: A Gaussian process library using TensorFlow*, J. Mach. Learn. Res. **18**(40), 1 (2017), arXiv:1610.08733.

[239] M. Blondel, Q. Berthet, M. Cuturi, *et al.*, *Efficient and modular implicit differentiation* (2021), arXiv:2105.15183.

[240] H.-Y. Huang, R. Kueng and J. Preskill, *Predicting many properties of a quantum system from very few measurements*, Nat. Phys. **16**(10), 1050 (2020), doi:10.1038/s41567-020-0932-7.

[241] C. E. Rasmussen and C. K. I. Williams, *Gaussian Processes for Machine Learning*, Adaptive Computation and Machine Learning. MIT Press, doi:10.7551/mitpress/3206.001.0001 (2005).

[242] R. V. Krems, *Bayesian machine learning for quantum molecular dynamics*, Phys. Chem. Chem. Phys. **21**(25), 13392 (2019), doi:10.1039/c9cp01883b.

[243] R. A. Vargas-Hernández and R. V. Krems, *Physical Extrapolation of Quantum Observables by Generalization with Gaussian Processes*, pp. 171–194, Springer International Publishing, Cham, doi:10.1007/978-3-030-40245-7_9 (2020).

[244] H.-Y. Huang, R. Kueng, G. Torlai, V. V. Albert and J. Preskill, *Provably efficient machine learning for quantum many-body problems*, Science **377**(6613) (2022), doi:10.1126/science.abk3333.

[245] P. A. M. Dirac and R. H. Fowler, *Quantum mechanics of many-electron systems*, Proc. R. Soc. A: Math. Phys. Eng. Sci. **123**(792), 714 (1929), doi:10.1098/rspa.1929.0094.

[246] G. Carleo and M. Troyer, *Solving the quantum many-body problem with artificial neural networks*, Science **355**(6325), 602 (2017), doi:10.1126/science.aag2302.

[247] K. Choo, A. Mezzacapo and G. Carleo, *Fermionic neural-network states for ab-initio electronic structure*, Nat. Commun. **11**(1), 2368 (2020), doi:10.1038/s41467-020-15724-9.

[248] H. Saito, *Solving the Bose–Hubbard model with machine learning*, J. Phys. Soc. Jpn. **86**(9), 093001 (2017), doi:10.7566/jpsj.86.093001.

[249] S. R. White, *Density matrix formulation for quantum renormalization groups*, Phys. Rev. Lett. **69**, 2863 (1992), doi:10.1103/PhysRevLett.69.2863.

[250] U. Schollwöck, *The density-matrix renormalization group in the age of matrix product states*, Ann. Phys. (N.Y.) **326**(1), 96 (2011), doi:10.1016/j.aop.2010.09.012.

[251] R. Orús, *A practical introduction to tensor networks: Matrix product states and projected entangled pair states*, Ann. Phys. (N.Y.) **349**, 117 (2014), doi:10.1016/j.aop.2014.06.013.

[252] M. V. den Nest, *Simulating quantum computers with probabilistic methods* (2010), arXiv:0911.1624.

[253] W. K. Hastings, *Monte Carlo sampling methods using Markov chains and their applications*, Biometrika **57**(1), 97 (1970), doi:10.2307/2334940.

[254] R. Jastrow, *Many-body problem with strong forces*, Phys. Rev. **98**, 1479 (1955), doi:10.1103/PhysRev.98.1479.

[255] E. Manousakis, *The spin-½ Heisenberg antiferromagnet on a square lattice and its application to the cuprous oxides*, Rev. Mod. Phys. **63**, 1 (1991), doi:10.1103/RevModPhys.63.1.

[256] T. Brown, B. Mann, N. Ryder, *et al.*, *Language models are few-shot learners*, In *NeurIPS 2020 – Adv. Neural Inf. Process. Syst.* (2020), arXiv:2005.14165.

[257] M. Caron, H. Touvron, I. Misra, *et al.*, *Emerging Properties in Self-Supervised Vision Transformers*, In *Proc. IEEE Int. Conf. Comput. Vis.*, pp. 9650–9660, doi:10.1109/ICCV48922.2021.00951 (2021).

[258] A. Nichol, P. Dhariwal, A. Ramesh, *et al.*, *GLIDE: Towards photorealistic image generation and editing with text-guided diffusion models* (2021), arXiv:2112.10741.
[259] A. Barra, A. Bernacchia, E. Santucci and P. Contucci, *On the equivalence of Hopfield networks and Boltzmann machines*, Neural Netw. **34**, 1 (2012), doi:10.1016/j.neunet.2012.06.003.
[260] G. Montufar, *Restricted Boltzmann machines: Introduction and review* (2018), arXiv:1806.07066.
[261] D.-L. Deng, X. Li and S. Das Sarma, *Quantum entanglement in neural network states*, Phys. Rev. X **7**(2) (2017), doi:10.1103/PhysRevX.7.021021.
[262] J. Chen, S. Cheng, H. Xie, L. Wang and T. Xiang, *Equivalence of restricted Boltzmann machines and tensor network states*, Phys. Rev. B **97**(8) (2018), doi:10.1103/PhysRevB.97.085104.
[263] X. Gao and L.-M. Duan, *Efficient representation of quantum many-body states with deep neural networks*, Nat. Commun. **8**(1) (2017), doi:10.1038/s41467-017-00705-2.
[264] D. Luo, G. Carleo, B. K. Clark and J. Stokes, *Gauge equivariant neural networks for quantum lattice gauge theories*, Phys. Rev. Lett. **127**, 276402 (2021), doi:10.1103/PhysRevLett.127.276402.
[265] A. Bansal, X. Chen, B. Russell, A. Gupta and D. Ramanan, *PixelNet: Representation of the pixels, by the pixels, and for the pixels* (2017), arXiv:1702.06506.
[266] M. Hibat-Allah, M. Ganahl, L. E. Hayward, R. G. Melko and J. Carrasquilla, *Recurrent neural network wave functions*, Phys. Rev. Res. **2**(2), 023358 (2020), doi:10.1103/PhysRevResearch.2.023358.
[267] M. Schmitt and M. Heyl, *Quantum many-body dynamics in two dimensions with artificial neural networks*, Phys. Rev. Lett. **125**, 100503 (2020), doi:10.1103/PhysRevLett.125.100503.
[268] C. Roth and A. H. MacDonald, *Group convolutional neural networks improve quantum state accuracy* (2021), arXiv:2104.05085.
[269] I. Glasser, N. Pancotti, M. August, I. D. Rodriguez and J. I. Cirac, *Neural-network quantum states, string-bond states, and chiral topological states*, Phys. Rev. X **8**, 011006 (2018), doi:10.1103/PhysRevX.8.011006.
[270] O. Sharir, A. Shashua and G. Carleo, *Neural tensor contractions and the expressive power of deep neural quantum states*, Phys. Rev. B **106**, 205136 (2022), doi:10.1103/PhysRevB.106.205136.
[271] Y. Levine, O. Sharir, N. Cohen and A. Shashua, *Quantum entanglement in deep learning architectures*, Phys. Rev. Lett. **122**, 065301 (2019), doi:10.1103/PhysRevLett.122.065301.
[272] P. Calabrese and J. Cardy, *Entanglement entropy and quantum field theory*, J. Stat. Mech. **2004**, P06002 (2004), doi:10.1088/1742-5468/2004/06/p06002.
[273] J. Eisert, M. Cramer and M. B. Plenio, *Colloquium: Area laws for the entanglement entropy*, Rev. Mod. Phys. **82**, 277 (2010), doi:10.1103/RevModPhys.82.277.
[274] M. Hibat-Allah, E. M. Inack, R. Wiersema, R. G. Melko and J. Carrasquilla, *Variational neural annealing*, Nat. Mach. Intell. **3**, 952 (2021), doi:10.1038/s42256-021-00401-3.
[275] K. Choo, G. Carleo, N. Regnault and T. Neupert, *Symmetries and many-body excitations with neural-network quantum states*, Phys. Rev. Lett. **121**, 167204 (2018), doi:10.1103/PhysRevLett.121.167204.

[276] A. Valenti, E. Greplova, N. H. Lindner and S. D. Huber, *Correlation-enhanced neural networks as interpretable variational quantum states*, Phys. Rev. Res. **4**, L012010 (2022), doi:10.1103/PhysRevResearch.4.L012010.

[277] G. Carleo, Y. Nomura and M. Imada, *Constructing exact representations of quantum many-body systems with deep neural networks*, Nat. Commun. **9**, 5322 (2018), doi:10.1038/s41467-018-07520-3.

[278] R. Kaubruegger, L. Pastori and J. C. Budich, *Chiral topological phases from artificial neural networks*, Phys. Rev. B **97**, 195136 (2018), doi:10.1103/PhysRevB.97.195136.

[279] Y. Zheng, H. He, N. Regnault and B. A. Bernevig, *Restricted Boltzmann machines and matrix product states of one-dimensional translationally invariant stabilizer codes*, Phys. Rev. B **99**, 155129 (2019), doi:10.1103/PhysRevB.99.155129.

[280] S. Lu, X. Gao and L.-M. Duan, *Efficient representation of topologically ordered states with restricted Boltzmann machines*, Phys. Rev. B **99**, 155136 (2019), doi:10.1103/PhysRevB.99.155136.

[281] Y. Huang and J. E. Moore, *Neural network representation of tensor network and chiral states*, Phys. Rev. Lett. **127**, 170601 (2021), doi:10.1103/PhysRevLett.127.170601.

[282] C.-Y. Park and M. J. Kastoryano, *Geometry of learning neural quantum states*, Phys. Rev. Res. **2**, 023232 (2020), doi:10.1103/PhysRevResearch.2.023232.

[283] S.-H. Lin and F. Pollmann, *Scaling of neural-network quantum states for time evolution*, Phys. Status Solidi B **259**(5), 2100172 (2022), doi:10.1002/pssb.202100172.

[284] F. Vicentini, D. Hofmann, A. Szabó, *et al.*, *NetKet 3: Machine learning toolbox for many-body quantum systems* (2021), arXiv:2112.10526.

[285] X. Yuan, S. Endo, Q. Zhao, Y. Li and S. C. Benjamin, *Theory of variational quantum simulation*, Quantum **3**, 191 (2019), doi:10.22331/q-2019-10-07-191.

[286] G. Carleo, F. Becca, M. Schiro and M. Fabrizio, *Localization and glassy dynamics of many-body quantum systems*, Sci. Rep. **2**, 243 (2012), doi:10.1038/10.1038/srep00243.

[287] I. L. Gutiérrez and C. B. Mendl, *Real time evolution with neural-network quantum states*, Quantum **6**, 627 (2022), doi:10.22331/q-2022-01-20-627.

[288] D. Hofmann, G. Fabiani, J. Mentink, G. Carleo and M. Sentef, *Role of stochastic noise and generalization error in the time propagation of neural-network quantum states*, SciPost Phys. **12**(5) (2022), doi:10.21468/scipostphys.12.5.165.

[289] S. Sorella, *Green function Monte Carlo with stochastic reconfiguration*, Phys. Rev. Lett. **80**, 4558 (1998), doi:10.1103/PhysRevLett.80.4558.

[290] F. Becca and S. Sorella, *Quantum Monte Carlo Approaches for Correlated Systems*, Cambridge University Press, doi:10.1017/9781316417041 (2017).

[291] D. Hangleiter, I. Roth, D. Nagaj and J. Eisert, *Easing the Monte Carlo sign problem*, Sci. Adv. **6**, eabb8341 (2020), doi:10.1126/sciadv.abb8341.

[292] D. Luo and B. K. Clark, *Backflow transformations via neural networks for quantum many-body wave functions*, Phys. Rev. Lett. **122**, 226401 (2019), doi:10.1103/PhysRevLett.122.226401.

[293] J. Hermann, Z. Schätzle and F. Noé, *Deep-neural-network solution of the electronic Schrödinger equation*, Nat. Chem. **12**, 891 (2020), doi:10.1038/s41557-020-0544-y.

[294] D. Pfau, J. S. Spencer, A. G. Matthews and W. M. C. Foulkes, *Ab initio solution of the many-electron Schrödinger equation with deep neural networks*, Phys. Rev. Res. **2**, 033429 (2020), doi:10.1103/PhysRevResearch.2.033429.
[295] J. Hermann, J. Spencer, K. Choo, *et al.*, *Ab-initio quantum chemistry with neural-network wavefunctions* Nat. Rev. Chem. **7**, 692–709 (2023), doi:10.1038/s41570-023-00516-8.
[296] S. B. Bravyi and A. Y. Kitaev, *Fermionic quantum computation*, Ann. Phys. **298**, 210 (2002), doi:10.1006/aphy.2002.6254.
[297] P. Jordan and E. Wigner, *Über das Paulische Äquivalenzverbot*, Zeitschrift für Physik **47**, 631 (1928), doi:10.1007/BF01331938.
[298] E. Zohar and J. I. Cirac, *Eliminating fermionic matter fields in lattice gauge theories*, Phys. Rev. B **98**, 075119 (2018), doi:10.1103/PhysRevB.98.075119.
[299] U. Borla, R. Verresen, F. Grusdt and S. Moroz, *Confined phases of one-dimensional spinless fermions coupled to Z_2 gauge theory*, Phys. Rev. Lett. **124**, 120503 (2020), doi:10.1103/PhysRevLett.124.120503.
[300] J. Nys and G. Carleo, *Variational solutions to fermion-to-qubit mappings in two spatial dimensions*, Quantum **6**, 833 (2022), doi:10.22331/q-2022-10-13-833.
[301] T. D. Barrett, A. Malyshev and A. I. Lvovsky, *Autoregressive neural-network wavefunctions for ab initio quantum chemistry*, Nat. Mach. Intell. **4**(4), 351 (2022), doi:10.1038/s42256-022-00461-z.
[302] B. Jonsson, B. Bauer and G. Carleo, *Neural-network states for the classical simulation of quantum computing* (2018), arXiv:1808.05232.
[303] M. Medvidović and G. Carleo, *Classical variational simulation of the quantum approximate optimization algorithm*, npj Quantum Inf. **7**, 101 (2021), doi:10.1038/s41534-021-00440-z.
[304] E. Farhi, J. Goldstone and S. Gutmann, *A quantum approximate optimization algorithm* (2014), arXiv:1411.4028.
[305] M. P. Harrigan, K. J. Sung, M. Neeley, *et al.*, *Quantum approximate optimization of non-planar graph problems on a planar superconducting processor*, Nat. Phys. **17**, 332 (2021), doi:10.1038/s41567-020-01105-y.
[306] J. Carrasquilla, D. Luo, F. Pérez, *et al.*, *Probabilistic simulation of quantum circuits using a deep-learning architecture*, Phys. Rev. A **104**, 032610 (2021), doi:10.1103/PhysRevA.104.032610.
[307] A. Vaswani, N. Shazeer, N. Parmar, *et al.*, *Attention is all you need*, In *Adv. Neural. Inf. Process. Syst.* (2017), arXiv:1706.03762.
[308] H.-P. Breuer and F. Petruccione, *The Theory of Open Quantum Systems*, Oxford University Press, doi:10.1093/acprof:oso/9780199213900.001.0001 (2007).
[309] N. Yoshioka and R. Hamazaki, *Constructing neural stationary states for open quantum many-body systems*, Phys. Rev. B **99**, 214306 (2019), doi:10.1103/PhysRevB.99.214306.
[310] A. Nagy and V. Savona, *Variational quantum Monte Carlo method with a neural-network ansatz for open quantum systems*, Phys. Rev. Lett. **122**, 250501 (2019), doi:10.1103/PhysRevLett.122.250501.
[311] F. Vicentini, A. Biella, N. Regnault and C. Ciuti, *Variational neural-network ansatz for steady states in open quantum systems*, Phys. Rev. Lett. **122**, 250503 (2019), doi:10.1103/PhysRevLett.122.250503.
[312] M. J. Hartmann and G. Carleo, *Neural-network approach to dissipative quantum many-body dynamics*, Phys. Rev. Lett. **122**, 250502 (2019), doi:10.1103/PhysRevLett.122.250502.

[313] D. Luo, Z. Chen, J. Carrasquilla and B. K. Clark, *Autoregressive neural network for simulating open quantum systems via a probabilistic formulation*, Phys. Rev. Lett. **128**, 090501 (2022), doi:10.1103/PhysRevLett.128.090501.

[314] M. Reh, M. Schmitt and M. Gärttner, *Time-dependent variational principle for open quantum systems with artificial neural networks*, Phys. Rev. Lett. **127**, 230501 (2021), doi:10.1103/PhysRevLett.127.230501.

[315] F. Minganti, A. Biella, N. Bartolo and C. Ciuti, *Spectral theory of Liouvillians for dissipative phase transitions*, Phys. Rev. A **98**, 042118 (2018), doi:10.1103/PhysRevA.98.042118.

[316] O. Gühne and G. Tóth, *Entanglement detection*, Physics Reports **474**(1), 1 (2009), doi:10.1016/j.physrep.2009.02.004.

[317] M. P. da Silva, O. Landon-Cardinal and D. Poulin, *Practical characterization of quantum devices without tomography*, Phys. Rev. Lett. **107**, 210404 (2011), doi:10.1103/PhysRevLett.107.210404.

[318] A. Tavakoli, *Semi-device-independent certification of independent quantum state and measurement devices*, Phys. Rev. Lett. **125**, 150503 (2020), doi:10.1103/PhysRevLett.125.150503.

[319] M. Kliesch and I. Roth, *Theory of quantum system certification*, PRX Quantum **2**, 010201 (2021), doi:10.1103/PRXQuantum.2.010201.

[320] N. Friis, G. Vitagliano, M. Malik and M. Huber, *Entanglement certification from theory to experiment*, Nat. Rev. Phys. **1**(1), 72 (2019), doi:10.1038/s42254-018-0003-5.

[321] J. Eisert, D. Hangleiter, N. Walk, *et al.*, *Quantum certification and benchmarking*, Nat. Rev. Phys. **2**(7), 382 (2020), doi:10.1038/s42254-020-0186-4.

[322] O. M. Sotnikov, I. A. Iakovlev, A. A. Iliasov, M. I. Katsnelson, A. A. Bagrov and V. V. Mazurenko, *Certification of quantum states with hidden structure of their bitstrings*, npj Quantum Inf. **8**(1), 41 (2022), doi:10.1038/s41534-022-00559-7.

[323] S. Chen, J. Li, B. Huang and A. Liu, *Tight bounds for quantum state certification with incoherent measurements*, In *2022 IEEE 63rd Annual Symposium on Foundations of Computer Science (FOCS)*, pp. 1205–1213. IEEE Computer Society, Los Alamitos, CA, USA, doi:10.1109/FOCS54457.2022.00118 (2022).

[324] A. Gočanin, I. Šupić and B. Dakić, *Sample-efficient device-independent quantum state verification and certification*, PRX Quantum **3**, 010317 (2022), doi:10.1103/PRXQuantum.3.010317.

[325] E.-C. Boghiu, F. Hirsch, P.-S. Lin, M. T. Quintino and J. Bowles, *Device-independent and semi-device-independent entanglement certification in broadcast Bell scenarios*, SciPost Phys. Core **6**, 028 (2023), doi:10.21468/SciPostPhysCore.6.2.028.

[326] D. Hangleiter, M. Kliesch, M. Schwarz and J. Eisert, *Direct certification of a class of quantum simulations*, Quantum Sci. Technol. **2**(1), 015004 (2017), doi:10.1088/2058-9565/2/1/015004.

[327] I. Frérot, M. Fadel and M. Lewenstein, *Probing quantum correlations in many-body systems: A review of scalable methods*, Rep. Prog. Phys. **86**(11), 114001 (2023), doi:10.1088/1361-6633/acf8d7.

[328] U. Leonhardt, *Quantum-state tomography and discrete Wigner function*, Phys. Rev. Lett. **74**, 4101 (1995), doi:10.1103/PhysRevLett.74.4101.

[329] A. G. White, D. F. V. James, P. H. Eberhard and P. G. Kwiat, *Nonmaximally entangled states: Production, characterization, and utilization*, Phys. Rev. Lett. **83**, 3103 (1999), doi:10.1103/PhysRevLett.83.3103.

[330] C. F. Roos, G. P. T. Lancaster, M. Riebe, *et al.*, *Bell states of atoms with ultralong lifetimes and their tomographic state analysis*, Phys. Rev. Lett. **92**, 220402 (2004), doi:10.1103/PhysRevLett.92.220402.
[331] H. Häffner, W. Hänsel, C. F. Roos, *et al.*, *Scalable multiparticle entanglement of trapped ions*, Nature **438**(7068), 643 (2005), doi:10.1038/nature04279.
[332] D. Gross, Y.-K. Liu, S. T. Flammia, S. Becker and J. Eisert, *Quantum state tomography via compressed sensing*, Phys. Rev. Lett. **105**, 150401 (2010), doi:10.1103/PhysRevLett.105.150401.
[333] D. Gross, *Recovering low-rank matrices from few coefficients in any basis*, IEEE Transactions on Information Theory **57**(3), 1548 (2011), doi:10.1109/TIT.2011.2104999.
[334] G. Tóth, W. Wieczorek, D. Gross, R. Krischek, C. Schwemmer and H. Weinfurter, *Permutationally invariant quantum tomography*, Phys. Rev. Lett. **105**, 250403 (2010), doi:10.1103/PhysRevLett.105.250403.
[335] T. Moroder, P. Hyllus, G. Tóth, *et al.*, *Permutationally invariant state reconstruction*, New Journal of Physics **14**(10), 105001 (2012), doi:10.1088/1367-2630/14/10/105001.
[336] M. Cramer, M. B. Plenio, S. T. Flammia, *et al.*, *Efficient quantum state tomography*, Nat. Commun. **1**(1) (2010), doi:10.1038/ncomms1147.
[337] T. Baumgratz, D. Gross, M. Cramer and M. B. Plenio, *Scalable reconstruction of density matrices*, Phys. Rev. Lett. **111**, 020401 (2013), doi:10.1103/PhysRevLett.111.020401.
[338] B. P. Lanyon, C. Maier, M. Holzäpfel, *et al.*, *Efficient tomography of a quantum many-body system*, Nat. Phys. **13**(12), 1158 (2017), doi:10.1038/nphys4244.
[339] A. Palmieri, E. Kovlakov, F. Bianchi, *et al.*, *Experimental neural network enhanced quantum tomography*, npj Quantum Inf. **6**, 20 (2020), doi:10.1038/s41534-020-0248-6.
[340] C. Pan and J. Zhang, *Deep learning-based quantum state tomography with imperfect measurement*, Int. J. Theor. Phys. **61**(9) (2022), doi:10.1007/s10773-022-05209-4.
[341] D. Koutný, L. Motka, Z. c. v. Hradil, J. Řeháček and L. L. Sánchez-Soto, *Neural-network quantum state tomography*, Phys. Rev. A **106**, 012409 (2022), doi:10.1103/PhysRevA.106.012409.
[342] S. Ahmed, C. Sánchez Muñoz, F. Nori and A. F. Kockum, *Quantum state tomography with conditional generative adversarial networks*, Phys. Rev. Lett. **127**, 140502 (2021), doi:10.1103/PhysRevLett.127.140502.
[343] H. Ma, Z. Sun, D. Dong, C. Chen and H. Rabitz, *Attention-based transformer networks for quantum state tomography* (2023), arXiv:2305.05433.
[344] A. M. Palmieri, G. Müller-Rigat, A. K. Srivastava, M. Lewenstein, G. Rajchel-Mieldzioć and M. Płodzień, *Enhancing quantum state tomography via resource-efficient attention-based neural networks* Phys. Rev. Res. **6**, 033248 (2024), doi:10.1103/PhysRevResearch.6.033248.
[345] G. Torlai, G. Mazzola, J. Carrasquilla, M. Troyer, R. Melko and G. Carleo, *Neural-network quantum state tomography*, Nat. Phys. **14**, 447 (2018), doi:10.1038/s41567-018-0048-5.
[346] A. Szabó and C. Castelnovo, *Neural network wave functions and the sign problem*, Phys. Rev. Res. **2**, 033075 (2020), doi:10.1103/PhysRevResearch.2.033075.
[347] T. Schmale, M. Reh and M. Gärttner, *Efficient quantum state tomography with convolutional neural networks*, npj Quantum Inf. **8**(1), 115 (2022), doi:10.1038/s41534-022-00621-4.

[348] G. Torlai, B. Timar, E. P. L. van Nieuwenburg, *et al.*, *Integrating neural networks with a quantum simulator for state reconstruction*, Phys. Rev. Lett. **123**, 230504 (2019), doi:10.1103/PhysRevLett.123.230504.

[349] S. Lohani, B. T. Kirby, M. Brodsky, O. Danaci and R. T. Glasser, *Machine learning assisted quantum state estimation*, Mach. Learn.: Sci. Technol. **1**(3), 035007 (2020), doi:10.1088/2632-2153/ab9a21.

[350] S. Lohani, T. A. Searles, B. T. Kirby and R. T. Glasser, *On the experimental feasibility of quantum state reconstruction via machine learning*, IEEE Trans. Quantum Eng. **2**, 1 (2021), doi:10.1109/TQE.2021.3106958.

[351] S. Lohani, J. M. Lukens, D. E. Jones, T. A. Searles, R. T. Glasser and B. T. Kirby, *Improving application performance with biased distributions of quantum states*, Phys. Rev. Res. **3**, 043145 (2021), doi:10.1103/PhysRevResearch.3.043145.

[352] S. Lohani, J. M. Lukens, R. T. Glasser, T. A. Searles and B. T. Kirby, *Data-centric machine learning in quantum information science*, Mach. Learn.: Sci. Technol. **3**(4), 04LT01 (2022), doi:10.1088/2632-2153/ac9036.

[353] O. Danaci, S. Lohani, B. T. Kirby and R. T. Glasser, *Machine learning pipeline for quantum state estimation with incomplete measurements*, Mach. Learn.: Sci. Technol. **2**(3), 035014 (2021), doi:10.1088/2632-2153/abe5f5.

[354] S. Aaronson, *Shadow tomography of quantum states*, In *Proc. 50th Annu. ACM SIGACT Symp. Theory Comput.*, STOC 2018, pp. 325–338. Association for Computing Machinery, New York, NY, USA, doi:10.1145/3188745.3188802 (2018).

[355] S. Aaronson and G. N. Rothblum, *Gentle measurement of quantum states and differential privacy*, In *Proc. 51st Annu. ACM SIGACT Symp. Theory Comput.*, STOC 2019, pp. 322–333. Association for Computing Machinery, New York, NY, USA, doi:10.1145/3313276.3316378 (2019).

[356] J. B. Altepeter, D. F. James and P. G. Kwiat, *4 qubit quantum state tomography*, In M. Paris and J. Řeháček, eds., *Quantum State Estimation*, pp. 113–145. Springer, Berlin, Heidelberg (2004).

[357] R. O'Donnell and J. Wright, *Efficient quantum tomography*, In *Proc. 48th Annu. ACM Symp. Theory Comput.*, STOC '16, p. 899–912. Association for Computing Machinery, New York, NY, USA, doi:10.1145/2897518.2897544 (2016).

[358] D. E. Koh and S. Grewal, *Classical Shadows with Noise*, Quantum **6**, 776 (2022), doi:10.22331/q-2022-08-16-776.

[359] A. Elben, R. Kueng, H.-Y. R. Huang, *et al.*, *Mixed-state entanglement from local randomized measurements*, Phys. Rev. Lett. **125**, 200501 (2020), doi:10.1103/PhysRevLett.125.200501.

[360] A. Elben, S. T. Flammia, H.-Y. Huang, *et al.*, *The randomized measurement toolbox*, Nat. Rev. Phys. **5**(1), 9 (2023), doi:10.1038/s42254-022-00535-2.

[361] K. Donatella, Z. Denis, A. Le Boité and C. Ciuti, *Dynamics with autoregressive neural quantum states: Application to critical quench dynamics*, Phys. Rev. A **108**(2), 022210 (2023).

[362] A. Sinibaldi, C. Giuliani, G. Carleo and F. Vicentini, *Unbiasing time-dependent Variational Monte Carlo by projected quantum evolution*, Quantum **7**, 1131 (2023), doi:10.22331/q-2023-10-10-1131.

[363] M. Kitagawa and M. Ueda, *Squeezed spin states*, Phys. Rev. A **47**, 5138 (1993), doi:10.1103/PhysRevA.47.5138.

[364] D. J. Wineland, J. J. Bollinger, W. M. Itano and D. J. Heinzen, *Squeezed atomic states and projection noise in spectroscopy*, Phys. Rev. A **50**, 67 (1994), doi:10.1103/PhysRevA.50.67.

[365] M. Płodzień, M. Kościelski, E. Witkowska and A. Sinatra, *Producing and storing spin-squeezed states and Greenberger-Horne-Zeilinger states in a one-dimensional optical lattice*, Phys. Rev. A **102**, 013328 (2020), doi:10.1103/PhysRevA.102.013328.

[366] M. Płodzień, M. Lewenstein, E. Witkowska and J. Chwedeńczuk, *One-axis twisting as a method of generating many-body Bell correlations*, Phys. Rev. Lett. **129**, 250402 (2022), doi:10.1103/PhysRevLett.129.250402.

[367] M. Płodzień, T. Wasak, E. Witkowska, M. Lewenstein and J. Chwedeńczuk, *Generation of scalable many-body Bell correlations in spin chains with short-range two-body interactions* Phys. Rev. Research **6**, 023050 (2024), doi:10.1103/PhysRevResearch.6.023050.

[368] T. Hernández Yanes, M. Płodzień, M. Mackoit Sinkevičienė, G. Žlabys, G. Juzeliūnas and E. Witkowska, *One- and two-axis squeezing via laser coupling in an atomic Fermi-Hubbard model*, Phys. Rev. Lett. **129**, 090403 (2022), doi:10.1103/PhysRevLett.129.090403.

[369] M. Dziurawiec, T. H. Yanes, M. Płodzień, M. Gajda, M. Lewenstein and E. Witkowska, *Accelerating many-body entanglement generation by dipolar interactions in the Bose-Hubbard model*, Phys. Rev. A **107**(1) (2023), doi:10.1103/PhysRevA.107.013311.

[370] T. Hernández Yanes, G. Žlabys, M. Płodzień, *et al.*, *Spin squeezing in open Heisenberg spin chains*, Phys. Rev. B **108**, 104301 (2023), doi:10.1103/PhysRevB.108.104301.

[371] C. Adams, G. Carleo, A. Lovato and N. Rocco, *Variational Monte Carlo calculations of $A \leq 4$ nuclei with an artificial neural-network correlator ansatz*, Phys. Rev. Lett. **127**, 022502 (2021), doi:10.1103/PhysRevLett.127.022502.

[372] J. Bausch and F. Leditzky, *Quantum codes from neural networks*, New J. Phys. **22**(2), 023005 (2020), doi:10.1088/1367-2630/ab6cdd.

[373] F. Vicentini, *Machine learning toolbox for quantum many body physics*, Nat. Rev. Phys. **3**, 156 (2021), doi:10.1038/s42254-021-00285-7.

[374] G. Carleo, *Beijing lecture notes and code*, Lecture Notes (2017).

[375] A. A. Melnikov, H. P. Nautrup, M. Krenn, *et al.*, *Active learning machine learns to create new quantum experiments*, Proc. Natl. Acad. Sci. U.S.A. **115**(6), 1221 (2018), doi:10.1073/pnas.1714936115.

[376] A. Fawzi, M. Balog, A. Huang, *et al.*, *Discovering faster matrix multiplication algorithms with reinforcement learning*, Nature **610**(7930), 47 (2022), doi:10.1038/s41586-022-05172-4.

[377] D. J. Mankowitz, A. Michi, A. Zhernov, *et al.*, *Faster sorting algorithms discovered using deep reinforcement learning*, Nature **618**(7964), 257 (2023), doi:10.1038/s41586-023-06004-9.

[378] S. R. Sutton and A. G. Barto, *Reinforcement Learning: An Introduction*, Bradford Book, doi:10.5555/980651.980663 (2018).

[379] R. S. Sutton, *Learning to predict by the methods of temporal differences*, Mach. Learn. **3**, 9 (1988), doi:10.1007/BF00115009.

[380] G. A. Rummery and M. Niranjan, *On-line Q-learning using connectionist systems*, CUED/F-INFENG/TR 166, University of Cambridge, Department of Engineering (1994).

[381] H. van Seijen, A. R. Mahmood, P. M. Pilarski, M. C. Machado and R. S. Sutton, *True online temporal-difference learning*, J. Mach. Learn. Res. **17**, 1 (2016), doi:10.48550/arXiv.1512.04087.

[382] G. H. John, *When the best move isn't optimal: Q-learning with exploration*, In *Proc. 12th Nat. Conf. Artif. Intell. (Vol. 2)* (1994).

[383] C. J. C. H. Watkins and P. Dayan, *Q-learning*, Mach. Learn. **8**, 279 (1992), doi:10.1007/BF00992698.

[384] J. E. Smith and R. L. Winkler, *The optimizer's curse: Skepticism and postdecision surprise in decision analysis*, Manag. Sci. **52**, 311 (2006), doi:10.1287/mnsc.1050.0451.

[385] S. Thrun and A. Schwartz, *Issues in using function approximation for reinforcement learning*, In *Proc. 4th Connectionist Models Summer School* (1993).

[386] H. Van Hasselt, *Double Q-learning*, In *NIPS 2010 – Adv. Neural Inf. Process. Syst.* (2010).

[387] L.-J. Lin, *Reinforcement Learning for Robots Using Neural Networks*, Ph.D. thesis, Carnegie Mellon University, USA, UMI Order No. GAX93-22750 (1992).

[388] H. Van Hasselt, A. Guez and D. Silver, *Deep reinforcement learning with double Q-learning*, In *Proc. AAAI Conf. Artif. Intell.* (2016), arXiv:1509.06461.

[389] P. Marbach and J. N. Tsitsiklis, *Simulation-based optimization of Markov reward processes*, IEEE Trans. Automat. Contr. **46**, 191 (2001), doi:10.1109/9.905687.

[390] R. S. Sutton, D. McAllester, S. Singh and Y. Mansour, *Policy gradient methods for reinforcement learning with function approximation*, In *NIPS 1999 – Adv. Neural Inf. Process. Syst.* (1999).

[391] R. J. Williams, *Simple statistical gradient-following algorithms for connectionist reinforcement learning*, Mach. Learn. **8**, 229 (1992), doi:10.1007/BF00992696.

[392] S. J. Rennie, E. Marcheret, Y. Mroueh, J. Ross and V. Goel, *Self-critical sequence training for image captioning*, In *Proc. IEEE Conf. Comput. Vision and Pattern Recognition*, doi:10.1109/CVPR.2017.131 (2017).

[393] J. Schulman, P. Moritz, S. Levine, M. Jordan and P. Abbeel, *High-dimensional continuous control using generalized advantage estimation*, In *ICLR 2016 – Int. Conf. Learn. Represent.* (2016), arXiv:1506.02438.

[394] A. G. Barto, R. S. Sutton and C. W. Anderson, *Neuronlike adaptive elements that can solve difficult learning control problems*, IEEE Trans. Syst. Man Cybern. Syst. **5**, 834 (1983), doi:10.1109/TSMC.1983.6313077.

[395] V. Konda and J. Tsitsiklis, *Actor-critic algorithms*, In *Adv. Neural Inf. Process. Syst.* (1999).

[396] T. Degris, M. White and R. S. Sutton, *Off-policy actor-critic*, In *ICML 2012 – Conf. Mach. Learn.*, pp. 179–186. Omnipress (2012), arXiv:1205.4839.

[397] V. Mnih, A. P. Badia, M. Mirza, *et al.*, *Asynchronous methods for deep reinforcement learning*, In *ICML 2016 – 33th Int. Conf. Mach. Learn.*, vol. 48, pp. 1928–1937 (2016), arXiv:1602.01783.

[398] S.-i. Amari, *Natural gradient works efficiently in learning*, Neural Comput. **10**(2), 251 (1998), doi:10.1162/089976698300017746.

[399] S. M. Kakade, *A natural policy gradient*, In *NIPS 2001 – Adv. Neural Inf. Process. Syst.* (2001).

[400] J. Peters and S. Schaal, *Natural actor-critic*, Neurocomputing **71**(7), 1180 (2008), doi:10.1016/j.neucom.2007.11.026.

[401] S. Bhatnagar, R. S. Sutton, M. Ghavamzadeh and M. Lee, *Natural actor–critic algorithms*, Automatica **45**(11), 2471 (2009), doi:10.1016/j.automatica.2009.07.008.

[402] J. Schulman, S. Levine, P. Abbeel, M. Jordan and P. Moritz, *Trust region policy optimization*, In *ICML 2015 – Int. Conf. Mach. Learn.* (2015), arXiv:1502.05477.

[403] Y. Wu, E. Mansimov, R. B. Grosse, S. Liao and J. Ba, *Scalable trust-region method for deep reinforcement learning using Kronecker-factored approximation*, In *NIPS 2017 – Adv. Neural Inf. Process. Syst.* (2017), arXiv:1708.05144.
[404] J. Schulman, F. Wolski, P. Dhariwal, A. Radford and O. Klimov, *Proximal policy optimization algorithms* (2017), arXiv:1707.06347.
[405] H. J. Briegel and G. De las Cuevas, *Projective simulation for artificial intelligence*, Sci. Rep. **2**(1), 1 (2012), doi:10.1038/srep00400.
[406] J. Mautner, A. Makmal, D. Manzano, M. Tiersch and H. J. Briegel, *Projective simulation for classical learning agents: A comprehensive investigation*, New Gener. Comput. **33**(1), 69 (2015), doi:10.1007/s00354-015-0102-0.
[407] A. A. Melnikov, A. Makmal and H. J. Briegel, *Benchmarking projective simulation in navigation problems*, IEEE Access **6**, 64639 (2018), doi:10.1109/ACCESS.2018.2876494.
[408] S. Jerbi, L. M. Trenkwalder, H. P. Nautrup, H. J. Briegel and V. Dunjko, *Quantum enhancements for deep reinforcement learning in large spaces*, PRX Quantum **2**(1), 010328 (2021), doi:10.1103/PRXQuantum.2.010328.
[409] W. L. Boyajian, J. Clausen, L. M. Trenkwalder, V. Dunjko and H. J. Briegel, *On the convergence of projective-simulation–based reinforcement learning in Markov decision processes*, Quantum Mach. Intell. **2**(2), 1 (2020), doi:10.1007/s42484-020-00023-9.
[410] A. A. Melnikov, A. Makmal, V. Dunjko and H. J. Briegel, *Projective simulation with generalization*, Sci. Rep. **7**(1), 1 (2017), doi:10.1038/s41598-017-14740-y.
[411] B. Eva, K. Ried, T. Müller, *et al.*, *How a minimal learning agent can infer the existence of unobserved variables in a complex environment Minds Mach.* **33**, 185–219 (2023), doi:10.1007/s11023-022-09619-5.
[412] M. Campbell, A. J. Hoane Jr and F.-h. Hsu, *Deep blue*, Artif. Intell. **134**(1–2), 57 (2002), doi:10.1016/S0004-3702(01)00129-1.
[413] A. Yee and M. Alvarado, *Pattern recognition and Monte-Carlo tree search for Go gaming better automation*, In *IBERAMIA 2012 – Adv. Artif. Intell.*, doi:10.1007/978-3-642-34654-5_2 (2012).
[414] R. Coulom, *Efficient selectivity and backup operators in Monte-Carlo tree search*, In *Int. Conf. Comput. Games*, pp. 72–83. Springer, doi:10.1007/978-3-540-75538-8_7 (2006).
[415] D. Silver, J. Schrittwieser, K. Simonyan, *et al.*, *Mastering the game of Go without human knowledge*, Nature **550**(7676), 354 (2017), doi:10.1038/nature24270.
[416] D. Silver, T. Hubert, J. Schrittwieser, *et al.*, *A general reinforcement learning algorithm that masters chess, shogi, and Go through self-play*, Science **362**(6419), 1140 (2018), doi:10.1126/science.aar6404.
[417] OpenAI, C. Berner, G. Brockman, *et al.*, *Dota 2 with large scale deep reinforcement learning* (2019), arXiv:1912.06680.
[418] W. H. Guss, C. Codel, K. Hofmann, *et al.*, *The MineRL 2019 competition on sample efficient reinforcement learning using human priors* (2019), arXiv:1904.10079.
[419] J. Schrittwieser, I. Antonoglou, T. Hubert, *et al.*, *Mastering Atari, Go, chess and shogi by planning with a learned model*, Nature **588**(7839), 604 (2020), doi:10.1038/s41586-020-03051-4.
[420] F. Marquardt, *Machine learning and quantum devices*, SciPost Phys. Lect. Notes p. 29 (2021), doi:10.21468/SciPostPhysLectNotes.29.

[421] R. Porotti, A. Essing, B. Huard and F. Marquardt, *A deep reinforcement learning for quantum state preparation with weak nonlinear measurements* Quantum **6**, 747 (2022), doi:10.22331/q-2022-06-28-747.
[422] T. Fösel, P. Tighineanu, T. Weiss and F. Marquardt, *Reinforcement learning with neural networks for quantum feedback*, Phys. Rev. X **8**, 031084 (2018), doi:10.1103/PhysRevX.8.031084.
[423] S. Borah, B. Sarma, M. Kewming, G. J. Milburn and J. Twamley, *Measurement-based feedback quantum control with deep reinforcement learning for a double-well nonlinear potential*, Phys. Rev. Lett. **127**, 190403 (2021), doi: 10.1103/PhysRevLett.127.190403.
[424] V. Nguyen, S. B. Orbell, D. T. Lennon, *et al.*, *Deep reinforcement learning for efficient measurement of quantum devices*, npj Quantum Inf. **7**(1), 100 (2021), doi:10.1038/s41534-021-00434-x.
[425] J. Preskill, *Quantum computing in the NISQ era and beyond*, Quantum **2**, 79 (2018), doi:10.22331/q-2018-08-06-79.
[426] T. Fösel, M. Yuezhen Niu, F. Marquardt and L. Li, *Quantum circuit optimization with deep reinforcement learning* (2021), arXiv:2103.07585.
[427] W. K. Wootters and W. H. Zurek, *A single quantum cannot be cloned*, Nature **299**(5886), 802 (1982), doi:10.1038/299802a0.
[428] P. W. Shor, *Scheme for reducing decoherence in quantum computer memory*, Phys. Rev. A **52**, R2493 (1995), doi:10.1103/PhysRevA.52.R2493.
[429] A. M. Steane, *Error correcting codes in quantum theory*, Phys. Rev. Lett. **77**, 793 (1996), doi:10.1103/PhysRevLett.77.793.
[430] D. Gottesman, *An introduction to quantum error correction and fault-tolerant quantum computation* (2009), arXiv:0904.2557.
[431] R. Sweke, M. S. Kesselring, E. P. L. van Nieuwenburg and J. Eisert, *Reinforcement learning decoders for fault-tolerant quantum computation*, Mach. Learn.: Sci. Technol. **2**(2), 025005 (2021), doi:10.1088/2632-2153/abc609.
[432] P. Andreasson, J. Johansson, S. Liljestrand and M. Granath, *Quantum error correction for the toric code using deep reinforcement learning*, Quantum **3**, 183 (2019), doi:10.22331/q-2019-09-02-183.
[433] D. Fitzek, M. Eliasson, A. F. Kockum and M. Granath, *Deep Q-learning decoder for depolarizing noise on the toric code*, Phys. Rev. Res. **2**, 023230 (2020), doi:10.1103/PhysRevResearch.2.023230.
[434] H. Théveniaut and E. van Nieuwenburg, *A NEAT quantum error decoder*, SciPost Phys. **11**, 5 (2021), doi:10.21468/SciPostPhys.11.1.005.
[435] M. Erhard, M. Krenn and A. Zeilinger, *Advances in high-dimensional quantum entanglement*, Nat. Rev. Phys. **2**(7), 365 (2020), doi:10.1038/s42254-020-0193-5.
[436] M. Krenn, M. Malik, R. Fickler, R. Lapkiewicz and A. Zeilinger, *Automated search for new quantum experiments*, Phys. Rev. Lett. **116**, 090405 (2016), doi:10.1103/PhysRevLett.116.090405.
[437] M. Krenn, J. S. Kottmann, N. Tischler and A. Aspuru-Guzik, *Conceptual understanding through efficient automated design of quantum optical experiments*, Phys. Rev. X **11**, 031044 (2021), doi:10.1103/PhysRevX.11.031044.
[438] M. Krenn, M. Erhard and A. Zeilinger, *Computer-inspired quantum experiments*, Nat. Rev. Phys. **2**, 649 (2020), doi:10.1038/s42254-020-0230-4.
[439] A. Peres, *Separability criterion for density matrices*, Phys. Rev. Lett. **77**(8), 1413 (1996), doi:10.1103/PhysRevLett.77.1413.

[440] B. Requena, G. Muñoz Gil, M. Lewenstein, V. Dunjko and J. Tura, *Certificates of quantum many-body properties assisted by machine learning*, Phys. Rev. Res. **5**, 013097 (2023), doi:10.1103/PhysRevResearch.5.013097.
[441] M. Bukov, A. G. R. Day, D. Sels, P. Weinberg, A. Polkovnikov and P. Mehta, *Reinforcement learning in different phases of quantum control*, Phys. Rev. X **8**, 031086 (2018), doi:10.1103/PhysRevX.8.031086.
[442] M. Y. Niu, S. Boixo, V. N. Smelyanskiy and H. Neven, *Universal quantum control through deep reinforcement learning*, npj Quantum Inf. **5**(33), 1 (2019), doi:10.1038/s41534-019-0141-3.
[443] K. A. McKiernan, E. Davis, M. S. Alam and C. Rigetti, *Automated quantum programming via reinforcement learning for combinatorial optimization* (2019), arXiv:1908.08054.
[444] Y.-H. Zhang, P.-L. Zheng, Y. Zhang and D.-L. Deng, *Topological quantum compiling with reinforcement learning*, Phys. Rev. Lett. **125**, 170501 (2020), doi:10.1103/PhysRevLett.125.170501.
[445] Y. Baum, M. Amico, S. Howell, *et al.*, *Experimental deep reinforcement learning for error-robust gate-set design on a superconducting quantum computer*, PRX Quantum **2**, 040324 (2021), doi:10.1103/PRXQuantum.2.040324.
[446] C. Cao, Z. An, S.-Y. Hou, D. L. Zhou and B. Zeng, *Quantum imaginary time evolution steered by reinforcement learning*, Commun. Phys. **5**(57), 1 (2022), doi:10.1038/s42005-022-00837-y.
[447] F. Metz and M. Bukov, *Self-correcting quantum many-body control using reinforcement learning with tensor networks*, Nat. Mach. Intell. **5**(7), 780 (2023), doi:10.1038/s42256-023-00687-5.
[448] Y. Qiu, M. Zhuang, J. Huang and C. Lee, *Efficient and robust entanglement generation with deep reinforcement learning for quantum metrology*, New J. Phys. **24**(8), 083011 (2022), doi:10.1088/1367-2630/ac8285.
[449] D. Silver, G. Lever, N. Heess, T. Degris, D. Wierstra and M. Riedmiller, *Deterministic policy gradient algorithms*, In *ICML 2014 – Int. Conf. Mach. Learn.*, doi:10.5555/3044805.3044850 (2014).
[450] S. Levine, *Reinforcement learning and control as probabilistic inference: Tutorial and review* (2018), arXiv:1805.00909.
[451] T. Haarnoja, A. Zhou, P. Abbeel and S. Levine, *Soft actor-critic: Off-policy maximum entropy deep reinforcement learning with a stochastic actor*, In *ICML 2018 – Int. Conf. Mach. Learn.* (2018), arXiv:1801.01290.
[452] A. Abdolmaleki, J. T. Springenberg, Y. Tassa, *et al.*, *Maximum a posteriori policy optimisation* (2018), arXiv:1806.06920.
[453] J. Degrave, F. Felici, J. Buchli, *et al.*, *Magnetic control of tokamak plasmas through deep reinforcement learning*, Nature **602**(7897), 414 (2022), doi:10.1038/s41586-021-04301-9.
[454] V. V. Sivak, A. Eickbusch, H. Liu, B. Royer, I. Tsioutsios and M. H. Devoret, *Model-free quantum control with reinforcement learning*, Phys. Rev. X **12**, 011059 (2022), doi:10.1103/PhysRevX.12.011059.
[455] V. V. Sivak, A. Eickbusch, B. Royer, *et al.*, *Real-time quantum error correction beyond break-even*, Nature **616**(7955), 50 (2023), doi:10.1038/s41586-023-05782-6.
[456] A. Karpathy, *Software 2.0*, Medium, Accessed: 04-08-2022 (2017).
[457] C. D. Schuman, T. E. Potok, R. M. Patton, *et al.*, *A survey of neuromorphic computing and neural networks in hardware* (2017), arXiv:1808.05232.

[458] K. Roy, A. Jaiswal and P. Panda, *Towards spike-based machine intelligence with neuromorphic computing*, Nature **575**(7784), 607 (2019), doi:10.1038/s41586-019-1677-2.

[459] M. Innes, A. Edelman, K. Fischer, *et al.*, *A differentiable programming system to bridge machine learning and scientific computing* (2019), arXiv:1907.07587.

[460] S. G. Johnson, *Notes on adjoint methods for 18.335*, Tech. rep., MIT, Introduction to Numerical Methods (2021).

[461] R. T. Chen, Y. Rubanova, J. Bettencourt and D. K. Duvenaud, *Neural ordinary differential equations*, In *NeurIPS 2018 – Adv. Neural Inf. Process. Syst.* (2018), arXiv:1806.07366.

[462] H.-J. Liao, J.-G. Liu, L. Wang and T. Xiang, *Differentiable programming tensor networks*, Phys. Rev. X **9**(3), 031041 (2019), doi:10.1103/PhysRevX.9.031041.

[463] B.-B. Chen, Y. Gao, Y.-B. Guo, *et al.*, *Automatic differentiation for second renormalization of tensor networks*, Phys. Rev. B **101**, 220409 (2020), doi:10.1103/PhysRevB.101.220409.

[464] G. Torlai, J. Carrasquilla, M. T. Fishman, R. G. Melko and M. P. A. Fisher, *Wavefunction positivization via automatic differentiation*, Phys. Rev. Res. **2**, 032060 (2020), doi:10.1103/PhysRevResearch.2.032060.

[465] J. Ingraham, A. Riesselman, C. Sander and D. Marks, *Learning protein structure with a differentiable simulator*, In *ICLR 2018 – Int. Conf. Learn. Represent.* (2018).

[466] S. S. Schoenholz and E. D. Cubuk, *JAX, M.D.: A framework for differentiable physics*, In *NeurIPS 2020 – Adv. Neural Inf. Process. Syst.* (2020), arXiv:1912.04232.

[467] T. Tamayo-Mendoza, C. Kreisbeck, R. Lindh and A. Aspuru-Guzik, *Automatic differentiation in quantum chemistry with applications to fully variational Hartree–Fock*, ACS Cent. Sci. **4**(5), 559 (2018), doi:10.1021/acscentsci.7b00586.

[468] L. Zhao and E. Neuscamman, *Excited state mean-field theory without automatic differentiation*, J. Chem. Phys. **152**(20), 204112 (2020), doi:10.1063/5.0003438.

[469] L. Li, S. Hoyer, R. Pederson, *et al.*, *Kohn-Sham equations as regularizer: Building prior knowledge into machine-learned physics*, Phys. Rev. Lett. **126**(3), 036401 (2021), doi:10.1103/PhysRevLett.126.036401.

[470] M. F. Kasim and S. M. Vinko, *Learning the exchange-correlation functional from nature with fully differentiable density functional theory*, Phys. Rev. Lett. **127**, 126403 (2021), doi:10.1103/PhysRevLett.127.126403.

[471] A. S. Abbott, B. Z. Abbott, J. M. Turney and H. F. Schaefer, *Arbitrary-order derivatives of quantum chemical methods via automatic differentiation*, J. Phys. Chem. Lett. **12**(12), 3232 (2021), doi:10.1021/acs.jpclett.1c00607.

[472] M. F. Kasim, S. Lehtola and S. M. Vinko, *DQC: A Python program package for differentiable quantum chemistry*, J. Chem. Phys. **156**(8), 084801 (2022), doi:10.1063/5.0076202.

[473] V. Bergholm, J. Izaac, M. Schuld, *et al.*, *PennyLane: Automatic differentiation of hybrid quantum-classical computations* (2020), arXiv:1811.04968.

[474] X. Zhang and G. K.-L. Chan, *Differentiable quantum chemistry with PySCF for molecules and materials at the mean-field level and beyond*, J. Chem. Phys. **157**(20) (2022), doi:10.1063/5.0118200, 204801.

[475] N. Yoshikawa and M. Sumita, *Automatic differentiation for the direct minimization approach to the Hartree–Fock method*, J. Phys. Chem. A **126**(45), 8487 (2022), doi:10.1021/acs.jpca.2c05922, PMID: 36346835.

[476] R. A. Vargas–Hernández, K. Jorner, R. Pollice and A. Aspuru–Guzik, *Inverse molecular design and parameter optimization with Hückel theory using automatic differentiation*, J. Chem. Phys. **158**(10) (2023), doi:10.1063/5.0137103.
[477] N. Khaneja, T. Reiss, C. Kehlet, T. Schulte-Herbrüggen and S. J. Glaser, *Optimal control of coupled spin dynamics: Design of NMR pulse sequences by gradient ascent algorithms*, J. Magn. Reson. **172**(2), 296 (2005), doi:10.1016/j.jmr.2004.11.004.
[478] N. Leung, M. Abdelhafez, J. Koch and D. Schuster, *Speedup for quantum optimal control from automatic differentiation based on graphics processing units*, Phys. Rev. A **95**(4) (2017), doi:10.1103/PhysRevA.95.042318.
[479] M. Abdelhafez, D. I. Schuster and J. Koch, *Gradient-based optimal control of open quantum systems using quantum trajectories and automatic differentiation*, Phys. Rev. A **99**, 052327 (2019), doi:10.1103/PhysRevA.99.052327.
[480] H. Jirari, *Optimal population inversion of a single dissipative two-level system*, Eur. Phys. J. B **92**(12), 265 (2019), doi:10.1140/epjb/e2019-100378-x.
[481] H. Jirari, *Time-optimal bang-bang control for the driven spin-boson system*, Phys. Rev. A **102**, 012613 (2020), doi:10.1103/PhysRevA.102.012613.
[482] F. Schäfer, M. Kloc, C. Bruder and N. Lörch, *A differentiable programming method for quantum control*, Mach. Learn.: Sci. Technol. **1**(3), 035009 (2020), doi:10.1088/2632-2153/ab9802.
[483] R. A. Vargas-Hernández, R. T. Q. Chen, K. A. Jung and P. Brumer, *Inverse design of dissipative quantum steady-states with implicit differentiation* (2020), arXiv:2011.12808.
[484] R. A. Vargas-Hernández, R. T. Q. Chen, K. A. Jung and P. Brumer, *Fully differentiable optimization protocols for non-equilibrium steady states*, New J. Phys. **23**(12), 123006 (2021), doi:10.1088/1367-2630/ac395e.
[485] I. Khait, J. Carrasquilla and D. Segal, *Optimal control of quantum thermal machines using machine learning*, Phys. Rev. Res. **4**, L012029 (2022), doi:10.1103/PhysRevResearch.4.L012029.
[486] L. Coopmans, D. Luo, G. Kells, B. K. Clark and J. Carrasquilla, *Protocol discovery for the quantum control of majoranas by differentiable programming and natural evolution strategies*, PRX Quantum **2**(2), 020332 (2021), doi:10.1103/PRXQuantum.2.020332.
[487] F. Schäfer, P. Sekatski, M. Koppenhöfer, C. Bruder and M. Kloc, *Control of stochastic quantum dynamics by differentiable programming*, Mach. Learn.: Sci. Technol. **2**(3), 035004 (2021), doi:10.1088/2632-2153/abec22.
[488] M. H. Goerz, S. C. Carrasco and V. S. Malinovsky, *Quantum optimal control via semi-automatic differentiation*, Quantum **6**, 871 (2022), doi:10.22331/q-2022-12-07-871.
[489] X.-Z. Luo, J.-G. Liu, P. Zhang and L. Wang, *Yao. jl: Extensible, efficient framework for quantum algorithm design*, Quantum **4**, 341 (2020), doi:10.22331/q-2020-10-11-341.
[490] O. Kyriienko, A. E. Paine and V. E. Elfving, *Solving nonlinear differential equations with differentiable quantum circuits*, Phys. Rev. A **103**(5), 052416 (2021), doi:10.1103/PhysRevA.103.052416.
[491] P. Huembeli and A. Dauphin, *Characterizing the loss landscape of variational quantum circuits*, Quantum Sci. Technol. **6**(2), 025011 (2021), doi:10.1088/2058-9565/abdbc9.

[492] A. G. Baydin, B. A. Pearlmutter, A. A. Radul and J. M. Siskind, *Automatic differentiation in machine learning: A survey*, J. Mach. Learn. Res. **18**(1), 5595–5637 (2018), doi:10.5555/3122009.3242010.

[493] R. E. Wengert, *A simple automatic derivative evaluation program*, Commun. ACM **7**(8), 463 (1964), doi:10.1145/355586.364791.

[494] A. Griewank and A. Walther, *Evaluating Derivatives: Principles and Techniques of Algorithmic Differentiation*, Society for Industrial and Applied Mathematics, Philadelphia, PA, doi:10.1137/1.9780898717761 (2008).

[495] S. Linnainmaa, *The representation of the cumulative rounding error of an algorithm as a Taylor expansion of the local rounding errors*, Ph.D. thesis, Univ. Helsinki, Finland, Master's Thesis (1970).

[496] A. Griewank, *Who invented the reverse mode of differentiation*, Documenta Math., Accessed: 04-01-2022 (2012).

[497] C. Rackauckas, *Parallel computing and scientific machine learning*, 18.337J/6.338J Lecture notes, MIT Lecture (2020).

[498] Y. Ma, V. Dixit, M. Innes, X. Guo and C. Rackauckas, *A comparison of automatic differentiation and continuous sensitivity analysis for derivatives of differential equation solutions* (2021), arXiv:1812.01892.

[499] L. Wang, *Implementation of an inverse Schrödinger problem in JAX*, Available as Google Colab Notebook: https://colab.research.google.com/drive/1e1NFA-E1Th7nN_9-DzQjAaglH6bwZtVU?usp=sharing (2021).

[500] H. Xie, J.-G. Liu and L. Wang, *Automatic differentiation of dominant eigensolver and its applications in quantum physics*, Phys. Rev. B **101**, 245139 (2020), doi:10.1103/PhysRevB.101.245139.

[501] L. Wang, *Implementation of a quantum optimal control problem in JAX*, Accessed: 03-11-2022. Available as Google Colab Notebook (2021).

[502] Q. Wang, R. Hu and P. Blonigan, *Least squares shadowing sensitivity analysis of chaotic limit cycle oscillations*, J. Comput. Phys. **267**, 210 (2014), doi:10.1016/j.jcp.2014.03.002.

[503] L. Metz, C. D. Freeman, S. S. Schoenholz and T. Kachman, *Gradients are not all you need* (2021), arXiv:2111.05803.

[504] W. S. Moses and V. Churavy, *Instead of rewriting foreign code for machine learning, automatically synthesize fast gradients*, In *NeurIPS 2020 – Adv. Neural Inf. Process. Syst.* (2020), arXiv:2010.01709.

[505] J.-G. Liu and T. Zhao, *Differentiate everything with a reversible embedded domain-specific language* (2020), arXiv:2003.04617.

[506] B. Silverman, *Density Estimation for Statistics and Data Analysis*, Chapman & Hall/CRC Monographs on Statistics & Applied Probability. Taylor & Francis, Boca Raton doi:10.1201/9781315140919 (1998).

[507] Z. Hradil, *Quantum-state estimation*, Phys. Rev. A **55**, R1561 (1997), doi:10.1103/PhysRevA.55.R1561.

[508] M. Paris and J. Rehacek, *Quantum State Estimation*, Lecture Notes in Physics. Springer-Verlag, Berlin/Heidelberg, doi:10.1007/b98673 (2004).

[509] Y. S. Teo, *Introduction to Quantum-State Estimation*, World Scientific, Singapore, doi:10.1142/9617 (2015).

[510] S. Kullback and R. A. Leibler, *On information and sufficiency*, Ann. Math. Stat. **22**(1), 79 (1951), doi:10.1214/aoms/1177729694.

[511] D. C. Hackett, C.-C. Hsieh, M. S. Albergo, *et al.*, *Flow-based sampling for multimodal distributions in lattice field theory* (2021), arXiv:2107.00734.

[512] K. A. Nicoli, C. Anders, L. Funcke, *et al.*, *Machine learning of thermodynamic observables in the presence of mode collapse* (2021), arXiv:2111.11303.
[513] R. G. Melko, G. Carleo, J. Carrasquilla and J. I. Cirac, *Restricted Boltzmann machines in quantum physics*, Nat. Phys. **15**(9), 887 (2019), doi:10.1038/s41567-019-0545-1.
[514] A. Van den Oord, N. Kalchbrenner, L. Espeholt, O. Vinyals, A. Graves and K. Kavukcuoglu, *Conditional image generation with PixelCNN decoders*, In *NIPS 2016: Adv. Neural Inf. Process. Syst.* (2016), arXiv:1606.05328.
[515] D. P. Kingma and M. Welling, *An introduction to variational autoencoders*, Found. Trends Mach. Learn. **12**(4), 307 (2019), doi:10.1561/2200000056.
[516] I. Goodfellow, J. Pouget-Abadie, M. Mirza, *et al.*, *Generative adversarial nets*, In *NIPS 2014 – Adv. Neural Inf. Process. Syst.* (2014), arXiv:1406.2661.
[517] A. Creswell, T. White, V. Dumoulin, K. Arulkumaran, B. Sengupta and A. A. Bharath, *Generative adversarial networks: An overview*, IEEE Signal Process. Mag. **35**(1), 53 (2018), doi:10.1109/MSP.2017.2765202.
[518] E. G. Tabak and E. Vanden-Eijnden, *Density estimation by dual ascent of the log-likelihood*, Commun. Math. Sci. **8**(1), 217 (2010), doi:10.4310/CMS.2010.v8.n1.a11.
[519] G. Papamakarios, E. Nalisnick, D. J. Rezende, S. Mohamed and B. Lakshminarayanan, *Normalizing flows for probabilistic modeling and inference*, J. Mach. Learn. Res. **22**(57), 1 (2021), arXiv:1912.02762.
[520] I. Kobyzev, S. Prince and M. Brubaker, *Normalizing flows: An introduction and review of current methods*, IEEE Trans. Pattern Anal. Mach. Intell. **11**, 3964 (2021), doi:10.1109/TPAMI.2020.2992934.
[521] J. Ho, A. Jain and P. Abbeel, *Denoising diffusion probabilistic models*, In *NeurIPS 2020 – Adv. Neural Inf. Process Syst.* (2020), arXiv:2006.11239.
[522] L. Yang, Z. Zhang and S. Hong, *Diffusion models: A comprehensive survey of methods and applications* (2022), arXiv:2209.00796.
[523] F. Noé, S. Olsson, J. Köhler and H. Wu, *Boltzmann generators: Sampling equilibrium states of many-body systems with deep learning*, Science **365**(6457), eaaw1147 (2019), doi:10.1126/science.aaw1147.
[524] K. A. Nicoli, C. J. Anders, L. Funcke, *et al.*, *Estimation of thermodynamic observables in lattice field theories with deep generative models*, Phys. Rev. Lett. **126**(3), 032001 (2021), doi:10.1103/PhysRevLett.126.032001.
[525] M. S. Albergo, G. Kanwar and P. E. Shanahan, *Flow-based generative models for Markov chain Monte Carlo in lattice field theory*, Phys. Rev. D **100**, 034515 (2019), doi:10.1103/PhysRevD.100.034515.
[526] M. Gabrié, G. M. Rotskoff and E. Vanden-Eijnden, *Adaptive Monte Carlo augmented with normalizing flows*, Proc. Natl. Acad. Sci. U.S.A. **119**(10), e2109420119 (2022), doi:10.1073/pnas.2109420119.
[527] G. E. Hinton, *A practical guide to training restricted Boltzmann machines*, In G. Montavon, G. B. Orr and K.-R. Müller, eds., *Neural Networks: Tricks of the Trade: Second Edition*, Lecture Notes in Computer Science, pp. 599–619. Springer, Berlin, Heidelberg, doi:10.1007/978-3-642-35289-8_32 (2012).
[528] M. Gabrié, E. W. Tramel and F. Krzakala, *Training restricted Boltzmann machines via the Thouless-Anderson-Palmer free energy*, In *NIPS 2015 – Adv. Neural Inf. Process. Syst.* (2015), arXiv:1506.02914.
[529] P. Ramachandran, T. L. Paine, P. Khorrami, *et al.*, *Fast generation for convolutional autoregressive models* (2017), arXiv:1704.06001.

[530] A. van den Oord, N. Kalchbrenner and K. Kavukcuoglu, *Pixel recurrent neural networks*, In *ICML 2016 – Int. Conf. Mach. Learn.* (2016), arXiv:1601.06759.

[531] M. Cristoforetti, G. Jurman, A. I. Nardelli and C. Furlanello, *Towards meaningful physics from generative models* (2019), arXiv:1705.09524.

[532] G. Kanwar, M. S. Albergo, D. Boyda, *et al.*, *Equivariant flow-based sampling for lattice gauge theory*, Phys. Rev. Lett. **125**(12), 121601 (2020), doi:10.1103/PhysRevLett.125.121601.

[533] K. A. Nicoli, *Deep generative models for thermodynamics of spin systems and field theories*, Ph.D. thesis, Technische Universität Berlin, Fakultät IV, Maschinelles Lernen, doi:10.14279/depositonce-17052 (2023).

[534] L. Dinh, D. Krueger and Y. Bengio, *NICE: Non-linear independent components estimation*, In *ICLR 2015 – Int. Conf. Learn. Represent.* (2015), arXiv:1410.8516.

[535] L. Dinh, J. Sohl-Dickstein and S. Bengio, *Density estimation using real NVP*, In *ICLR 2017 – Int. Conf. Learn. Represent.* (2017), arXiv:1605.08803.

[536] D. P. Kingma and P. Dhariwal, *Glow: Generative flow with invertible 1x1 convolutions*, In *NeurIPS 2018 – Adv. Neural Inf. Process Syst.* (2018), arXiv:1807.03039.

[537] C. Durkan, A. Bekasov, I. Murray and G. Papamakarios, *Neural spline flows*, In *NeurIPS 2019 – Adv. Neural Inf. Process Syst.* (2019), arXiv:1906.04032.

[538] L. Grenioux, A. Durmus, É. Moulines and M. Gabrié, *On sampling with approximate transport maps* In ICML 2023 – 40th Int. Conf. Mach. Learn., vol. 202, pp. 11698–11733. PMLR (2023), arXiv:2302.04763.

[539] T. Müller, B. Mcwilliams, F. Rousselle, M. Gross and J. Novák, *Neural importance sampling*, ACM Trans. Graph. **38**(5), 1 (2019), doi:10.1145/3341156.

[540] S. Bacchio, P. Kessel, S. Schaefer and L. Vaitl, *Learning trivializing gradient flows for lattice gauge theories*, Phys. Rev. D **107**, L051504 (2023), doi:10.1103/PhysRevD.107.L051504.

[541] L. Del Debbio, J. M. Rossney and M. Wilson, *Machine learning trivializing maps: A first step towards understanding how flow-based samplers scale up* (2021), arXiv:2112.15532.

[542] L. Del Debbio, J. Marsh Rossney and M. Wilson, *Efficient modeling of trivializing maps for lattice ϕ^4 theory using normalizing flows: A first look at scalability*, Phys. Rev. D **104**, 094507 (2021), doi:10.1103/PhysRevD.104.094507.

[543] R. Abbott, M. S. Albergo, A. Botev, *et al.*, *Aspects of scaling and scalability for flow-based sampling of lattice QCD*, Eur. Phys. J. A **59**, 257 (2023), doi:10.1140/epja/s10050-023-01154-w.

[544] G. Kanwar, M. S. Albergo, D. Boyda, *et al.*, *Equivariant flow-based sampling for lattice gauge theory*, Phys. Rev. Lett. **125**, 121601 (2020), doi:10.1103/PhysRevLett.125.121601.

[545] D. Boyda, G. Kanwar, S. Racanière, *et al.*, *Sampling using* $\mathrm{SU}(n)$ *gauge equivariant flows*, Phys. Rev. D **103**, 074504 (2021), doi:10.1103/PhysRevD.103.074504.

[546] J. Köhler, L. Klein and F. Noé, *Equivariant flows: Exact likelihood generative learning for symmetric densities*, In *ICML 2020 – Int. Conf. Mach. Learn.* (2020), arXiv:2006.02425.

[547] V. G. Satorras, E. Hoogeboom, F. B. Fuchs, I. Posner and M. Welling, *$E(n)$ equivariant normalizing flows*, In *NeurIPS 2021 – Adv. Neural Inf. Process Syst.* (2021), arXiv:2105.09016.

[548] G. Jerfel, S. Wang, C. Wong-Fannjiang, K. A. Heller, Y. Ma and M. I. Jordan, *Variational refinement for importance sampling using the forward Kullback-Leibler divergence*, In *PLMR 2021 – Proc. Mach. Learn. Res.* (2021), arXiv:2106.15980.

[549] K. A. Nicoli, C. J. Anders, T. Hartung, K. Jansen, P. Kessel and S. Nakajima, *Detecting and mitigating mode-collapse for flow-based sampling of lattice field theories*, Phys. Rev. D **108**, 114501 (2023), doi:10.1103/PhysRevD.108.114501.
[550] L. Vaitl, K. A. Nicoli, S. Nakajima and P. Kessel, *Gradients should stay on path: Better estimators of the reverse- and forward KL divergence for normalizing flows*, Mach. Learn.: Sci. Technol. **3**(4), 045006 (2022), doi:10.1088/2632-2153/ac9455.
[551] L. Vaitl, K. A. Nicoli, S. Nakajima and P. Kessel, *Path-gradient estimators for continuous normalizing flows*, In *PLMR 2022 – Proc. Mach. Learn. Res.* (2022), arXiv:2206.09016.
[552] M. Arbel, A. Matthews and A. Doucet, *Annealed flow transport Monte Carlo*, In *PLMR 2021 – Proc. Mach. Learn. Res.* (2021), arXiv:2102.07501.
[553] L. I. Midgley, V. Stimper, G. N. Simm, B. Schölkopf and J. M. Hernández-Lobato, *Flow annealed importance sampling bootstrap* (2022), arXiv:2208.01893.
[554] A. Matthews, M. Arbel, D. J. Rezende and A. Doucet, *Continual repeated annealed flow transport Monte Carlo*, In *PLMR 2022 – Proc. Mach. Learn. Res.* (2022), arXiv:2201.13117.
[555] L. Wang, *Generative models for physicists*, Tech. rep., Institute of Physics, Chinese Academy of Sciences, GitHub.io (2018).
[556] M. S. Albergo, G. Kanwar, S. Racanière, *et al.*, *Flow-based sampling for fermionic lattice field theories*, Phys. Rev. D **104**, 114507 (2021), doi:10.1103/PhysRevD.104.114507.
[557] E. Lustig, O. Yair, R. Talmon and M. Segev, *Identifying topological phase transitions in experiments using manifold learning*, Phys. Rev. Lett. **125**(12), 127401 (2020), doi:10.1103/PhysRevLett.125.127401.
[558] E. Greplova, C. Gold, B. Kratochwil, *et al.*, *Fully automated identification of two-dimensional material samples*, Phys. Rev. Appl. **13**(6), 064017 (2020), doi:10.1103/PhysRevApplied.13.064017.
[559] R. Durrer, B. Kratochwil, J. V. Koski, *et al.*, *Automated tuning of double quantum dots into specific charge states using neural networks*, Phys. Rev. Appl. **13**(5), 054019 (2020), doi:10.1103/PhysRevApplied.13.054019.
[560] E. Greplova and S. Huber group, GitHub repository to "Fully automated search for 2D material samples" (2019).
[561] P. Mostosi, H. Schindelin, P. Kollmannsberger and A. Thorn, *Haruspex: A neural network for the automatic identification of oligonucleotides and protein secondary structure in cryo-electron microscopy maps*, Angew. Chem. Int. Ed. **59**(35), 14788 (2020), doi:10.1002/anie.202000421.
[562] K. Nolte, Y. Gao, S. Stäb, P. Kollmannsberger and A. Thorn, *Detecting ice artefacts in processed macromolecular diffraction data with machine learning*, Acta Crystallogr. D **78**(2), 187 (2022), doi:10.1107/S205979832101202X.
[563] A. U. Lode, R. Lin, M. Büttner, *et al.*, *Optimized observable readout from single-shot images of ultracold atoms via machine learning*, Phys. Rev. A **104**(4), L041301 (2021), doi:10.1103/PhysRevA.104.L041301.
[564] R. Lin, C. Georges, J. Klinder, *et al.*, *Mott transition in a cavity-boson system: A quantitative comparison between theory and experiment*, SciPost Phys. **11**(2), 030 (2021), doi:10.21468/SciPostPhys.11.2.030.
[565] R. Lin, P. Molignini, L. Papariello, *et al.*, *MCTDH-X: The multiconfigurational time-dependent Hartree method for indistinguishable particles software*, Quantum Sci. Technol. **5**(2), 024004 (2020), doi:10.1088/2058-9565/ab788b.

[566] J. Zhang, G. Pagano, P. W. Hess, *et al.*, *Observation of a many-body dynamical phase transition with a 53-qubit quantum simulator*, Nature **551**(7682), 601 (2017), doi:10.1038/nature24654.

[567] H. Bernien, S. Schwartz, A. Keesling, *et al.*, *Probing many-body dynamics on a 51-atom quantum simulator*, Nature **551**(7682), 579 (2017), doi:10.1038/nature24622.

[568] B. Chiaro, C. Neill, A. Bohrdt, *et al.*, *Direct measurement of nonlocal interactions in the many-body localized phase*, Phys. Rev. Res. **4**, 013148 (2022), doi:10.1103/PhysRevResearch.4.013148.

[569] M. Rispoli, A. Lukin, R. Schittko, *et al.*, *Quantum critical behaviour at the many-body localization transition*, Nature **573**(7774), 385 (2019), doi:10.1038/s41586-019-1527-2.

[570] A. Valenti, G. Jin, J. Léonard, S. D. Huber and E. Greplova, *Scalable Hamiltonian learning for large-scale out-of-equilibrium quantum dynamics*, Phys. Rev. A **105**, 023302 (2022), doi:10.1103/PhysRevA.105.023302.

[571] A. Gresch, L. Bittel and M. Kliesch, *Scalable approach to many-body localization via quantum data* (2022), arXiv:2202.08853.

[572] V. Gebhart, R. Santagati, A. A. Gentile, *et al.*, *Learning quantum systems*, Nat. Rev. Phys. **5**(3), 141 (2023), doi:10.1038/s42254-022-00552-1.

[573] A. Cervera-Lierta, M. Krenn and A. Aspuru-Guzik, *Design of quantum optical experiments with logic artificial intelligence*, Quantum **6**, 836 (2022), doi:10.22331/q-2022-10-13-836.

[574] D. Flam-Shepherd, T. Wu, X. Gu, A. Cervera-Lierta, M. Krenn and A. Aspuru-Guzik, *Learning interpretable representations of entanglement in quantum optics experiments using deep generative models*, Nat. Mach. Intell. **4**, 544–554 (2022), doi:10.1038/s42256-022-00493-5.

[575] R. D. King, J. Rowland, S. G. Oliver, *et al.*, *The automation of science*, Science **324**(5923), 85 (2009), doi:10.1126/science.1165620.

[576] F. Häse, L. M. Roch and A. Aspuru-Guzik, *Next-generation experimentation with self-driving laboratories*, Trends Chem. **1**(3), 282 (2019), doi:10.1016/j.trechm.2019.02.007.

[577] A. A. Gentile, B. Flynn, S. Knauer, *et al.*, *Learning models of quantum systems from experiments*, Nat. Phys. **17**(7), 837 (2021), doi:10.1038/s41567-021-01201-7.

[578] T. E. O'Brien, L. B. Ioffe, Y. Su, *et al.*, *Quantum computation of molecular structure using data from challenging-to-classically-simulate nuclear magnetic resonance experiments*, PRX Quantum **3**, 030345 (2022), doi:10.1103/PRXQuantum.3.030345.

[579] N. M. van Esbroeck, D. T. Lennon, H. Moon, *et al.*, *Quantum device fine-tuning using unsupervised embedding learning*, New J. Phys. **22**(9), 095003 (2020), doi:10.1088/1367-2630/abb64c.

[580] B. Severin, D. T. Lennon, L. C. Camenzind, *et al.*, *Cross-architecture tuning of silicon and SiGe-based quantum devices using machine learning*, Sci. Rep. **14**, 17281 (2024), doi:10.1038/s41598-024-67787-z.

[581] J. P. Zwolak, T. McJunkin, S. S. Kalantre, *et al.*, *Ray-based framework for state identification in quantum dot devices*, PRX Quantum **2**, 020335 (2021), doi:10.1103/PRXQuantum.2.020335.

[582] A. Dawid, N. Bigagli, D. W. Savin and S. Will, *Automated graph-based detection of quantum control schemes: Application to molecular laser cooling*, Phys. Rev. Research **7**, 013135 (2025), doi:10.1103/PhysRevResearch.7.013135.

[583] N. Wiebe, A. Kapoor and K. M. Svore, *Quantum algorithms for nearest-neighbor methods for supervised and unsupervised learning*, Quantum Inf. Comput. **15**(3–4), 316 (2015), doi:10.26421/QIC15.3-4-7.
[584] P. Raccuglia, K. C. Elbert, P. D. F. Adler, *et al.*, *Machine-learning-assisted materials discovery using failed experiments*, Nature **533**(7601), 73 (2016), doi:10.1038/nature17439.
[585] L. Zdeborová, *Understanding deep learning is also a job for physicists*, Nat. Phys. **16**(6), 602 (2020), doi:10.1038/s41567-020-0929-2.
[586] T. M. Cover, *Geometrical and statistical properties of systems of linear inequalities with applications in pattern recognition*, IEEE Trans. Comput. **EC-14**(3), 326 (1965), doi:10.1109/PGEC.1965.264137.
[587] E. Gardner, *Maximum storage capacity in neural networks*, EPL **4**(4), 481 (1987), doi:10.1209/0295-5075/4/4/016.
[588] D. J. Amit, H. Gutfreund and H. Sompolinsky, *Storing infinite numbers of patterns in a spin-glass model of neural networks*, Phys. Rev. Lett. **55**(14), 1530 (1985), doi:10.1103/PhysRevLett.55.1530.
[589] B. Derrida, E. Gardner and A. Zippelius, *An exactly solvable asymmetric neural network model*, EPL **4**(2), 167 (1987), doi:10.1209/0295-5075/4/2/007.
[590] B. Derrida and J. P. Nadal, *Learning and forgetting on asymmetric, diluted neural networks*, J. Stat. Phys. **49**(5-6), 993 (1987), doi:10.1007/BF01017556.
[591] C. Peterson, *A mean field theory learning algorithm for neural networks*, Complex Syst. **1**, 995 (1987).
[592] W. Krauth and M. Mézard, *Storage capacity of memory networks with binary couplings*, J. Phys. **50**(20), 3057 (1989), doi:10.1051/jphys:0198900500200305700.
[593] G. Györgyi, *First-order transition to perfect generalization in a neural network with binary synapses*, Phys. Rev. A **41**(12), 7097 (1990), doi:10.1103/PhysRevA.41.7097.
[594] M. Opper and D. Haussler, *Generalization performance of Bayes optimal classification algorithm for learning a perceptron*, Phys. Rev. Lett. **66**(20), 2677 (1991), doi:10.1103/PhysRevLett.66.2677.
[595] D. Sherrington and S. Kirkpatrick, *Solvable model of a spin-glass*, Phys. Rev. Lett. **35**(26), 1792 (1975), doi:10.1103/PhysRevLett.35.1792.
[596] M. Mézard, G. Parisi and M. A. Virasoro, *SK model: The replica solution without replicas*, EPL **1**(2), 77 (1986), doi:10.1209/0295-5075/1/2/006.
[597] F. Barahona, *On the computational complexity of Ising spin glass models*, J. Phys. A: Math. Gen. **15**, 3241 (1982), doi:10.1088/0305-4470/15/10/028.
[598] C. Fan, M. Shen, Z. Nussinov, *et al. Searching for spin glass ground states through deep reinforcement learning*. Nat. Commun. **14**, 725 (2023), doi:10.1038/s41467-023-36363-w.
[599] S. F. Edwards and P. W. Anderson, *Theory of spin glasses*, J. Phys. F: Met. Phys. **5**, 965 (1975), doi:10.1088/0305-4608/5/5/017.
[600] D. J. Thouless, P. W. Anderson and R. G. Palmer, *Solution of 'Solvable model of a spin glass'*, Philos. Mag. **35**(3), 593 (1977), doi:10.1080/14786437708235992.
[601] M. Talagrand, *The Parisi formula*, Ann. Math. **163**(1), 221 (2006), doi:10.4007/annals.2006.163.221.
[602] D. Panchenko, *Introduction to the SK model*, Curr. Dev. Math. **2014**(1), 231 (2014), doi:10.4310/cdm.2014.v2014.n1.a4.
[603] T. Castellani and A. Cavagna, *Spin-glass theory for pedestrians*, J. Stat. Mech. p. P05012 (2005), doi:10.1088/1742-5468/2005/05/P05012.

[604] M. Gabrié, *Mean-field inference methods for neural networks*, J. Phys. A: Math. Theor. **53**(22), 223002 (2020), doi:10.1088/1751-8121/ab7f65.
[605] J. Barbier, F. Krzakala, N. Macris, L. Miolane and L. Zdeborová, *Optimal errors and phase transitions in high-dimensional generalized linear models*, Proc. Natl. Acad. Sci. U.S.A. **116**(12), 5451 (2019), doi:10.1073/pnas.1802705116.
[606] S. Rangan, *Generalized approximate message passing for estimation with random linear mixing*, In *2011 IEEE Int. Symp. Inf. Theory Proc.*. IEEE, doi:10.1109/isit.2011.6033942 (2011).
[607] L. Zdeborová and F. Krzakala, *Statistical physics of inference: Thresholds and algorithms*, Adv. Phys. **65**(5), 453 (2016), doi:10.1080/00018732.2016.1211393.
[608] R. Monasson and R. Zecchina, *Learning and generalization theories of large committee machines*, Mod. Phys. Lett. B **09**(30), 1887 (1995), doi:10.1142/s0217984995001868.
[609] B. Aubin, A. Maillard, J. Barbier, F. Krzakala, N. Macris and L. Zdeborová, *The committee machine: Computational to statistical gaps in learning a two-layers neural network*, J. Stat. Mech. Theor. Exp. **2019**(12), 124023 (2019), doi:10.1088/1742-5468/ab43d2.
[610] A. Rahimi and B. Recht, *Random features for large-scale kernel machines*, In *NIPS 2007 – Adv. Neural Inf. Process. Syst.* (2007).
[611] A. Rahimi and B. Recht, *Weighted sums of random kitchen sinks: Replacing minimization with randomization in learning*, In *NIPS 2008 – Adv. Neural Inf. Process. Syst.* (2008).
[612] H. Schwarze and J. Hertz, *Generalization in fully connected committee machines*, EPL **21**(7), 786 (1993), doi:10.1209/0295-5075/21/7/012.
[613] H. Schwarze, *Learning a rule in a multilayer neural network*, J. Phys. A: Math. Gen. **26**(21), 5781 (1993), doi:10.1088/0305-4470/26/21/017.
[614] F. Gerace, B. Loureiro, F. Krzakala, M. Mézard and L. Zdeborová, *Generalisation error in learning with random features and the hidden manifold model*, J. Stat. Mech. **2021**(12), 124013 (2021), doi:10.1088/1742-5468/ac3ae6.
[615] S. D'Ascoli, M. Gabrié, L. Sagun and G. Biroli, *More data or more parameters? Investigating the effect of data structure on generalization*, In *NeurIPS 2021 – Adv. Neural Inf. Process. Syst.* (2021), arXiv:2103.05524.
[616] S. Goldt, M. S. Advani, A. M. Saxe, F. Krzakala and L. Zdeborová, *Dynamics of stochastic gradient descent for two-layer neural networks in the teacher-student setup*, J. Stat. Mech. Theor. Exp. **2020**(12), 1 (2020), doi:10.1088/1742-5468/abc61e.
[617] M. Biehl and P. Riegler, *On-line learning with a perceptron*, EPL **28**(7), 525 (1994), doi:10.1209/0295-5075/28/7/012.
[618] M. Biehl and H. Schwarze, *Learning by on-line gradient descent*, J. Phys. A: Math. Gen. **28**(3), 643 (1995), doi:10.1088/0305-4470/28/3/018.
[619] S. Goldt, B. Loureiro, G. Reeves, F. Krzakala, M. Mézard and L. Zdeborová, *The Gaussian equivalence of generative models for learning with shallow neural networks* (2020), arXiv:2006.14709.
[620] F. Mignacco, F. Krzakala, P. Urbani and L. Zdeborová, *Dynamical mean-field theory for stochastic gradient descent in Gaussian mixture classification*, In *NeurIPS 2020 – Adv. Neural Inf. Process. Syst.* (2020), arXiv:2006.06098.
[621] M. Lewenstein, *Quantum perceptrons*, J. Mod. Opt. **41**(12), 2491 (1994), doi:10.1080/09500349414552331.
[622] A. Gratsea, V. Kasper and M. Lewenstein, *Storage properties of a quantum perceptron*, Phys. Rev. E 110(2), 024127 (2021), doi:10.1103/PhysRevE.110.024127.

[623] M. Lewenstein, A. Gratsea, A. Riera-Campeny, A. Aloy, V. Kasper and A. Sanpera, *Storage capacity and learning capability of quantum neural networks*, Quantum Sci. Technol. **6**(4), 045002 (2021), doi:10.1088/2058-9565/ac070f.

[624] A. Gratsea and P. Huembeli, *Exploring quantum perceptron and quantum neural network structures with a teacher-student scheme*, Quantum Mach. Intell. **4**(1), 2 (2022), doi:10.1007/s42484-021-00058-6.

[625] Y. Feng and Y. Tu, *Phases of learning dynamics in artificial neural networks in the absence or presence of mislabeled data*, Mach. Learn.: Sci. Technol. **2**(4), 043001 (2021), doi:10.1088/2632-2153/abf5b9.

[626] A. Decelle, C. Furtlehner and B. Seoane, *Equilibrium and non-equilibrium regimes in the learning of restricted Boltzmann machines*, In *NeurIPS 2021 – Adv. Neural Process. Syst.* (2021), arXiv:2103.05524.

[627] H.-S. Zhong, H. Wang, Y.-H. Deng, *et al.*, *Quantum computational advantage using photons*, Science **370**(6523), 1460 (2020), doi:10.1126/science.abe8770.

[628] F. Arute, K. Arya, R. Babbush, *et al.*, *Quantum supremacy using a programmable superconducting processor*, Nature **574**(7779), 505 (2019), doi:10.1038/s41586-019-1666-5.

[629] C. D. Bruzewicz, J. Chiaverini, R. McConnell and J. M. Sage, *Trapped-ion quantum computing: Progress and challenges*, Appl. Phys. Rev. **6**(2), 021314 (2019), doi:10.1063/1.5088164.

[630] M. Saffman, *Quantum computing with atomic qubits and Rydberg interactions: Progress and challenges*, J. Phys. B: At. Mol. Opt. Phys. **49**(20), 202001 (2016), doi:10.1088/0953-4075/49/20/202001.

[631] L. Henriet, L. Beguin, A. Signoles, *et al.*, *Quantum computing with neutral atoms*, Quantum **4**, 327 (2020), doi:10.22331/q-2020-09-21-327.

[632] L. S. Madsen, F. Laudenbach, M. F. Askarani, *et al.*, *Quantum computational advantage with a programmable photonic processor*, Nature **606**(7912), 75 (2022), doi:10.1038/s41586-022-04725-x.

[633] Y. Liu, S. Arunachalam and K. Temme, *A rigorous and robust quantum speed-up in supervised machine learning*, Nat. Phys. **17**(9), 1013 (2021), doi:10.1038/s41567-021-01287-z.

[634] J. Herrmann, S. M. Llima, A. Remm, *et al.*, *Realizing quantum convolutional neural networks on a superconducting quantum processor to recognize quantum phases*, Nat. Commun. **13**(1), 4144 (2022), doi:10.1038/s41467-022-31679-5.

[635] H.-Y. Huang, M. Broughton, J. Cotler, *et al.*, *Quantum advantage in learning from experiments*, Science **376**(6598), 1182 (2022), doi:10.1126/science.abn7293.

[636] M. Gong, H.-L. Huang, S. Wang, *et al.*, *Quantum neuronal sensing of quantum many-body states on a 61-qubit programmable superconducting processor*, Sci. Bull. **68**(9), 906 (2023), doi:10.1016/j.scib.2023.04.003.

[637] L. Hales and S. Hallgren, *An improved quantum Fourier transform algorithm and applications*, In *FOCS 2000 – 41st Annu. IEEE Symp. Found. Comput. Sci.*, pp. 515–525, doi:10.1109/SFCS.2000.892139 (2000).

[638] P. Shor, *Algorithms for quantum computation: Discrete logarithms and factoring*, In *FOCS 1994 – 35th Annu. IEEE Symp. Found. Comput. Sci.*, pp. 124–134, doi:10.1109/SFCS.1994.365700 (1994).

[639] A. Y. Kitaev, *Quantum measurements and the Abelian stabilizer problem* (1995), arXiv:quant-ph/9511026.

[640] A. W. Harrow, A. Hassidim and S. Lloyd, *Quantum algorithm for linear systems of equations*, Phys. Rev. Lett. **103**, 150502 (2009), doi:10.1103/PhysRevLett.103.150502.

[641] P. Rebentrost, M. Mohseni and S. Lloyd, *Quantum support vector machine for big data classification*, Phys. Rev. Lett. **113**, 130503 (2014), doi:10.1103/PhysRevLett.113.130503.

[642] Z. Li, X. Liu, N. Xu and J. Du, *Experimental realization of a quantum support vector machine*, Phys. Rev. Lett. **114**, 140504 (2015), doi:10.1103/PhysRevLett.114.140504.

[643] N. Wiebe, D. Braun and S. Lloyd, *Quantum algorithm for data fitting*, Phys. Rev. Lett. **109**, 050505 (2012), doi:10.1103/PhysRevLett.109.050505.

[644] A. Gilyén, S. Lloyd and E. Tang, *Quantum-inspired low-rank stochastic regression with logarithmic dependence on the dimension* (2018), arXiv:1811.04909.

[645] A. Ekert and R. Jozsa, *Quantum computation and Shor's factoring algorithm*, Rev. Mod. Phys. **68**, 733 (1996), doi:10.1103/RevModPhys.68.733.

[646] C. M. Dawson and M. A. Nielsen, *The Solovay-Kitaev algorithm* (2005), arXiv:quant-ph/0505030.

[647] Y. Y. Atas, J. Zhang, R. Lewis, A. Jahanpour, J. F. Haase and C. A. Muschik, *SU(2) hadrons on a quantum computer via a variational approach*, Nat. Commun. **12**(1) (2021), doi:10.1038/s41467-021-26825-4.

[648] K. Temme, S. Bravyi and J. M. Gambetta, *Error mitigation for short-depth quantum circuits*, Phys. Rev. Lett. **119**, 180509 (2017), doi:10.1103/PhysRevLett.119.180509.

[649] S. Endo, S. C. Benjamin and Y. Li, *Practical quantum error mitigation for near-future applications*, Phys. Rev. X **8**(3) (2018), doi:10.1103/PhysRevX.8.031027.

[650] L. Funcke, T. Hartung, K. Jansen, S. Kühn, P. Stornati and X. Wang, *Measurement error mitigation in quantum computers through classical bit-flip correction*, Phys. Rev. A **105**, 062404 (2022), doi:10.1103/PhysRevA.105.062404.

[651] A. D. Córcoles, E. Magesan, S. J. Srinivasan, *et al.*, *Demonstration of a quantum error detection code using a square lattice of four superconducting qubits*, Nat. Commun. **6**(1), 6979 (2015), doi:10.1038/ncomms7979.

[652] V. Havlíček, A. D. Córcoles, K. Temme, *et al.*, *Supervised learning with quantum-enhanced feature spaces*, Nature **567**(7747), 209 (2019), doi:10.1038/s41586-019-0980-2.

[653] M. Schuld and N. Killoran, *Quantum machine learning in feature Hilbert spaces*, Phys. Rev. Lett. **122**, 040504 (2019), doi:10.1103/PhysRevLett.122.040504.

[654] C. M. Wilson, J. S. Otterbach, N. Tezak, *et al.*, *Quantum kitchen sinks: An algorithm for machine learning on near-term quantum computers* (2018), arXiv:1806.08321.

[655] J. Dai and R. V. Krems, *Quantum Gaussian process model of potential energy surface for a polyatomic molecule*, J. Chem. Phys. **156**(18), 184802 (2022), doi:10.1063/5.0088821.

[656] *Quantum feature maps and kernels*, chapter of the Qiskit textbook "Introduction to Quantum Computing," Accessed: 03-02-2022.

[657] E. Tang, *Quantum principal component analysis only achieves an exponential speedup because of its state preparation assumptions*, Phys. Rev. Lett. **127**(6), 060503 (2021), doi:10.1103/PhysRevLett.127.060503.

[658] J. M. Kübler, S. Buchholz and B. Schölkopf, *The inductive bias of quantum kernels*, In *NeurIPS 2021 – Adv. Neural Inf. Process. Syst.*, pp. 12661–12673 (2021), arXiv:2106.03747.

[659] T. Haug, C. N. Self and M. Kim, *Quantum machine learning of large datasets using randomized measurements*, Mach. Learn.: Sci. Technol. **4**(1), 015005 (2023), doi:10.1088/2632-2153/acb0b4.
[660] J. Liu, F. Tacchino, J. R. Glick, L. Jiang and A. Mezzacapo, *Representation learning via quantum neural tangent kernels*, PRX Quantum **3**, 030323 (2022), doi:10.1103/PRXQuantum.3.030323.
[661] E. Torabian and R. V. Krems, *Compositional optimization of quantum circuits for quantum kernels of support vector machines*, Phys. Rev. Res. **5**, 013211 (2023), doi:10.1103/PhysRevResearch.5.013211.
[662] A. Peruzzo, J. McClean, P. Shadbolt, *et al.*, *A variational eigenvalue solver on a photonic quantum processor*, Nat. Comm. **5**(1) (2014), doi:10.1038/ncomms5213.
[663] L. Funcke, T. Hartung, K. Jansen, S. Kühn and P. Stornati, *Dimensional expressivity analysis of parametric quantum circuits*, Quantum **5**, 422 (2021), doi:10.22331/q-2021-03-29-422.
[664] A. Gresch and M. Kliesch, *Guaranteed efficient energy estimation of quantum many-body Hamiltonians using ShadowGrouping* Nat. Comm. **16**, 689 (2025), doi:10.1038/s41467-024-54859-x.
[665] J. Li, X. Yang, X. Peng and C.-P. Sun, *Hybrid quantum-classical approach to quantum optimal control*, Phys. Rev. Lett. **118**, 150503 (2017), doi:10.1103/PhysRevLett.118.150503.
[666] A. Pérez-Salinas, A. Cervera-Lierta, E. Gil-Fuster and J. I. Latorre, *Data re-uploading for a universal quantum classifier*, Quantum **4**, 226 (2020), doi:10.22331/q-2020-02-06-226.
[667] I. Cong, S. Choi and M. D. Lukin, *Quantum convolutional neural networks*, Nat. Phys. **15**(12), 1273 (2019), doi:10.1038/s41567-019-0648-8.
[668] M. Schuld, *Supervised quantum machine learning models are kernel methods* (2021), arXiv:2101.11020.
[669] S. Jerbi, L. J. Fiderer, H. Poulsen Nautrup, J. M. Kübler, H. J. Briegel and V. Dunjko, *Quantum machine learning beyond kernel methods*, Nat. Commun. **14**(1), 517 (2023), doi:10.1038/s41467-023-36159-y.
[670] J. Bausch, *Recurrent quantum neural networks*, In *NeurIPS 2020 – Adv. Neural Inf. Process. Syst.* (2020), arXiv:2006.14619.
[671] S. Jerbi, C. Gyurik, S. Marshall, H. Briegel and V. Dunjko, *Parametrized quantum policies for reinforcement learning*, In *NeurIPS 2021 – Adv. Neural Inf. Process. Syst.*, pp. 28362–28375 (2021), arXiv:2103.05577.
[672] A. Skolik, S. Jerbi and V. Dunjko, *Quantum agents in the Gym: A variational quantum algorithm for deep Q-learning*, Quantum **6**, 720 (2022), doi:10.22331/q-2022-05-24-720.
[673] G. Brockman, V. Cheung, L. Pettersson, *et al.*, *OpenAI Gym* (2016), arXiv:1606.01540.
[674] J. Romero, J. P. Olson and A. Aspuru-Guzik, *Quantum autoencoders for efficient compression of quantum data*, Quantum Sci. Technol. **2**(4), 045001 (2017), doi:10.1088/2058-9565/aa8072.
[675] D. Bondarenko and P. Feldmann, *Quantum autoencoders to denoise quantum data*, Phys. Rev. Lett. **124**(13) (2020), doi:10.1103/PhysRevLett.124.130502.
[676] D. F. Locher, L. Cardarelli and M. Müller, *Quantum Error Correction with Quantum Autoencoders*, Quantum **7**, 942 (2023), doi:10.22331/q-2023-03-09-942.

[677] G. Verdon, J. Marks, S. Nanda, S. Leichenauer and J. Hidary, *Quantum Hamiltonian-based models and the variational quantum thermalizer algorithm* (2019), arXiv:1910.02071.

[678] S. Lloyd and C. Weedbrook, *Quantum generative adversarial learning*, Phys. Rev. Lett. **121**, 040502 (2018), doi:10.1103/PhysRevLett.121.040502.

[679] P.-L. Dallaire-Demers and N. Killoran, *Quantum generative adversarial networks*, Phys. Rev. A **98**, 012324 (2018), doi:10.1103/PhysRevA.98.012324.

[680] J.-G. Liu and L. Wang, *Differentiable learning of quantum circuit Born machines*, Phys. Rev. A **98**, 062324 (2018), doi:10.1103/PhysRevA.98.062324.

[681] B. Coyle, D. Mills, V. Danos and E. Kashefi, *The Born supremacy: Quantum advantage and training of an Ising Born machine*, npj Quantum Inf. **6**(1) (2020), doi:10.1038/s41534-020-00288-9.

[682] M. Benedetti, D. Garcia-Pintos, O. Perdomo, V. Leyton-Ortega, Y. Nam and A. Perdomo-Ortiz, *A generative modeling approach for benchmarking and training shallow quantum circuits*, npj Quantum Inf. **5**(1) (2019), doi:10.1038/s41534-019-0157-8.

[683] L. Viola and S. Lloyd, *Dynamical suppression of decoherence in two-state quantum systems*, Phys. Rev. A **58**, 2733 (1998), doi:10.1103/PhysRevA.58.2733.

[684] F. Kleißler, A. Lazariev and S. Arroyo-Camejo, *Universal, high-fidelity quantum gates based on superadiabatic, geometric phases on a solid-state spin-qubit at room temperature*, npj Quantum Inf. **4**(1) (2018), doi:10.1038/s41534-018-0098-7.

[685] M. Taherkhani, M. Willatzen, E. V. Denning, I. E. Protsenko and N. Gregersen, *High-fidelity optical quantum gates based on type-II double quantum dots in a nanowire*, Phys. Rev. B **99**, 165305 (2019), doi:10.1103/PhysRevB.99.165305.

[686] E. Zahedinejad, J. Ghosh and B. C. Sanders, *High-fidelity single-shot toffoli gate via quantum control*, Phys. Rev. Lett. **114**, 200502 (2015), doi:10.1103/PhysRevLett.114.200502.

[687] D. Yu, H. Wang, D. Ma, X. Zhao and J. Qian, *Adiabatic and high-fidelity quantum gates with hybrid Rydberg-Rydberg interactions*, Opt. Express **27**(16), 23080 (2019), doi:10.1364/OE.27.023080.

[688] F. Haddadfarshi and F. Mintert, *High fidelity quantum gates of trapped ions in the presence of motional heating*, New J. Phys. **18**(12), 123007 (2016), doi:10.1088/1367-2630/18/12/123007.

[689] S. Li, J. Xue, T. Chen and Z. Xue, *High-fidelity geometric quantum gates with short paths on superconducting circuits*, Adv. Quantum Technol. **4**(5), 2000140 (2021), doi:10.1002/qute.202000140.

[690] J. R. McClean, S. Boixo, V. N. Smelyanskiy, R. Babbush and H. Neven, *Barren plateaus in quantum neural network training landscapes*, Nat. Commun. **9**(1), 1 (2018), doi:10.1038/s41467-018-07090-4.

[691] M. Cerezo, A. Sone, T. Volkoff, L. Cincio and P. J. Coles, *Cost function dependent barren plateaus in shallow parametrized quantum circuits*, Nat. Commun. **12**(1), 1 (2021), doi:10.1038/s41467-021-21728-w.

[692] L. Bittel and M. Kliesch, *Training variational quantum algorithms is NP-hard*, Phys. Rev. Lett. **127**, 120502 (2021), doi:10.1103/PhysRevLett.127.120502.

[693] S. Sim, P. D. Johnson and A. Aspuru-Guzik, *Expressibility and entangling capability of parameterized quantum circuits for hybrid quantum-classical algorithms*, Adv. Quantum Technol. **2**(12), 1900070 (2019), doi:10.1002/qute.201900070.

[694] L. Bittel, S. Gharibian and M. Kliesch, *The optimal depth of variational quantum algorithms is QCMA-hard to approximate*, In *38th Comput. Complexity Conf. (CCC 2023)*, vol. 264, pp. 34:1–34:24, doi:10.4230/LIPIcs.CCC.2023.34 (2023), arXiv:2211.12519.
[695] A. Y. Kitaev, *Quantum computations: Algorithms and error correction*, Russ. Math. Surv. **52**(6), 1191 (1997), doi:10.1070/rm1997v052n06abeh002155.
[696] M. S. Rudolph, S. Sim, A. Raza, *et al.*, *ORQVIZ: Visualizing high-dimensional landscapes in variational quantum algorithms* (2021), arXiv:2111.04695.
[697] J. Eisert, *Entangling power and quantum circuit complexity*, Phys. Rev. Lett. **127**, 020501 (2021), doi:10.1103/PhysRevLett.127.020501.
[698] H.-Y. Huang, R. Kueng and J. Preskill, *Information-theoretic bounds on quantum advantage in machine learning*, Phys. Rev. Lett. **126**, 190505 (2021), doi:10.1103/PhysRevLett.126.190505.
[699] K. Bharti, A. Cervera-Lierta, T. H. Kyaw, *et al.*, *Noisy intermediate-scale quantum algorithms*, Rev. Mod. Phys. **94**(1), 015004 (2022), doi:10.1103/RevModPhys.94.015004.
[700] G. Munoz-Gil, G. Volpe, M. A. Garcia-March, *et al.*, *Objective comparison of methods to decode anomalous diffusion*, Nat. Commun. **12**(1), 6253 (2021), doi:10.1038/s41467-021-26320-w.
[701] G. Muñoz-Gil, M. A. Garcia-March, C. Manzo, J. D. Martín-Guerrero and M. Lewenstein, *Single trajectory characterization via machine learning*, New J. Phys. **22**(1), 013010 (2020), doi:10.1088/1367-2630/ab6065.
[702] G. Munoz-Gil, C. Romero-Aristizabal, N. Mateos, *et al.*, *Particle flow modulates growth dynamics and nanoscale-arrested growth of transcription factor condensates in living cells*, bioRxiv (2022), doi:10.1101/2022.01.11.475940.
[703] H. B. Moss and R.-R. Griffiths, *Gaussian process molecule property prediction with FlowMO* (2020), arXiv:2010.01118.
[704] A. Glielmo, Y. Rath, G. Csányi, A. De Vita and G. H. Booth, *Gaussian process states: A data-driven representation of quantum many-body physics*, Phys. Rev. X **10**, 041026 (2020), doi:10.1103/PhysRevX.10.041026.
[705] K. Choo, T. Neupert and G. Carleo, *Two-dimensional frustrated J_1–J_2 model studied with neural network quantum states*, Phys. Rev. B **100**(12) (2019), doi:10.1103/PhysRevB.100.125124.
[706] M. Secor, A. V. Soudackov and S. Hammes-Schiffer, *Artificial neural networks as propagators in quantum dynamics*, J. Phys. Chem. Lett. **12**(43), 10654 (2021), doi:10.1021/acs.jpclett.1c03117.
[707] V. Havlicek, *Amplitude ratios and neural network quantum states*, Quantum **7**, 938 (2023), doi:10.22331/q-2023-03-02-938.
[708] J. Yao, L. Lin and M. Bukov, *Reinforcement learning for many-body ground-state preparation inspired by counterdiabatic driving*, Phys. Rev. X **11**, 031070 (2021), doi:10.1103/PhysRevX.11.031070.
[709] H. P. Nautrup, N. Delfosse, V. Dunjko, H. J. Briegel and N. Friis, *Optimizing quantum error correction codes with reinforcement learning*, Quantum **3**, 215 (2019), doi:10.22331/q-2019-12-16-215.
[710] P. Peng, X. Huang, C. Yin, L. Joseph, C. Ramanathan and P. Cappellaro, *Deep reinforcement learning for quantum Hamiltonian engineering*, Phys. Rev. Appl. **18**, 024033 (2022), doi:10.1103/PhysRevApplied.18.024033.
[711] J. Jumper, R. Evans, A. Pritzel, *et al.*, *Highly accurate protein structure prediction with AlphaFold*, Nature **596**(7873), 583 (2021), doi:10.1038/s41586-021-03819-2.

[712] M. Varadi, S. Anyango, M. Deshpande, *et al.*, *AlphaFold protein structure database: Massively expanding the structural coverage of protein-sequence space with high-accuracy models*, Nucleic Acids Res. **50**(D1), D439 (2021), doi:10.1093/nar/gkab1061.

[713] A. Davies, P. Veličković, L. Buesing, *et al.*, *Advancing mathematics by guiding human intuition with AI*, Nature **600**(7887), 70 (2021), doi:10.1038/s41586-021-04086-x.

[714] T. Kriváchy, Y. Cai, D. Cavalcanti, A. Tavakoli, N. Gisin and N. Brunner, *A neural network oracle for quantum nonlocality problems in networks*, npj Quantum Inf. **6**, 70 (2020), doi:10.1038/s41534-020-00305-x.

[715] A. Pozas-Kerstjens, N. Gisin and M.-O. Renou, *Proofs of network quantum nonlocality in continuous families of distributions*, Phys. Rev. Lett. **130**, 090201 (2023), doi:10.1103/PhysRevLett.130.090201.

[716] A. Pozas-Kerstjens, G. Muñoz-Gil, E. Piñol, *et al.*, *Efficient training of energy-based models via spin-glass control*, Mach. Learn.: Sci. Technol. **2**(2), 025026 (2021), doi:10.1088/2632-2153/abe807.

[717] L. G. Wright, T. Onodera, M. M. Stein, *et al.*, *Deep physical neural networks trained with backpropagation*, Nature **601**(7894), 549 (2022), doi:10.1038/s41586-021-04223-6.

[718] K. Wagner and D. Psaltis, *Optical neural networks: An introduction by the feature editors*, Appl. Opt. **32**(8), 1261 (1993), doi:10.1364/AO.32.001261.

[719] Y. Zuo, B. Li, Y. Zhao, *et al.*, *All-optical neural network with nonlinear activation functions*, Optica **6**(9), 1132 (2019), doi:10.1364/OPTICA.6.001132.

[720] X. Sui, Q. Wu, J. Liu, Q. Chen and G. Gu, *A review of optical neural networks*, IEEE Access **8**, 70773 (2020), doi:10.1109/ACCESS.2020.2987333.

[721] H. Zhang, M. Gu, X. D. Jiang, *et al.*, *An optical neural chip for implementing complex-valued neural network*, Nat. Commun. **12**(1), 457 (2021), doi:10.1038/s41467-020-20719-7.

[722] Hui Zhang, J. Thompson, M. Gu, *et al.*, *Efficient on-chip training of optical neural networks using genetic algorithm*, ACS Photonics **8**(6), 1662 (2021), doi:10.1021/acsphotonics.1c00035.

[723] X. Xu, M. Tan, B. Corcoran, *et al.*, *11 TOPS photonic convolutional accelerator for optical neural networks*, Nature **589**(7840), 44 (2021), doi:10.1038/s41586-020-03063-0.

[724] J. Liu, Q. Wu, X. Sui, *et al.*, *Research progress in optical neural networks: Theory, applications and developments*, PhotoniX **2**(1), 5 (2021), doi:10.1186/s43074-021-00026-0.

[725] T. Wang, S.-Y. Ma, L. G. Wright, T. Onodera, B. C. Richard and P. L. McMahon, *An optical neural network using less than 1 photon per multiplication*, Nat. Commun. **13**(1), 123 (2022), doi:10.1038/s41467-021-27774-8.

[726] H. Xu, S. Ghosh, M. Matuszewski and T. C. Liew, *Universal self-correcting computing with disordered exciton-polariton neural networks*, Phys. Rev. Appl. **13**, 064074 (2020), doi:10.1103/PhysRevApplied.13.064074.

[727] D. Ballarini, A. Gianfrate, R. Panico, *et al.*, *Polaritonic neuromorphic computing outperforms linear classifiers*, Nano Lett. **20**(5), 3506 (2020), doi:10.1021/acs.nanolett.0c00435.

[728] M. Matuszewski, A. Opala, R. Mirek, *et al.*, *Energy-efficient neural network inference with microcavity exciton polaritons*, Phys. Rev. Appl. **16**, 024045 (2021), doi:10.1103/PhysRevApplied.16.024045.

[729] R. Mirek, A. Opala, P. Comaron, *et al.*, *Neuromorphic binarized polariton networks*, Nano Lett. **21**(9), 3715 (2021), doi:10.1021/acs.nanolett.0c04696.

[730] D. Zvyagintseva, H. Sigurdsson, V. K. Kozin, *et al.*, *Machine learning of phase transitions in nonlinear polariton lattices*, Commun. Phys. **5**(1), 8 (2022), doi:10.1038/s42005-021-00755-5.

[731] J. J. Hopfield, *Neural networks and physical systems with emergent collective computational abilities*, Proc. Natl. Acad. Sci. U.S.A. **79**(8), 2554 (1982), doi:10.1073/pnas.79.8.2554.

[732] P. Rotondo, M. Marcuzzi, J. P. Garrahan, I. Lesanovsky and M. Müller, *Open quantum generalisation of Hopfield neural networks*, J. Phys. A: Math. Theor. **51**(11), 115301 (2018), doi:10.1088/1751-8121/aaabcb.

[733] K. B. Petersen and M. S. Pedersen, *The matrix cookbook*, Mathematics - Waterloo University, Accessed: 03-04-2022 (2012).

[734] A. Morvan, B. Villalonga, X. Mi, et al., Phase transitions in random circuit sampling. Nature 634(8033), 328–333 (2024). https://doi.org/10.1038/s41586-024-07998-6.

[735] D. Bluvstein, S. J. Evered, A. A. Geim, et al., Logical quantum processor based on reconfigurable atom arrays. Nature 626(7997), 58–65 (2024). https://doi.org/10.1038/s41586-023-06927-3.

Index